LOG QUALITY CONTROL

Richard M. Bateman

PRESIDENT, VIZILOG, INC.

International Human Resources Development Corporation

BOSTON

 For information address: IHRDC, Publishers, 137 Newbury Street, Boston, MA 02116.

Library of Congress Cataloging in Publication Data

Bateman, Richard M., 1940–
Log quality control.

1. Oil well logging—Quality control. I. Title.
TN871.35.B43 1984 622'.3328 84–6687
ISBN 0-934634-89-0

Printed in the United States of America

LOG QUALITY CONTROL

CONTENTS

PREFACE

Accurate logs are essential for valid formation evaluation. With this in mind, this book has been prepared to help you judge the quality of field logs.

Logging check lists tend to become burdensome, lengthy, and somewhat dictatorial toward the service company personnel; although all the information is without question important to logging operations. Thus, the problem is: how to apply quality control while taking advantage of the service company personnel's experience, and with the least amount of effort from the operator's representative?

The responsibility for good quality logs must rest with you, the operator's representative at the well. This book provides the detailed information you will need. Included for each current log type are the following:

Tools Available
Measurement Principle
Uses and Limitations
Calibration and Quality Control
Field Examples

Computer control of log data gathering and processing at the well-site is an established and growing reality. Therefore, this manual also includes quality control criteria for these types of logs and processed presentations.

ACKNOWLEDGMENTS

Vizilog, Inc. gratefully acknowledges the help and assistance, as well as the kind permission of

Amoco Production Company,
Birdwell Division, Seismograph Service Corporation,
Chevron Oil Field Research Company,
Dresser Industries, Inc.,
Gearhart Industries,
Petrophysical Data Consultants, Inc.,
Schlumberger Well Services, and
Welex, a Halliburton Company,

to reproduce log examples, and other previously published articles and/or figures without which this publication would have been impossible. Particular thanks is given to Mr. Al Jageler and Mr. Bob Murphy of Amoco and Mr. Chuck Neuman of Chevron, whose wisdom and encyclopedic knowledge of log-related matters were indispensable.

CHAPTER ONE

GENERAL LOG QUALITY CONTROL

INTRODUCTION

Logging Objectives

Ideally, a reservoir should be defined by drilling as few wells as possible. In those wells, enough data should be gathered to extrapolate reservoir parameters fieldwide to arrive at realistic figures for both the economic evaluation of the reservoir and the planning of the optimum recovery method.

Logging offers a way of gathering the data needed for both economic analysis and production planning.

What then are the parameters the manager, the geologist, the geophysicist, and the reservoir and production engineers need and which of those can logging reasonably be expected to provide?

The geologist needs to know the stratigraphy, the structural and sedimentary features, and the mineralogy of the formations through which the well was drilled.

The geophysicist needs to relate seismic references to specific horizons encountered in the drilled section.

The reservoir engineer needs to know the vertical and lateral extent of the reservoir, its porosity (and type of porosity) and permeability, and its fluid content and recoverability.

The production engineer needs to know the rock properties; be aware of overpressure if it exists; and be able to assess sanding and associated problems and the need for secondary recovery efforts or pressure maintenance. Once the well is on production, the production engineer will also need to know the dynamic behavior of the well under production conditions and must be able to diagnose problems as the well ages. He may also need to know formation injectivity and residual oil saturations in order to plan waterflooding and monitor the progress of the waterflood when it is operational.

The manager needs to know the vital inputs to an economic study. These are the original hydrocarbon in place; recoverability; cost of development; and, based on those factors, the profitability of producing the reservoir.

Log measurements can give the majority of the parameters required. Specifically, logs provide either a direct measurement or an indication of:

1. porosity, both primary and secondary (fractures and vugs)
2. permeability
3. water saturation and hydrocarbon movability

4. hydrocarbon type (oil, gas, or condensate)
5. lithology
6. formation dip and structure
7. sedimentary environment
8. travel times of elastic waves in the formation

From these data, estimates can be made of the reservoir size and the hydrocarbons in place.

Logging techniques in cased holes can provide much of the data needed to monitor primary production and also to gauge the applicability of waterflooding and monitor its progress when installed. In producing wells, logging can provide measurements of:

1. flow rates
2. fluid type
3. pressure
4. temperature
5. oil and/or gas saturation

From these measurements, dynamic well behavior can be understood better, remedial work planned, and secondary or tertiary recovery proposals evaluated and monitored.

In summary, logging, when properly applied, can answer a great many questions from a wide spectrum of special interest groups on topics ranging from basic geology to economics.

Measurement Principles

With a few specialized exceptions, well logs are continuous records of various measurements plotted versus depth. For any measurement to have meaning, it must be related to generally accepted standard units by a valid and specified system of comparison. All distances on Earth, for example, are related to the standard meter, which is defined as 1,650,763.73 wavelengths in vacuo of the radiation in the unperturbed transition between levels $2p_{10d}$ and $5d_5$ in krypton 86. This primary standard is unhandy for most needs, so we resort to secondary standards such as yardsticks (which are, theoretically, 0.9144 meter long). Any yardstick or other distance-measuring device must have two fundamental calibration marks, one of which corresponds to zero distance and the other to a distance of one yard (or whatever). We can subdivide the distance any way we want to, so long as our subdivisions are consistent, but the end points are fixed by definition.

The calibrations of logging tools follow, where possible, a comparable method. A sensing system is designed which produces consistent and repeatable indications over the range of measurements for which the tool is intended. The indications usually are designed to follow either a linear or a logarithmic function of the measured parameter. Calibration consists of adjusting this system to make it read the known values of the two calibration standards: a low value (zero, if possible) and a high value at or near the maximum expected measurement. In some types of logging tools, such as the sonic and sidewall neutron porosity tools, intermediate readings are made during the calibration. These are not calibration readings, but functional checks that verify the integrity of the measuring systems.

The guiding principles of any sound calibration philosophy are (1) to use calibration standards that duplicate the actual parameters to be measured, within the limits of practicality; and (2) to select calibration values that bracket the anticipated range of measurements, again subject to practical considerations. Each compromise with reality in these matters introduces a new area of potential errors, and such errors can be cumulative.

To illustrate the significance of these principles, we can continue the yardstick analogy. You could use a yardstick to measure volume, or even weight (requiring an assumption of density); but principle (1) tells us it is better to use the yardstick for distance, a volumetrically calibrated container for volume, and a scale for weight. You could use the yardstick for miles or millimeters of distance, but according to principle (2) the accuracy is best when you use it for small multiples or major subdivisions of a yard.

Some tools, particularly radiation-measuring devices, require massive calibration installations which cannot reasonably be taken to the field. Such tools are operationally calibrated by the use of portable secondary field calibration standards. All field and (where applicable) shop calibrations are routinely recorded on film and presented as a part of the optical log. In this manual the significance of these calibration films is explained, the primary and secondary calibration standards for each type of tool are specified, and typical examples of each type of log are given. The presentation and scaling—and sometimes the galvanometer assignments—will vary from those shown, according to local needs. However, the function and scaling of each trace are always shown on the log heading, as illustrated. It would be impractical to present examples of every log in every lithological environment. The example logs are presented in a final form and will differ from field prints in that the scale inserts and inserts labeled *Repeat Section* and *Calibration Record* will not necessarily be presented.

The log calibration records that are now part of every field log are an objective verification of the tool's correct functioning. To be able to use these records effectively, you should *fully understand* the calibration of each tool, and the corresponding significance of each step of the calibration record.

Modern Logging Tools

The actual running of a log involves as much the tool on the end of the logging cable as the cable itself and the surface controlling and recording apparatus. So before discussing downhole tools, the common elements of all logs will be presented.

Figure 1.1 illustrates the basic components of any logging system. A sensor, incorporated in a sonde together with its associated electronics, is suspended in the hole by a multiconductor cable. The sensor is

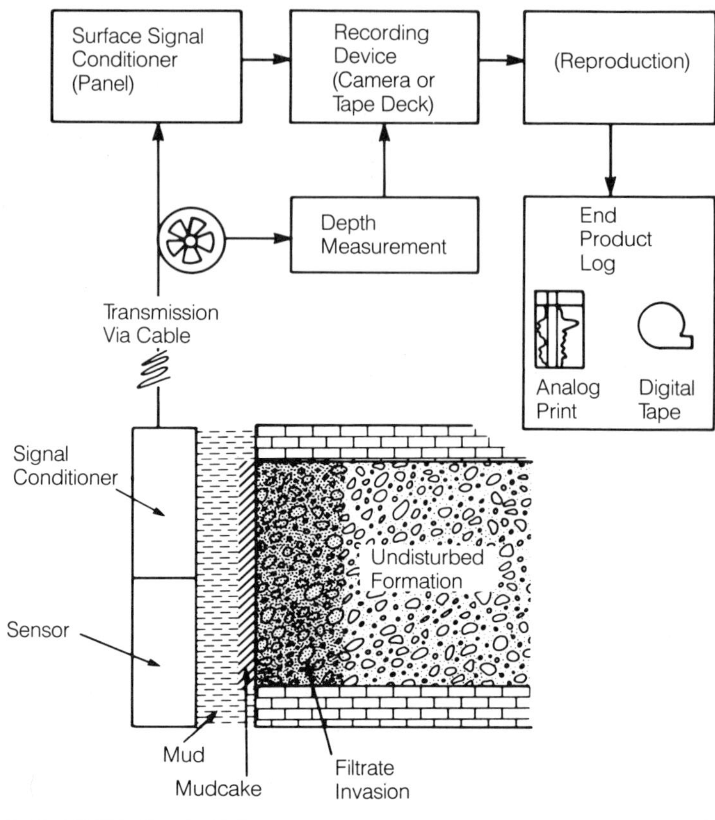

FIGURE 1.1 *Basic Components of a Logging System.*

separated from the virgin formation by a portion of the mud column; by mudcake; and, more often than not, by an invaded zone in the surrounding formations. The signals from the sensor are conditioned by the electronics for transmission up the cable to the control panel; and this, in turn, conditions the signals for the recorder. As the cable is raised or lowered, it activates some depth-measuring device, a sheave wheel for example, which activates a recording device, either an optical camera (making a film) or tape deck (making a digital recording on magnetic tape). Finally, some form of reproduction takes place to provide a hard copy of the recorded data.

In general, well logging jargon distinguishes among the *logging survey*, a *logging tool*, a *log*, and a *curve*. There is frequently some confusion about these terms in the minds of oil company engineers and logging engineers when logging matters are discussed.

Making a logging survey involves running into a well with a logging tool, or combination of tools, and recording the measurements of wellbore and near wellbore properties as the tool is withdrawn from the well. For example, on a logging order form, it may be noted that at

5000 ft it is planned to run an IES—induction electrical survey. The logging company personnel then know to take to the well an IRT—induction resistivity tool; and the record produced by the logging operation will be an *induction log*. (Unfortunately, this log is often referred to as an IES log, which is not strictly true.) An induction log may contain not only the induction resistivity curve but, on the same presentation, other curves as well, such as the gamma ray curve, spontaneous potential (SP) curve, and a spherically focused resistivity curve. It is usually assumed that when someone says he has an induction log from a well, that log will, in fact, have other standard curves recorded on it in addition to the induction curve.

To recap: A *logging survey* is taken to mean the act of providing a tool and making a survey. A *log* means a strip of paper with several curves recorded on it. A *logging tool* may be made up with a number of distinct measuring devices all run at the same time. The following logging tools are in use in the 1980s:

Resistivity Tools	*Cased-Hole Tools*	*Porosity–Lithology Tools*
induction	pulsed neutron	sonic
laterolog	production logging devices	density and lithologic density
microresistivity		neutron

These basic devices will answer 90% of the questions about the formation. Omitted from the list are various types of old logs (such as electric logs); some newcomers such as the electromagnetic propagation tool and the carbon/oxygen tool; and some standard auxiliary devices which, while important, do not rate as separate tools since they always piggy-back along with one or other of the basic tools. These *auxiliary tools* include spontaneous potential (SP), gamma ray, and caliper devices.

PLANNING AND CONDUCTING A WELL LOGGING OPERATION

Mud Programs

Planning a competent job must begin before the well is spudded. Such variables as mud and casing programs, bit sizes, allowable deviation, and drilling speed should be chosen with full awareness of their bearing on a successful and reliable logging job. *Successful* means that all the logging tools went to bottom without great difficulty and ran logs over the desired intervals without sticking. *Reliable* means that the tools operated properly, of course; but it also means that wellbore conditions were favorable to the gathering of data undistorted by environmental effects.

From the time a logging suite is chosen, certain limits are automatically imposed on the drilling program. The physical requirements of various logging tools must be known and provided for: some tools can be run only in open hole; some require mud in the hole; some require bigger hole sizes than others; etc. More subtle are the variables that can alter the tool responses: mud and drilling-speed combinations that cause rugose or out-of-round holes; mud additives that affect the logs; and, most common, the mud-filtrate loss that governs depth of invasion of the formations. Invasion affects all logs, and its influence is often hard to nail down. By no means all negative, invasion effects are much used in log analysis. The parameters R_{xo} and S_{xo} (resistivity and filtrate saturation in flushed zone) are necessary in computer programs that evaluate hydrocarbon density, because they allow more accurate and comprehensive solutions to the overall evaluation problem.

In general, invasion should be kept shallow but finite, and this requires good-quality, low-water-loss muds throughout the drilling. Thus, the money-saving practice of mudding up only after most of the hole is drilled is likely to be a false economy. (Misleading logs can result in overlooked pay zones, or conversely, setting pipe on dusters.)

No matter how good the mud, invasion cannot be shut off entirely by mudcake. Thus, since the longer a zone is exposed to drilling mud the deeper invasion will be, it may be advisable to make intermediate log runs on zones of potential interest as soon after penetration as possible.

Choosing a Logging Suite

The logging suite needed depends on the questions to be answered. Development wells in consistent geological facies may need no more than two log runs: a resistivity combination and a porosity combination. The latter may or may not include both neutron and density logs, and the former may include a sonic log. Wildcat wells will require a more comprehensive set of logs. After deciding what information is needed to analyze the well, it is a good idea to confer with the reservoir engineer, completion engineer, and geophysicist, who may need additional log information. Table 1.1 lists recommended logging programs for most logging situations.

Contacting the Service Companies

WELL DATA. After the logging program has been determined, the logging-service company should be notified. If complete and accurate instructions as to the logging requirements are given to the service company at least 24 hours ahead of time, many hours of rig time can be saved. Some of the more important items of instruction to the service company during this "will call" operation concern the well data and the log data. These items are listed below and should be transmitted to the service company prior to logging time:

1. Company name, name of representative, his office and home phone number.
2. Well name, number, and location data.

TABLE 1.1 *General Recommended Logging Programs*

Condition	Data Desired	Recommended Services	Remarks
Fresh mud $\frac{R_{mf}}{R_w} > 2$	Correlation and lithology in sand/shale	Dual Induction–SFL–GR/SP; Induction Electrical; Dual Laterolog–GR/SP (low porosity and/or high resistivities); Sidewall Cores	The dual induction and dual laterolog devices are superior to single induction and single laterolog in all cases Sidewall cores give lithology in sand/shale
	Porosity; water saturation; lithology in carbonates and evaporates; hydrocarbon type	Density and/or Neutron and/or Sonic and GR; Formation Tester	For shaly sands or simple mixed lithologies, density–neutron or neutron–sonic. For complex mixed lithologies use all three. For hydrocarbon type as above and/or Formation Tester
	Producible hydrocarbons and permeability indicators	Proximity-Microlog; Microlaterolog or Micro-SFL; Dual Resistivity device; Sidewall Cores; Formation Tester	Proximity log used for thick mudcake; Microlaterlog for thin mudcake; Micro-SFL available with Dual Laterolog. Sidewall cores give permeability estimate in shaly sands
	Hydrocarbon indication at the wellsite	Wellsite Computer Products, including: apparent formation water resistivity, R_{wa}; formation factor/resistivity overlay; and R_{xo}/R_t vs. SP overlay	Available with computer units only. Others available with both units
	Formation dip magnitude and directional	Dipmeter	Wellbore directional information is also available
Salt mud $\frac{R_{mf}}{R_w} < 2$	Correlation and lithology in sand/shale	Dual Laterolog–Micro-SFL–GR	In some areas of low porosity and/or mixed lithology the neutron or density–neutron is usable as a correlation log
	Porosity, water saturation, and lithology in carbonates and evaporates; hydrocarbon type	Density and/or Neutron and/or Sonic and GR; Formation Tester	See remarks under fresh mud
	Producible hydrocarbons and permeability indicators	Proximity–Microlog; Microlaterolog-Microlog; Dual Laterolog–Micro-SFL; Sidewall Core; Formation Tester	In very salty muds or when R_{xo} is less than R_{mc}, the microlog results may be unsatisfactory. See remarks under fresh mud

TABLE 1.1 *(continued)*

Condition	Data Desired	Recommended Services	Remarks
Salt mud (*continued*)	Hydrocarbon indication at the wellsite	Wellsite Computer Products; R_{wa}, R_o Overlay	See remarks under fresh mud. Apparent formation water resistivity may not be satisfactory in low and/or mixed lithologies
	Formation dip magnitude and directional	Dipmeter	Wellbore directional information is also available
Oil-base mud	Correlation and lithology in sand/shale	Dual Induction–GR; Induction–GR; Sidewall Cores	High temperature (350 to 400°F) will restrict the use of the dual induction and sidewall cores
	Porosity, water saturation, lithology in carbonates and evaporites; hydrocarbon type	Density and/or Neutron and/or Sonic; Formation Tester	See remarks under fresh mud. High temperature (350°F) and small holes (6 in.) restrict the use of the formation tester
	Producible hydrocarbons and permeability indicators	Dual Induction; Sidewall Cores; Formation Tester	Sidewall cores give permeability estimates in shaly sands
	Hydrocarbon indication at the wellsite	Wellsite Computer Products; R_{wa}, R_o Overlay	See remarks under fresh mud
	Formation dip magnitude and directional	Dipmeter	See remarks under fresh mud
Air- or Gas-filled hole	Resistivity, correlation, and lithology in sand/shale	Dual Induction–GR; Induction–GR	
	Porosity, water saturation, gas saturation, estimated lithology in carbonates and evaporates	Density and/or Neutron	Dual spacing neutron CNL cannot be used; must use gamma ray-neutron (GRN) or preferably sidewall neutron porosity (SNP)
	Permeability indicator	Temperature Log; Noise Log	Gas entry
	Hydrocarbon indication at the wellsite	Wellsite Computer Products; R_o Overlay, R_{wa}	See remarks under fresh mud
Fresh or unknown formation water	Resistivity; correlation and estimated lithology in sand/shale	See fresh mud and salt mud	See fresh mud and salt mud

	Porosity water saturation; estimated lithology in carbonates and evaporates; hydrocarbon type	Density and/or Neutron and/or Sonic and GR; Electromagnetic Propagation Tool; Inelastic Neutron Scattering and Capture Gamma Ray Spectroscopy; Formation Tester	Under conditions of no invasion the electromagnetic propagation tool will yield water saturation directly. See remarks under fresh mud
	Producible hydrocarbons and permeability indicators	Sidewall Cores; Formation Tester	
	Hydrocarbon indication at the wellsite	Wellsite Computer Products	See remarks under fresh mud. May use electromagnetic propagation tool
Cased hole	Fluid type and lithology	Pulsed Neutron Log; Inelastic Neutron Scattering and Capture Gamma Ray Spectroscopy; GR and Natural Gamma Ray Spectroscopy	
	Porosity and hydrocarbon type	Pulsed Neutron Log; Gamma Ray–Neutron; Density Formation Tester; Inelastic Neutron Scattering and Capture Gamma Ray Spectroscopy	Dual spacing neutron (CNL) cannot be used if hole is gas filled. Under favorable conditions the density may be used for porosity
	Permeability	Formation Tester	

3. Lease directions (always give explicit directions).
4. Approximate time the well will be ready.
5. Approximate depth and deviation of the well, casing size and depth, depth reference datum, and elevation information.
6. Hole size—advise the service company of possible large washed-out zones so that special borehole contact tools may be obtained that will reach out to the formations. Small holes will require special slim-hole tools.
7. Indicate the approximate bottomhole temperature and request that bottomhole temperatures be measured on each run. This gives the service company a chance to obtain special tools designed for logging high-temperature holes.
8. The logging services required. Table 1.1 will help select the appropriate logging program. Table 1.2 will help with the correct logging nomenclature for the particular service company.
9. If the well location is a considerable distance from the service-company office, or in a remote part of the country, request that spare tools be brought along whenever possible. This is especially important when roads are such that the tools may be shaken around in the truck during transportation. Most logging equipment is built to take great shocks in a vertical position, but will not stand much pounding in a horizontal position. Allow ample time for the truck to drive slowly.

LOGGING INTERVALS. For the first log run on a well, service companies assess an operation charge all the way to the surface even if the shallow portion of the well is cased and normally not logged. Therefore, if a radioactivity device (gamma ray, neutron, or density) is recorded on the first log run it should be recorded all the way back to the surface if rig time permits. The service charges always include both a small-scale or correlation (1 or 2 in./100 ft) and large-scale or detail (5 in./100 ft) log of all the footage logged. The 5-in./100-ft log should be closely examined to determine the zones that need further evaluation by the auxiliary logs (sonic, dipmeter, etc.). These auxiliary logs are assessed a depth charge and an operation charge for the footage recorded—the minimum operation charge is generally for 2000 to 3000 ft.

All subsequent log runs in the same well at a later date are assessed a depth charge and an operation charge for the footage recorded with a minimum operation charge. A 200-ft repeat section of the log should always be recorded over the primary zone of interest. In addition, all log runs should overlap the old hole by 200 ft. If the old hole has been cased off, the radioactivity logs (gamma ray, neutron, density) should still overlap the old hole. This is important for quality control. Request a 200-ft repeat, for quality control, of any zones exhibiting abnormal log response. If possible run a detailed (5 in./100 ft) log over any potentially productive zones not detailed on the previous logging run.

MUD DATA. Instruct the drilling crew to catch a circulated sample of the mud for measuring R_m, R_{mf}, and R_{mc} values. This sample should be taken just prior to coming out of the hole for the logging operation. Normally, too little attention is applied to catching a circulated mud sample. This can create serious problems in log interpretation because the value of the mud filtrate resistivity is most important in determining formation water resistivities and producible hydrocarbons. Mud resistivity values are needed to correct apparent resistivity values to true resistivity values. If it is necessary to obtain the mud sample from the pit, make certain the sample is obtained from the bottom of the pit close to the pick-up line.

DRILLSTEM TEST WATER. If a drillstem test has been run prior to the logging run, retain samples of all recovered water so that a resistivity measurement can be made by the logging engineer. (The last sample of water recovered is of particular importance.) This provides vital information for log interpretation without increasing the delay sometimes associated with chemical analysis of the water column.

LOG SCALES. Request that the logging engineer bring the logs that were previously recorded on the well. In addition, if the logs on an offset well are available over the proper intervals, request that the engineer bring such logs as a guide in selecting scales, and for quality control in the new well. If the engineer cannot supply these logs, try to provide them from office files.

In the absence of an offset log or previous experience in the area, watch the log going down the hole and set up the scales of the SP, sonic, density, neutron, and gamma ray curves to deflect at least 60% of the recording track or tracks. Avoid an excess of backup (off-scale) traces for these curves. Scale selection is generally predetermined for the logarithmic resistivity curves. For linear resistivity curves, use as sensitive a scale as possible to obtain readable values over the zone of interest.

DISPATCHING THE SERVICE COMPANY. When the well is ready, notify the logging company in plenty of time to allow for loading the truck, rounding up the crew, eating, and other preparation. In the continental United States, on an average, allow four minutes per mile from the time of notification. A minimum of 1½ hours is usually required for the logging crew to load the truck and shop check the instruments. The loading time must be added to travel time. When above-average hole temperatures for the area are expected, give 36 hours notice so that the hot-hole tools can be heat tested at the shop. Repeat the earlier instructions as to the type of logs to be run and any special equipment that is needed. Advise the logging company if the logging run is to be delayed an appreciable amount of time or cancelled.

TABLE 1.2 *Service Company Nomenclature*

Schlumberger	Gearhart	Dresser Atlas	Welex
Electrical Log (ES)	Electric Log	Electrolog	Electric Log
Induction Electric Log	Induction Electrical Log	Induction Electrolog	Induction Electric Log
Induction Spherically Focused Log			
Dual Induction Spherically Focused Log	Dual Induction-Laterolog	Dual Induction Focused Log	Dual Induction Log
Laterolog-3	Laterolog-3	Focused Log	Guard Log
Dual Laterolog	Dual Laterolog	Dual Laterolog	Dual Guard Log
Microlog	Micro-Electrical Log	Minilog	Contact Log
Microlaterolog	Microlaterolog	Microlaterolog	$F_o R_{xo}$Log
Proximity Log Microspherically Focused Log		Proximity Log	
Borehole Compensated Sonic Log	Borehole Compensated Sonic	Borehole Compensated Acoustilog	Acoustic Velocity Log
Long Spaced Sonic Log		Long Spacing BHC Acoustilog	
Cement Bond/Variable Density Log	Sonic Cement Bond System	Acoustic Cement Bond Log	Microseismogram
Gamma Ray Neutron	Gamma Ray Neutron	Gamma Ray Neutron	Gamma Ray Neutron

Sidewall Neutron Porosity Log	Sidewall Neutron Porosity Log	Sidewall Epithermal Neutron Log	Sidewall Neutron Log
Compensated Neutron Log	Compensated Neutron Log	Compensated Neutron Log	Dual Spaced Neutron
Thermal Neutron Decay Time Log		Neutron Lifetime Log	Thermal Multigate Decay
Dual Spacing TDT		Dual Detector Neutron	
Formation Density Log	Compensated Density Log	Compensated Densilog	Density Log
Litho-Density Log			
High Resolution Dipmeter	Four-Electrode Dipmeter	Diplog	Diplog
Formation Interval Tester		Formation Tester	Formation Test
Repeat Formation Tester	Selective Formation Tester	Formation Multi Tester	Multiset Tester
Sidewall Sampler	Sidewall Core Gun	Corgun	Sidewall Coring
Electromagnetic Propagation Log			
Borehole Geometry Tool	X-Y Caliper	Caliper Log	Caliper
Ultra Long Spacing Electric Log			
Natural Gamma Ray Spectrometry		Spectralog	Compensated Spectral Natural Gamma
General Spectroscopy Tool		Carbon/Oxygen Log	
Well Seismic Tool			
Fracture Identification Log	Fracture Detection Log		

LOG QUALITY CONTROL

General Log Quality Control

Service-company personnel expect an operator's representative to be available and will invite him into the logging unit during the logging operation. However, if possible, discuss the log with the logging engineer before and after the logging operation.

The most critical time during a logging operation is when the tool is within 1000 ft of the bottom of the well. Do not distract the logging engineer during this time. Even the best engineers may make mistakes when they are distracted. Give the engineer a chance to perform the operation with minimum interruption.

After each log is complete, discuss the logs with the engineer as thoroughly as possible. Ask for an explanation of any abnormal curve responses, equipment failure, or hole problems and *enter the information in the "remarks" column of the log heading*. This may be done after the first print is made but before any more prints are made. If there is any question about validity, request a rerun of the log before the crew rigs down. Generally, 200 ft of repeat in relatively smooth hole should be enough to verify the log. Everyone is reluctant to go back in the hole after rigging down; however, *once pipe is set, it will be impossible to get another resistivity survey of any type*.

The operator's representative at the well site has the final responsibility for obtaining competent logs. There are many ways to verify various aspects of the logs; this manual describes the principal methods. The effectiveness with which individuals use these techniques will vary, since experience and common sense are certainly important factors. No matter how competent and conscientious an observer may be, however, there are ways in which bad logs can defy detection at the wellsite. To this degree, log quality also depends on the competence and integrity of the logging company's engineer. An important objective is to develop relationships of mutual trust with the logging company personnel who perform the work.

A log consists of two kinds of data: the so-called fixed-heading information and varying measurements made with a downhole logging tool. The former includes tool calibration records as well as the familiar heading information and remarks. These calibrations are the only completely objective verifications of log quality available. Learn what they mean and how to use them. The depth-related log measurements will include one or more repeat sections, usually of about two hundred feet. They are a valuable, though not conclusive, indication of correct tool operation, and should be examined carefully on every log run.

Log Headings

It should be apparent that heading data, calibration data, and any remarks pertinent to log quality must be obtained at the wellsite and entered on each log. Some compromises are permissible for field prints; for example, the "additional services" block cannot be properly completed until the job is finished. However, even field-print headings should always be as complete as possible, including any remarks that pertain to log measurements or depths. Obviously much of this well information must come from the operator's representative, the drilling contractor, or other service-company people. Be sure this information is valid and accurate.

The following is an item-by-item discussion (based on the standard API log-heading format) of the fixed-heading portion of a log. Item numbers refer to the circled numbers on figures 1.2 and 1.3.

1. Well name and company: Check for correct spelling and designation of well name, well number, location, and company name.
2. Date of operation: The correct dates are particularly important where multiple runs are made on one well. Time between runs is sometimes important for evaluation of invasion effects.
3. Depth data: Driller's depths, casing point, and logger's depths are important for many reasons. In directionally drilled holes, include a true vertical depth in the remarks when possible.
4. Mud data: Mud type is very important to the response of most tools. Include the best available description and measured parameters of the mud. The mud resistivity should be measured in the mud cup rather than the filtrate cell, and correct procedures followed for measuring mud filtrate *and* mudcake. Changes in mud programs during the drilling operation should be noted on the log heading in the remarks column.
5. Bottomhole temperatures (BHT): Insist on a maximum-reading thermometer on every trip in the hole. Where conditions warrant, a group of three thermometers should be run; at least two will nearly always agree closely. Depending on the time after circulation, the temperature recorded on the first log might not be an indication of the true bottomhole temperature. This is why it is useful to run BHT thermometers on subsequent logging trips. If subsequent tools are to be run to the bottom of the well, without a cleanout trip, the same thermometers can be rerun. The logging company should be informed beforehand of these requirements.
6. Truck number and location: If questions arise, the location of the recording truck and the name of the engineer will be needed.
7. Time: Elapsed time for each log run is an important factor for many reasons.
8. Gauge: The bit size must be known for confident log evaluation. If bit size (of, e.g., the core bit) has been changed, make sure this is entered as a remark on the log heading.
9. Run number: The run number reflects the number of times this type of log has been recorded in this well.
10. Remarks column: Any information that can lead to a better interpretation of the log should be entered here: e.g., hole conditions; tool malfunctions; abnormal responses; and drilling problems, such as stuck drill pipe and fishing operations (depths and particulars). This is

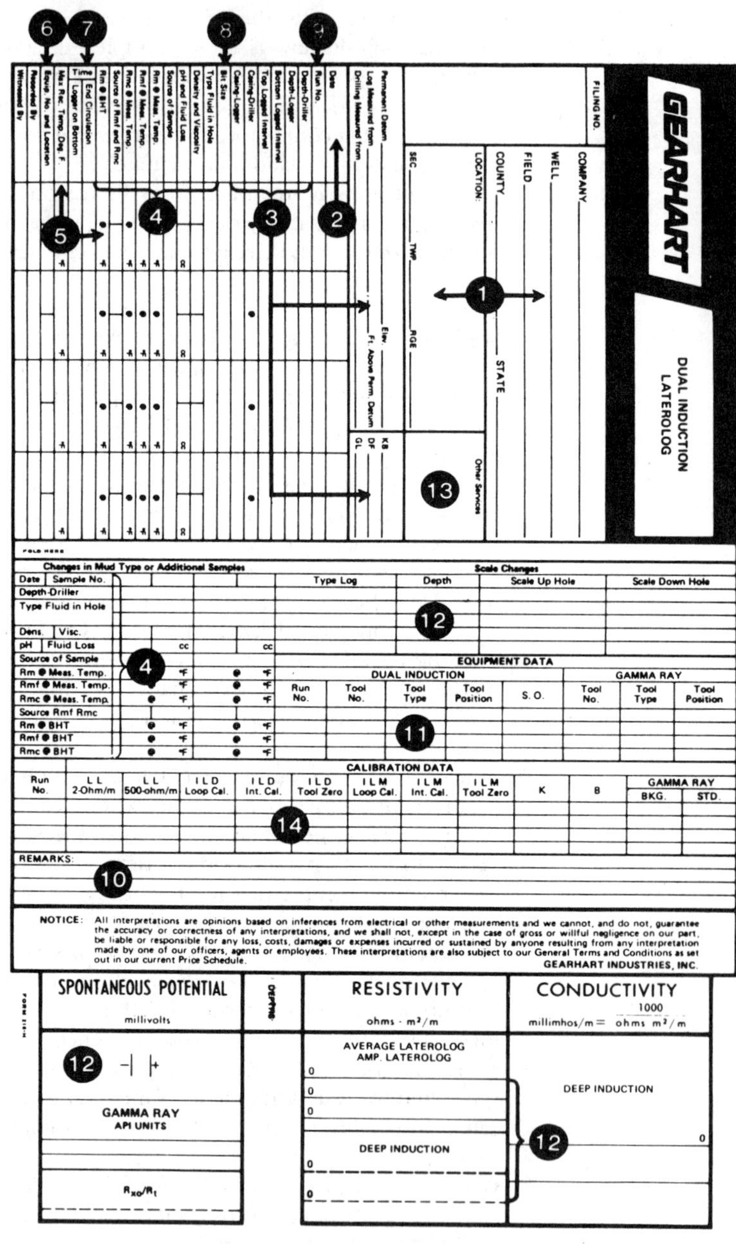

GEARHART

DUAL INDUCTION LATEROLOG

FILING NO.

COMPANY
WELL
FIELD
COUNTY STATE
LOCATION:
SEC TWP RGE
Other Services

Permanent Datum Elev.
Log Measured from Ft. Above Perm. Datum
Drilling Measured from
KB
DF
GL

Date
Run No.
Depth-Driller
Depth-Logger
Bottom Logged Interval
Top Logged Interval
Casing-Driller
Casing-Logger
Bit Size
Type Fluid in Hole
Density and Viscosity
pH and Fluid Loss
Source of Sample
Rm @ Meas. Temp.
Rmf @ Meas. Temp.
Rmc @ Meas. Temp.
Source of Rmf and Rmc
Rm @ BHT
Time: End Circulation
Time: Logger on Bottom
Max Rec. Temp. Deg. F
Equip. No. and Location
Recorded By
Witnessed By

FOLD HERE

Changes in Mud Type or Additional Samples			Scale Changes			
Date / Sample No.			Type Log	Depth	Scale Up Hole	Scale Down Hole
Depth-Driller						
Type Fluid in Hole						
Dens. / Visc.						
pH / Fluid Loss	cc	cc				
Source of Sample						
Rm @ Meas. Temp.	@ °F	@ °F				
Rmf @ Meas. Temp.	@ °F	@ °F				
Rmc @ Meas. Temp.	@ °F	@ °F				
Source Rmf Rmc						
Rm @ BHT	@ °F	@ °F				
Rmf @ BHT	@ °F	@ °F				
Rmc @ BHT	@ °F	@ °F				

EQUIPMENT DATA

DUAL INDUCTION					GAMMA RAY		
Run No.	Tool No.	Tool Type	Tool Position	S. O.	Tool No.	Tool Type	Tool Position

CALIBRATION DATA

Run No.	L L 2-Ohm/m	L L 500-ohm/m	I L D Loop Cal.	I L D Int. Cal.	I L D Tool Zero	I L M Loop Cal.	I L M Int. Cal.	I L M Tool Zero	K	B	GAMMA RAY BKG.	GAMMA RAY STD.

REMARKS:

NOTICE: All interpretations are opinions based on inferences from electrical or other measurements and we cannot, and do not, guarantee the accuracy or correctness of any interpretations, and we shall not, except in the case of gross or willful negligence on our part, be liable or responsible for any loss, costs, damages or expenses incurred or sustained by anyone resulting from any interpretation made by one of our officers, agents or employees. These interpretations are also subject to our General Terms and Conditions as set out in our current Price Schedule.
GEARHART INDUSTRIES, INC.

FORM 210-H

SPONTANEOUS POTENTIAL millivolts	DEPTHS	RESISTIVITY ohms · m²/m	CONDUCTIVITY millimhos/m = 1000 / ohms m²/m
− +		AVERAGE LATEROLOG AMP. LATEROLOG 0 0 0	DEEP INDUCTION 0
GAMMA RAY API UNITS		DEEP INDUCTION 0 0	
R_{xo}/R_t			

FIGURE 1.2 *Log Heading. Courtesy Gearhart Industries, Inc.*

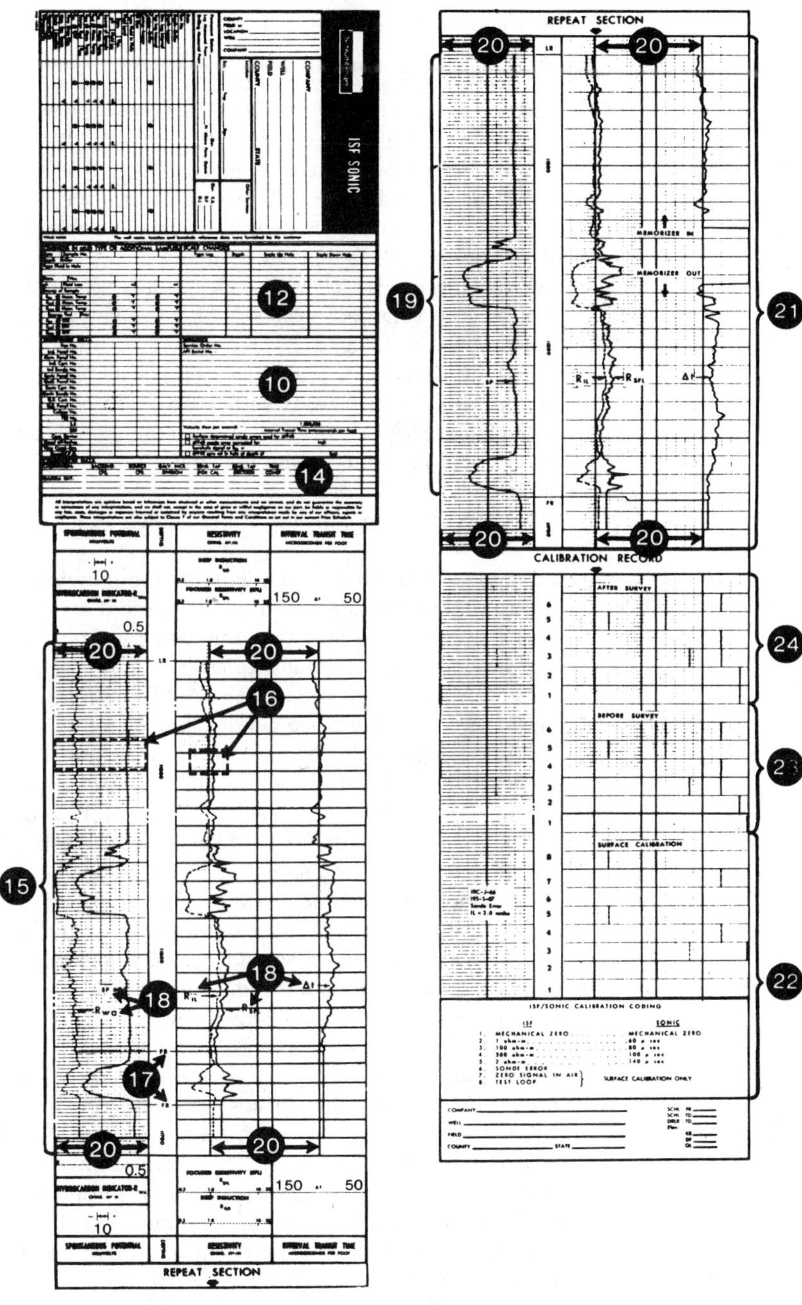

FIGURE 1.3 *Overall Format and Organization of the Log. Courtesy Schlumberger Well Services.*

the most overlooked, yet one of the most important, portions of the log heading.

11. Equipment type and serial number: Each major component of any logging tool is uniquely identified by number. For example, DIS-DB-149 means a *D*ual *I*nduction *S*onde, Model DB, Tool Number 149 (Schlumberger usage). It sometimes happens that subtle tool malfunctions will go undetected while several logs are run with the same equipment. When this happens, it is vital to be able to identify the logs run with this tool, so that any resulting errors can be rectified. Some logging companies use color codes to identify certain types of tools. This practice is not acceptable for log-heading tool identification.
12. Scale and scale changes: The log scale is *the* most important item on the log heading. If the scale is missing or incorrect, log evaluation is compromised. Correctness of the log scale must be monitored by observing the log and by checking the calibrations. Any scale changes should be noted *both* on the heading *and* at the point of change on the log.
13. Other services: Generally this is not known until the final operation is completed.
14. Calibration data: This is supplementary calibration data and should correspond to the calibration data presented on film. Although most service companies follow the same general procedures, the specifics may vary. Always ask for an explanation if there are any questions.

15–21. Curve presentations: After thoroughly checking the log heading, check for the proper placement of labeling of the curves. Log curves should be presented as continuous lines with no gaps or smears (15). Dashed lines, dotted lines, and solid lines used to represent different log curves should be legible, as should the depth lines (16). Reference points that show the depth at which each logging tool began recording should be clearly marked (17). Each logging curve should be clearly labeled (18). Time marks on the left side of Track 1 should be clearly visible (19). Each log curve should stay within its designated log track and the tracks should be easily differentiated from each other and should have the same number of divisions (same scale) from top to bottom of the logged interval (20). There should be at least 200 feet of repeat section (21).

22. Surface calibration: This is also sometimes referred to as *shop calibration*. This calibration film relates the signals produced by the wellsite calibration standard to some absolute standard. The response of any logging tool must be related to useful formation parameters by means of some such calibration standard. Primary calibration standards are usually too cumbersome and/or environmentally sensitive to be taken to the wellsite. Portable secondary calibration standards are built to bridge this gap. These, obviously, introduce the possibility of error. Surface calibration film allows you to verify that the wellsite calibration duplicates accurately the master shop calibrations. Individual logging tools can vary in their responses to identical calibration standards. Such variations (such as induction-tool sonde errors) are cancelled during calibration; but, obviously, the surface calibration

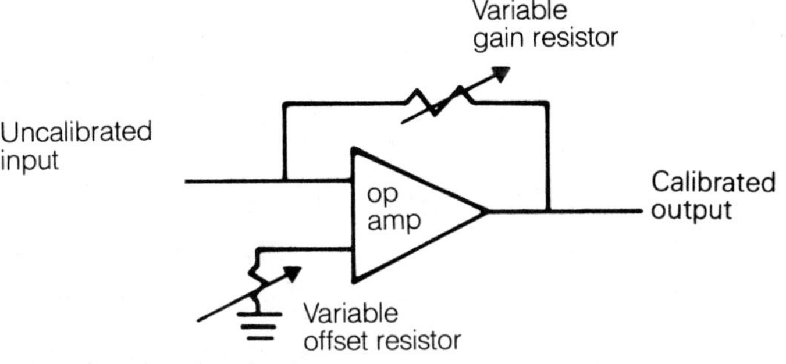

FIGURE 1.4 *Analog Calibration System. Courtesy Schlumberger Well Services.*

record must refer to the same equipment used to run the log. Tool serial number must therefore appear on both the log heading and the calibration—and match.

23. Before survey calibration: This calibration film is recorded, either on the surface or in the borehole, before the log is recorded. In addition to the values from the shop calibration, values for measurements that do not require shop calibrations are presented. These include caliper, sonic, dipmeter, etc. Be sure the values recorded are accurate and correspond to the values on the calibration tail.
24. After survey calibration: This calibration is recorded after the log is run. Sometimes drift in the values is acceptable. Tolerances are given within the sections discussing individual tools. It is vital to be sure that the after survey calibration is made *after* the log is recorded and not before. There is a propensity among logging engineers to make this record before running the log, in the interest of saving rig time. The easiest way to insure that the calibrations are recorded in the proper sequence is to look at the uncut film while it is drying or before it is cut and spliced.

Calibration Systems

The digital computer brought a revolution into logging techniques in the midseventies. Among its innovations was a completely new concept about how to record and scale the data. The older analog systems typically took the downhole signal, shifted its zero point and set its response slope; did certain other conditioning such as reciprocating, converting linear to logarithmic or vice versa, etc.; and then recorded this much-massaged signal (fig. 1.4). The computer-operated system, on the other hand, nearly always records the raw data on tape, just as it comes from the tool. Mathematical processing is then used to condition the signal according to calibration points, scales, etc. This conditioned signal is what goes on the film, and on a final tape record (fig. 1.5).

The later method has the major advantage of preserving the raw tool measurement—calibration or other conditioning can be done

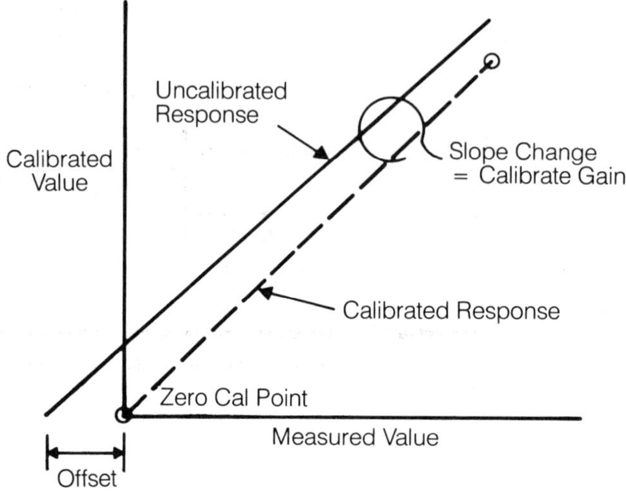

FIGURE 1.5 *Computer-Unit Calibration System. Courtesy Schlumberger Well Services.*

after the log is run, or at any other time. It also permits the use of precise digital values for calibration, thus ending dependency on the logging engineer's judgment in setting an optical galvanometer. Computer-unit calibrations are quite different from the earlier type. Since the industry standard is moving toward computer-operated logging units, this type of calibration is described in detail.

COMPUTER-UNIT CALIBRATION. The standard log format of an induction log for one logging company includes a calibration summary like the one in figure 1.6. Some of those accustomed to the earlier film calibration record have wondered why that same format could not be created by the computer units. The answer is—it could, easily; but it would be a little like the early automobile with a fake horse mounted in front. The figures listed on the calibration summary are the actual digital values used to calculate a calibrated value at each six-inch level of the log. Making an optical calibration record from these figures would entail a loss of accuracy, thus nullifying one of the significant advantages of the computer logging unit.

SHOP SUMMARY. The calibration record begins with a *shop summary* (fig. 1.6), though this does not mean that the shop calibration has to be run first.

Before any calibration is done, the computer assembles the various tool readings it needs. Remember that these are unaltered, raw data signals from the tool. Each downhole tool will produce somewhat

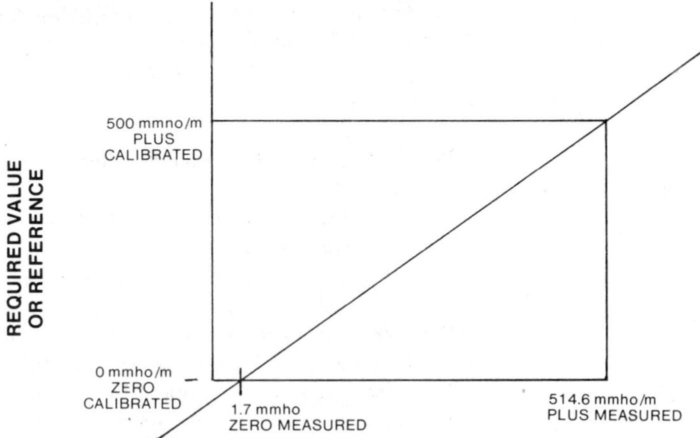

SHOP SUMMARY

PERFORMED: 78/08/07
PROGRAM FILE: SHOP (VERSION 10.2 78/ 6/27)

DITD ELECTRONICS CALIBRATION SUMMARY

	ELECTRONICS CALIBRATION				SONDE		
	MEASURED		CALIBRATED		ERROR	TEST	
	ZERO	PLUS	ZERO	PLUS	CORR.	LOOP	UNITS
ILD	1.7	514.6	0.0	499.9	5.33	511.2	MM/M
ILM	3.4	497.4	0.0	499.9	5.06	498.7	MM/M

(IS:119 , IC:141)

FIGURE 1.6 *Computer-Unit Surface Calibration Summary. Courtesy Schlumberger Well Services.*

different signals, so the individual tool serial numbers are a necessary part of the record. The signals recorded are:

zero measured	—sonde disconnected electrically
plus measured	—signal produced with the internal calibrate resistor in the circuit
sonde error corr.	—signal with the sonde suspended in a zero-conductivity area
test loop	—signal with the test loop in place; no other signal source

The *zero-calibrated* and *plus-calibrated* figures are the desired calibration values. The 499.9 value, instead of 500, is a software feature that does not affect the validity of the calibration. Once these values are in the computer's memory, the program logic constructs an algorithm that will be applied to every measurement made by the tool. The form of this algorithm is

$$y = mx + b,$$

where y = desired (calibrated) value,
x = tool signal,
m = calculated slope (= 500/(plus measured − zero measured), and
b = x intercept of the m slope line.

AFTER SURVEY TOOL-CHECK SUMMARY. Here we have a comparison of the zero-calibrated and plus-calibrated values before and after the log (fig. 1.7). These values are computed by use of the calibration algorithm with the tool in the sonde-disconnected and calibrate-resistor-on modes.

The difference, if any, between before and after values is normally caused by electronics drift or sonde distortions under pressure and temperature. Logging companies should specify the tolerances they consider allowable. Be familiar with these tolerances, and aware of the effect they could have on log calibrations. Any out-of-tolerance reading is an indication of a defective tool, a leaking cable, or some other malfunction. If the difference is not too great, then accept the log. The logging engineer may be able to reprocess the log using a linear interpolation between the before and after values as a basis for the logging algorithm.

BEFORE SURVEY CALIBRATION SUMMARY. Here the wellsite calibration values are presented for comparison with those of the shop summary (fig. 1.7). The zero-measured and plus-measured values should agree closely (though seldom perfectly) with those on the shop summary, and the sonde errors should be identical. An exception to this could be when the logging engineer sets his sonde errors in the borehole. This is sometimes a perfectly valid procedure, but it should not be used without the operator's approval; and the fact should be entered as a remark on the log heading.

The gamma-ray (GR) calibration shown in figure 1.7 is presented whenever the gamma ray is run as part of the combination tool. Its calibration will be described in a later chapter.

Note: As mentioned earlier, any of the computer-unit logs can be rerun later, using the raw data recorded on tape at the well. This is a powerful procedure through which logs can be corrected for miscalibration, depth discrepancies, inappropriate scaling, etc. However, unless precautions are taken, serious confusion can arise. The customer tape received at the well will contain, in addition to the raw log data, calibration factors, scales, and other information that corresponds to the graphic log. If different information is used when a second (rerun) film is made, it is possible to have a customer tape and

```
                AFTER SURVEY TOOL CHECK SUMMARY

  PERFORMED:      78/09/18
  PROGRAM FILE:   NUC      (VERSION    10.1       78/ 5/12)

DITD                    TOOL  CHECK

                   ZERO                       PLUS
          BEFORE         AFTER          BEFORE       AFTER      UNITS
   ILD      0.0            0.0          499.9        494.4      MM/M
   ILM      0.0           -0.0          499.9        494.8      MM/M
   SFL      0.0            0.1          499.9        496.7      MM/M

ILM  SONDE ERROR CORRECTION :           5.0    MM/M
ILD  SONDE ERROR CORRECTION :           8.5    MM/M


                BEFORE SURVEY CALIBRATION SUMMARY

  PERFORMED:      78/09/18
  PROGRAM FILE:   NUC      (VERSION    10.1       78/ 5/12)

DITD              ELECTRONICS CALIBRATION SUMMARY

             MEASURED                         CALIBRATED
          ZERO     PLUS                   ZERO        PLUS          UNITS
   ILD     2.7     514.3                   0.0        499.9         MM/M
   ILM     4.4     496.3                   0.0        499.9         MM/M
   SFL     0.1     539.0                   0.0        499.9         MM/M

ILM  SONDE ERROR CORRECTION :           4.5    MM/M
ILD  SONDE ERROR CORRECTION :           4.5    MM/M

SGTE               DETECTOR CALIBRATION SUMMARY

             MEASURED
          BKGD     JIG          CALIBRATED            UNITS
   GR      19      200              164               GAPI
```

Log	Curve	Tolerance Zero	Tolerance Plus	Units
IES	ILd	±2	±20	mm/m
	SFL	±2	±20	mm/m
Sm. Hole IES	ILd	±2	±50	mm/m
	16″ N	±0.08	±1.05	Ω.M
DIL	ILm	±2	±20	mm/m
	ILd	±2	±20	mm/m
	LL8	±2	±20	mm/m
DISF	ILm	±2	±20	mm/m
	ILd	±2	±20	mm/m
	SFL	±2	±20	mm/m

FIGURE 1.7 *Example of Field Calibration Films and Tolerance Limits. Courtesy Schlumberger Well Services.*

a graphic log that are not compatible. Be *sure* that the tape and the logs you receive reflect exactly the same data.

Acceptance Standards

A large percentage of all logs probably contains erroneous data of some nature, but all such logs are not necessarily worthless. Often a log can be corrected visually or mathematically. Sometimes the log must be rerun before valid conclusions can be made; however, the cost of rerunning the log might outweigh the importance of the error. When the log is not rerun, the error should be noted on the heading in the remarks column and also noted on the log opposite any zone of interest.

For the more serious errors, it is a mistake to think a bad log is better than none; a bad log may influence important decisions. Consequently, it is imperative that a bad log be rerun. The problem, of course, is to determine whether to accept or reject a questionable log. One reliable way to determine if a log should be rejected is to ask the question "Is the interpretation accurate?" When in doubt, rerun the log. Also ask "Can everyone who will use this log see the error and/or be able to make an accurate interpretation?" When in doubt, rerun the log. Remember that 95% of those who use the log are not log experts.

A definite criterion for acceptance or rejection of a log is difficult to establish; each situation will be somewhat different. Good judgment should outweigh written instructions when deciding whether to accept or rerun a log. The following guidelines should assist in making such decisions.

TECHNICAL QUALITY. The technical quality of the data may be affected by many things. The thought that occurs first is the question of equipment malfunction. Other possible causes of poor data are rugose borehole; sticking tools; logging engineer's errors; tool rotation; excess logging speed; deviated wells; poor centralization or eccentralization; and formation alteration.

Often an anomaly will suggest the possibility of a malfunction. This possibility should be clarified by repeating the log in that section. After all, the anomaly may be significant. It is interesting to recall that the SP was originally an anomaly that interfered with measurement of formation resistivity.

REPEATABILITY. Properly functioning resistivity tools, run under conditions that are within their capability, will nearly always repeat very well. As a functional check of the equipment, a repeat section of 200 ft or more is routinely run, and should be required except in unusual circumstances.

Sonic tools should also repeat well.

Microresistivity tools may follow a different borehole path on the repeat pass, which can produce considerable differences in the log. However, a third repeat run should coincide with one pass or the other.

Radioactivity tools are subject to statistical fluctuations that prevent perfect repeats. Some of these tools (density, sidewall neutron) are also subject to the problem of following different borehole paths. However, in general radioactivity logs should repeat well.

Aside from equipment failures, factors that could cause poor repeats include:

washed out holes, particularly those of extremely noncircular cross section

variable tool centering, particularly in large holes with fairly high mud conductivities
sensors following different paths in borehole (for pad-type tools)
presence of metallic fish in the borehole
comparing an up run with a down run (may be quite different with some types of equipment)

The repeatability of a log run may be affected by time-related phenomena. This is primarily apparent in invasion profiles. Invading filtrate can penetrate deeper, migrate vertically, accumulate as "annuli," or dissipate altogether with the passage of time. The log response, particularly of the shallower-reading devices, may continue to change for many days after the well is first logged. Though unusual, such changes can be very troublesome; but from the viewpoint of log quality they are usually recognizable. The changes occur only in the invaded sections, not in the shales or other impervious rocks.

OFFSET LOGS. If the well is in a developing field, or a consistent geological block, available offset logs are likely to be useful. This is especially true in an unfamiliar area.

ABSOLUTE LOG VALUES. Comparison of log readings with known absolute values is seldom possible, but when it can be done, this positive crosscheck should by all means be used. The necessary beds consist of pure, zero-porosity minerals such as halite, anhydrite, or limestone. Table 1.3 lists the absolute values for several of the more common tools.

Casing can sometimes be used as a check. All caliper tools should read the same in casing. The diameter indicated is usually slightly greater than that of new casing due to drillpipe wear. The two diame-

TABLE 1.3 *Log Values of Zero Porosity Formations*

Measurement		Units[a]	Halite	Anhydrite	Limestone
Resistivity		$\Omega \cdot m$	Infinite	Infinite	Infinite
Sonic		μs/ft	67.0	50.0	47.6
Bulk Density		g/cc	2.032	2.977	2.71
CNL[b]	Neutron tools	% porosity	−1.0	−0.5	0.0
SNP[b]		% porosity	+4.0	−0.5	0.0
CNS[c]		% porosity	+5.0	−0.2	0.0
SNL[d]		% porosity	0.0	0.0	0.0
CNL[d]		% porosity	0.0	0.0	0.0
SWN[d]		% porosity	1.0	0.0	0.0

[a]% porosity in limestone matrix units.
[b]Schlumberger.
[c]Gearhart.
[d]Dresser.

ters measured by a four-arm caliper should be equal. The sonic should read about 56 μs/ft in unbonded casing.

Depth Measurements

Measurement of depth is perhaps the logging company's most basic function, but one that tends to get lost among the more glamorous parameters. Absolute depth control is provided either by a calibrated sheave or by magnetic marks placed on the logging cable every 100 ft. In either case, the operational procedure for obtaining accurate depth control is rather rigorous and if followed properly will almost always result in accurate depth measurements. This is one of the places where it is advisable to be on terms of mutual trust with the logging engineer. It may be possible to detect evidence of inaccurate depth measurements, but absolute verification is very difficult. Compare logger's TD (total depth) and casing depth with depths reported by the driller; watch for excessive tie-in corrections with previous log runs, and check the apparent depths of known markers.

Relative depth control means insuring that all measurements are on depth with each other. All curves that are recorded on the same trip in the hole should be on depth with each other within ±6 in. In addition, each subsequent log should match the base log within 2 ft in straight holes and 4 ft in wells that are highly deviated (greater than 30°).

Depth control is particularly important for sidewall sampling. Sidewall samples shot off-depth are of no use. The depth tie-in should be recorded on film, and the samples checked against the log as they are pressed from the core bullets. *If there is any doubt as to where samples were shot, they should be resampled.*

Logging Speeds

As previously indicated the logging speed in feet per minute is indicated by gaps or ticks along the edge of the film track (fig. 1.3, item 19). Acceptable logging speeds depend on the type of log, the type of unit (computer or conventional), the intended use of the data, and type of formation being logged. Normal routine logging speeds are listed in table 1.4.

TABLE 1.4 *Recommended Logging Speeds*

Tool	Ft/min	Ft/hr	Remarks
Resistivity log	100	6000	
Resistivity + GR log	60	3600	GR for correlation
Radiation log	30	1800	
Sonic log	60	3600	Slower if the sonic is noisy
Dipmeter	60	3600	
Microresistivity	40	2400	

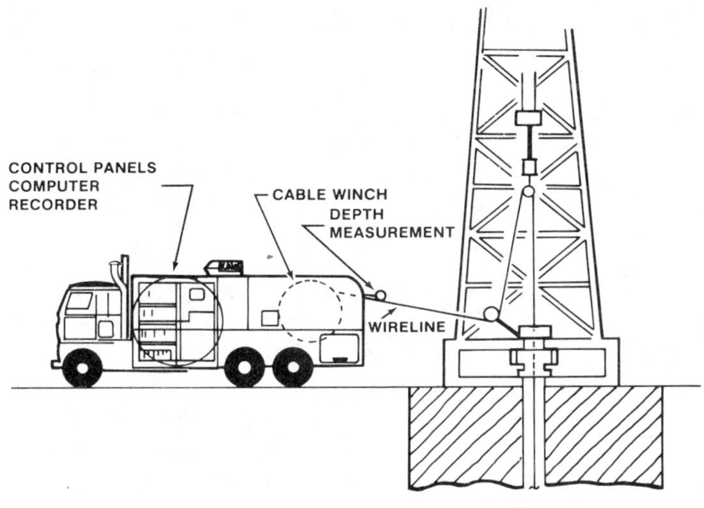

FIGURE 1.8 *Setup for a Logging Job. Courtesy Gearhart Industries, Inc.*

With the computer logging unit, speeds can be doubled under certain circumstances. If the well is not an exploratory or secondary/ tertiary recovery well, the speeds can generally be doubled—with the exception of the neutron and density logs.

THE LOGGING ENVIRONMENT

Rigging Up to Run a Log

Figure 1.8 shows a typical setup for a logging job. A logging truck is anchored 100 to 200 ft from the well. Two sheaves are mounted in the derrick with one suspended from the crown block and the other chained down near the rotary table. The logging cable from the truck winch is then passed around the sheaves and attached to the logging-tool string; the logging tool is then lowered into the hole.

Two mechanical details of this method of rigging up are worth noting. Between the top sheave and the elevators there is a tension device that measures the logging-cable strain and displays it in the logging truck (fig. 1.9). The tension on the elevators is twice the tension on the cable. The elevators should be securely latched and the traveling block braked and chained.

The tiedown chain for the lower sheave (fig. 1.9) is also of great importance. When it breaks or comes untied, the cable quite probably will break and the sheave can be catapulted several hundred feet away. The tiedown chain should go directly from the sheave to the anchor point and not be arranged in a V.

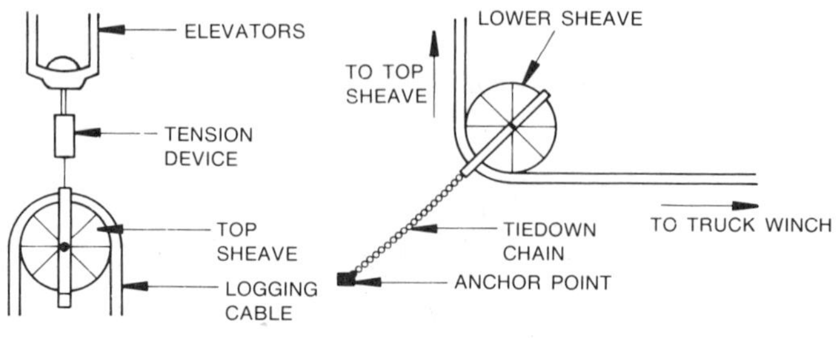

FIGURE 1.9 *Top and Bottom Sheave Wheel Arrangements.*

Logging Trucks

Logging-service companies offer a variety of logging units. Each type of unit has the following components:

logging cable
winch to raise and lower the cable in the well
self-contained 120-V AC generator
set of surface control panels
set of downhole tools (sondes and cartridges)
recording mechanism (tape and/or film)

Figure 1.10 shows a cutaway of a typical logging truck. Land units are mounted on a specially adapted truck chassis reinforced to bear the load of a full winch of cable (up to 30,000 ft long). The instrument and recorder cabs are usually cramped, noisy, too hot or too cold, and filled with ammonia fumes from an ozalid copier. Although not designed to do so, these conditions keep the merely curious onlooker away. Do not be deterred. If you have good reason to be in the logging unit, stay there and do what you must.

Offshore units, powered by diesel engines, are mounted on skids and bolted (or welded) to the deck of the drilling barge, vessel, or platform. Some logging units can be disassembled into many small components and flown into remote jungles suspended under helicopters. Nevertheless, all logging units are basically the same.

Good mechanical maintenance is required to avoid problems during logging operations. An engine that stops during logging operations will leave the logging tool dangling in the well. If it stays there too long without moving, it may become stuck. To prevent this, the traveling block from which the top sheave is suspended can be raised and lowered so that the tool is kept in motion. Another item that can cause grief is the 120-V generator. If it fails, all logging panels and tools go dead. Rig power may then be substituted.

Logging Cables

Modern logging cables are of two types: monoconductor and multiconductor. Monoconductor cables are used for completion ser-

FIGURE 1.10 *Cutaway View of a Logging Truck. Courtesy Schlumberger Well Services.*

vices—such as shooting perforating guns, setting wireline packers and plugs—and for production logging surveys—such as flowmeter and temperature logs—in producing wells. Multiconductor cables are used by most logging-service companies for recording openhole surveys.

Monoconductor cables are (approximately) ½ or ¼ in. in diameter—the smaller cable is used for jobs where high wellhead pressure is encountered. Multiconductor cables contain six or seven individual insulated conductors in the core. The outer sheath is composed of two counterwound layers of steel wire. Such a cable has a breaking strength of 14,000 to 18,000 lb and weighs 300 to 400 lb/1000 ft. It is quite elastic and has a stretch coefficient of around 10^{-6} ft/ft/lb.

The Head and the Weakpoint

The cable ends at the logging *head*. The head anchors the cable and attaches to the logging tool by means of a threaded ring. Thus, the head provides the electrical connection between the individual cable conductors and the various pins in the top of the tool. It also provides the mechanical connection. Built into the head is a *weakpoint*, which is a short length of aircraft cable designed to break at some given tension. The standard breaking point is 6000 lb, but deephole weak-

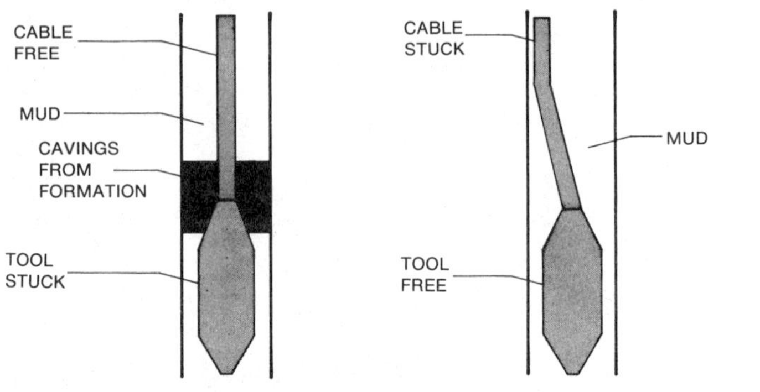

FIGURE 1.11 *Two Ways of Getting Stuck.*

points break at 3500 lb. The weakpoint is necessary to provide a means to free the cable from the tool in cases where the tool has become irrevocably stuck in the wellbore. To fully understand the function of the weakpoint, it is necessary to know the ways the tool can become stuck.

Getting Stuck

There are two ways of getting stuck. Either the tool will stick and the cable in the hole above the tool will remain free, or the tool will remain free and the cable will become *key seated* further up the hole above the tool. Figure 1.11 illustrates the difference. Once the equipment is stuck, the first thing to do is to determine whether it is the tool or the cable that is stuck. The standard procedure is to put normal logging tension on the cable and let it sit for a few minutes while the following data are gathered:

a. the present depth of the tool
b. the surface tension that was on the cable just before getting stuck
c. the cable type, size, etc.
d. the cable-head weakpoint rating

Once this data has been gathered, make certain that the tiedown chain on the lower sheave is secure, then mark the cable (using chalk or friction tape) at the rotary table. Securely position a T-bar clamp around the cable just above the rotary table. (If, in the subsequent tug of war, the cable breaks at the top sheave, this clamp will hold the cable at the surface and prevent it from snaking down the hole on top of the tool.) Now apply 1000 lb over the tension already on the cable and measure the distance the cable mark has moved. This will be the stretch produced in the elastic cable due to 1000 lb extra tension. Knowing this distance, the length of free cable can be estimated from a stretch chart or from knowledge of the stretch coefficient. If the length of free cable so determined proves to be the present logging

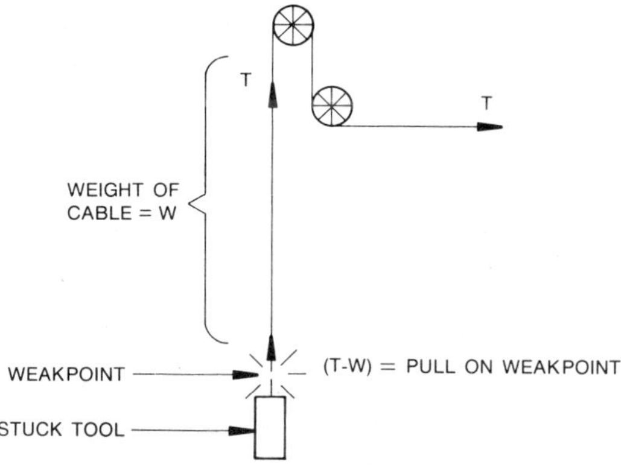

FIGURE 1.12 *Breaking the Cable at the Weakpoint.*

depth, then the tool is evidently stuck and the cable is free. On the other hand, if the calculated length of free cable is less than the present logging depth, the cable itself must be stuck higher up the hole. If the cable proves to be stuck, it is counterproductive to apply any further tension since this will merely compound the problem by aggravating the differential pressure sticking of the cable.

If it is the tool that is stuck, pulling on the cable will have one of three results. The tool will pop free, the weakpoint will break (leaving the tool in the hole but saving the cable), or the cable will break at the point of maximum tension at the top sheave. Of the three, the first is to be preferred. Of the other two, the breaking of the weakpoint is preferred. But which will occur first? Will the cable part at the surface before the weakpoint breaks? Figure 1.12 will help to explain the tensions involved.

If the cable weakpoint is 6000 lb, and 10,000 ft of cable weighing 3500 lb is in the hole, it will be necessary to apply 9500 lb (6000 + 3500) at the surface in order to apply 6000 lb at the weakpoint. At greater depths, the surface tension required to break the weakpoint will be even higher and may exceed the breaking strength (14,000 to 18,000 lb) of the cable itself. Thus, for deep holes a weakpoint of only 3500 lbs is preferred.

Differential pressure sticking of the cable is caused when the cable cuts through the mudcake. One side of the cable is exposed to formation pressure while the other side is exposed to the hydrostatic mud column. When hydrostatic pressure is significantly higher than formation pressure, the cable is forced against the formation and the resulting friction stops any further cable movement (fig. 1.13). This should be kept in mind when designing both mud and logging programs.

Fishing Alternatives

There are several alternatives to recovering the stuck tool and/or cable:

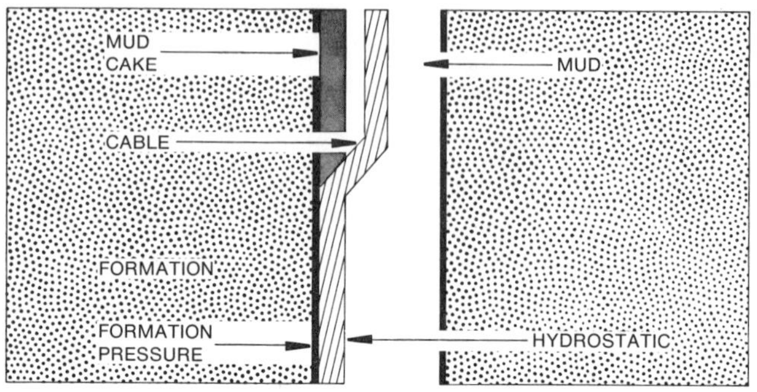

FIGURE 1.13 *Differential Pressure Sticking of the Cable.*

1. Leave the cable attached to the tool and run a side-door overshot.
2. Use the cut-and-thread technique.
3. Break the weakpoint; recover the cable; and fish for the logging tool with the drillpipe or push it to the bottom of the hole and mill it up.

Figure 1.14 illustrates these three methods. The side-door overshot is not recommended at depths greater than 3000 ft. The cut-and-thread technique is the surest way to recover a stuck logging tool.

Logging Tools

Logging tools are cylindrical tubes containing sensors and associated electronics. They are attached to the logging cable at the logging head. Although there are large variations in size and shape, a typical logging tool is 3⅝ inches in diameter and 10 to 30 ft long. They are built to withstand pressures up to 20,000 psi and temperatures of 300 to 400°F. The internal sensors and electronics are ruggedly built to withstand

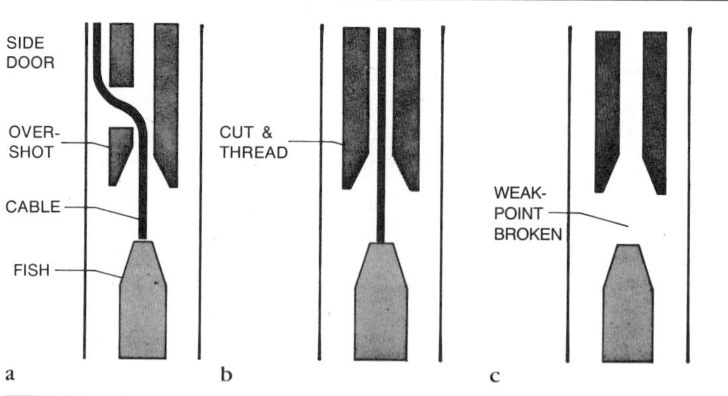

FIGURE 1.14 *Fishing Alternatives: (a) Side-Door Overshot, (b) Cut-and-Thread, and (c) Fishing with Drillpipe.*

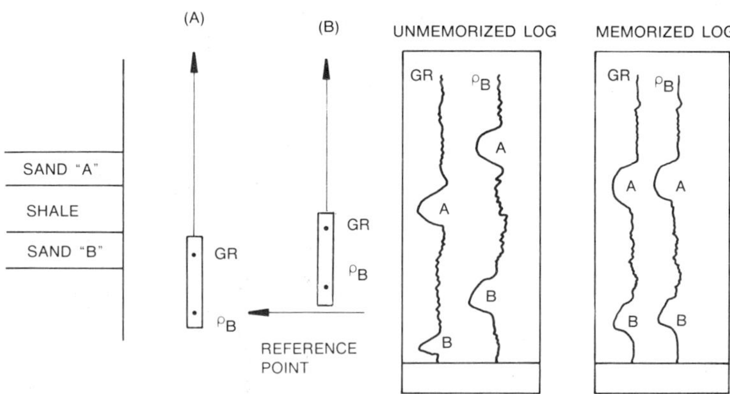

FIGURE 1.15 *Memorizing Logs.*

physical abuse. Modern tools are modularized to allow combination tool strings. By appropriate mixing and matching, various logging sensors can be connected together. The method has obvious limitations; for example, very long tools are difficult to handle and the capacity of the conductors in the cable is limited in their information-transmitting power.

Because the multiple sensors of a logging tool are located at different points along the axis of the tool, their respective measurements have to be memorized and placed at a common depth reference on the log. Thus, the signal from the sensor highest on the tool must be "remembered" until the signal from the lowest sensor arrives. See figure 1.15, where the reference point for the survey is the density skid recording the bulk-density measurement. Higher up the tool, the gamma ray sensor is recording a gamma ray curve. Without memorization, the gamma ray curve will be recorded off-depth and appear on the log to be deeper than the bulk-density curve by a distance equal to the spacing of the two sensors. It is important to assure that all curves recorded simultaneously are on-depth on the log.

Another depth problem arises when several surveys are recorded on different trips into the hole. Unless care is taken, these logs may not be on-depth with each other. The only method of assuring good depth control is to insist on a repeat section that passes a good marker bed. Each subsequent log should be placed on-depth using this repeat section before the main logging run is made.

The Borehole Environment

As the drill bit penetrates a permeable formation, an invasion process begins. Since the pressure in the mud column exceeds formation pressure, fluid from the drilling mud will move into the formation (provided it is porous and permeable) and deposit a mudcake on the

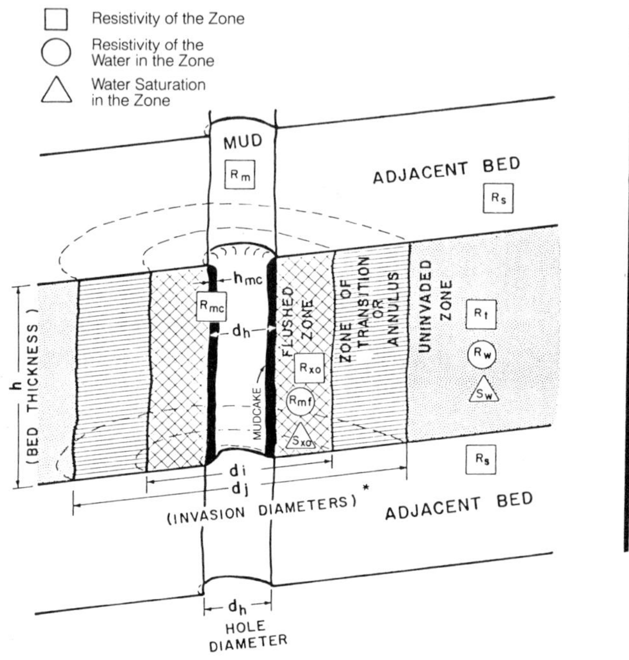

FIGURE 1.16 *Symbols Used in Log Interpretation. Courtesy Schlumberger Well Services.*

TABLE 1.5 *Logging Terms for Borehole Environment*

Name of Zone	Dimension	Fluid Content	Fluid/Water Resistivity	Rock Resistivity	Saturation
Bed	h	—	—	—	—
Adjacent bed	—	—	—	R_s	—
Mud	d_h	mud	R_m	—	—
Mudcake	h_{mc}	—	—	R_{mc}	—
Flushed or invaded	d_i	mud filtrate + residual oil	R_{mf}	R_{xo}	S_{xo}
Transition	d_j	mixed + hydrocarbons	—	—	—
Uninvaded (undisturbed)		connate water + hydrocarbons	R_w	R_t	S_w

borehole wall. Figure 1.16 illustrates the process and the names used in logging literature for the various zones that surround the borehole.

The terms used in Table 1.5 should be well known to everyone involved in the evaluation of well logs. It is important to distinguish between the resistivity of the fluid within the pore space and the resistivity of the rock itself. The flushed zone is important because it affects the readings of some logging tools and because it forms a "reservoir" of mud filtrate that will be recovered on a drillstem test before formation fluids are recovered.

CHAPTER TWO

RESISTIVITY LOGGING

GENERAL RESISTIVITY

Resistivity is the electrical resistance of a specific amount of material. In well logging, resistivity is the specific resistance of a volume of rock one meter square and one meter long. Resistivity can be measured from voltage or current variations (electrical devices) or from induced current variations (induction devices). This chapter includes devices for measuring both resistivity and conductivity—exemplified by laterolog devices and induction devices, respectively. In either case, the measurement is normally presented in resistivity units. The tools in common use generally make two or three simultaneous measurements, which are presented on the same resisitivity grid. In addition, an SP, gamma ray, and/or caliper measurement may be recorded. Resistivity tools and the measurements they record are summarized in table 2.1. With rare exceptions, the only conductive material in subsurface formations is water containing dissolved salts. All resistivity logs are therefore measurements of the water-filled and electrically connected pore space within the tool's effective range. Exceptions to this rule are rare; but mention should be made of galena and pyrite, both metallic ores that conduct electricity in the "dry" state.

Old Electric Logs

Basically all electrical resistivity measuring devices are similar and only the designs of the electrode systems are different. Several electrode configurations are available such that an array of resistivity devices can provide short normal, long normal, long lateral, laterolog, microlateral, micronormal, microlaterolog, proximity, and dipmeter curves.

The electrical log (ES) consists of an SP, a short normal, a long normal (or limestone lateral), and a long lateral curve. The simplicity of the sonde permits numerous electrode spacings. The ES log has been replaced with the induction log; therefore, the theory of electrode spacings will not be discussed fully here. However, enough material will be included to provide a guide for certain situations where an ES is run for a specific requirement (e.g., an ES drillstem sonde run through drillpipe, ES surveys in core holes, salt–mud correlation runs, and partnership arrangements).

ELECTRODE ARRANGEMENTS. Most ES sondes are four-electrode devices with the electrodes arranged as illustrated in figure 2.1. Electrode spacings are measured from the current electrode to the primary measuring electrode for normal devices (AM) and from the current

TABLE 2.1 *Tools for Deep Resistivity Measurements*

Tool Name	Service Company Abbreviation	Curves Recorded	Service Company[a]	Remarks
Dual Induction Spherically Focused Log	DISF	R_{ILd},R_{ILm},R_{SFL},SP	S	
Dual Laterolog	DLL	R_{LLd},R_{LLs},SP	S	GR generally preferred
Induction Spherically Focused Log	ISF	R_{ILd},R_{SFL},C_{ILd},SP	S	Being replaced by the DISF
Induction Electric Survey	IES	R_{ILd},$R_{16in.}$,C_{ILd},SP	S	Specialty tool for small holes and high temperature
Ultra Long Spacing Electric Log	ULSEL	R_{75ft},R_{150ft},R_{600ft},R_{1000ft}	S	Salt-dome profiling
Drillstem Electric Log	ES	$R_{16in.}$,R_{64ft}, Normal, $R_{18ft18in.}$ Lateral, SP	S	Pump down through drill pipe
Dual Induction Laterolog	DIL	R_{ILd},R_{ILm},R_{LL3},SP	GI	
Dual Laterolog	DLL	R_{LLd},R_{LLs},SP	GI	GR generally preferred
Induction Electrical Log	IEL	R_{ILd},$R_{16in.}$,SP	GI	
Laterolog	LL3	R_{LL3},GR	GI	
Dual Induction Focused Log	DIFL	R_{ILd},R_{ILm},R_{FL},SP	DA	
Dual Laterolog	DLL	R_{LLd},R_{LLs},SP	DA	GR generally preferred
Induction Electrolog	IEL	R_{ILd},$R_{16in.}$,SP	DA	
Dual Induction Log		R_{ILd},R_{ILm},SP	W	
Dual Guard Log		SHALLOW GUARD		
		DEEP GUARD	W	
Induction Electrical Log	IEL	R_{ILd},$R_{18in.}$,C_{ILd},SP	W	
Guard Log		DEEP GUARD	W	

[a]S, Schlumberger; GI, Gearhart Industries; DA, Dresser Atlas; W, Welex

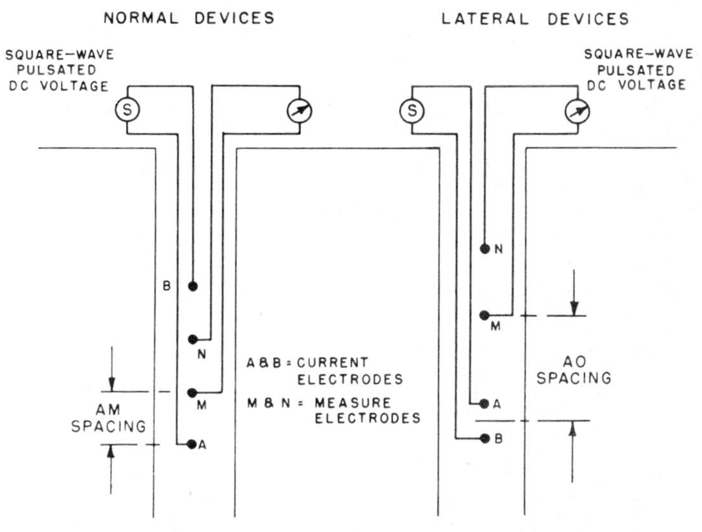

FIGURE 2.1 *Four Electrode Arrangements for Electric Logs.*

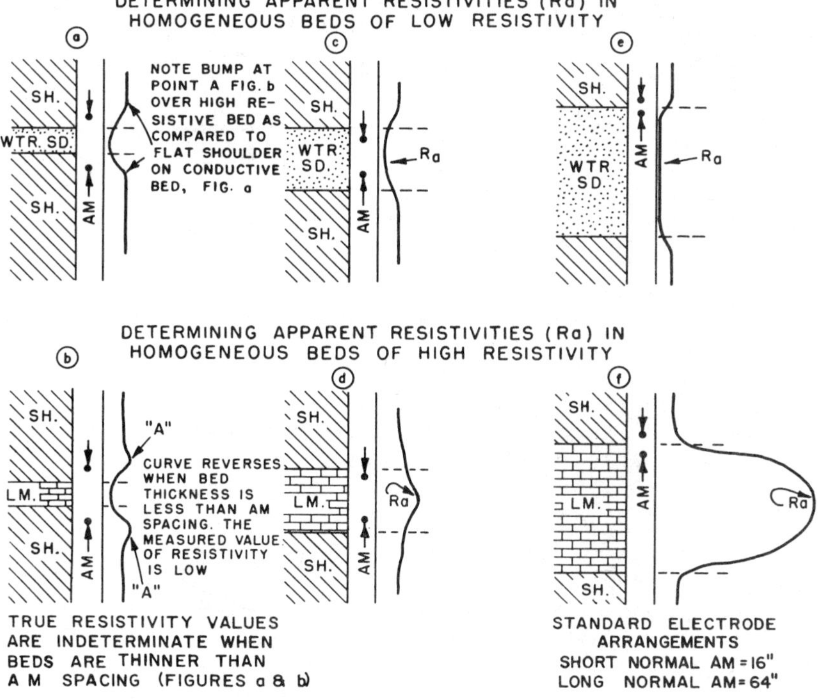

FIGURE 2.2 *Bed Thickness Effects for Normal Devices.*

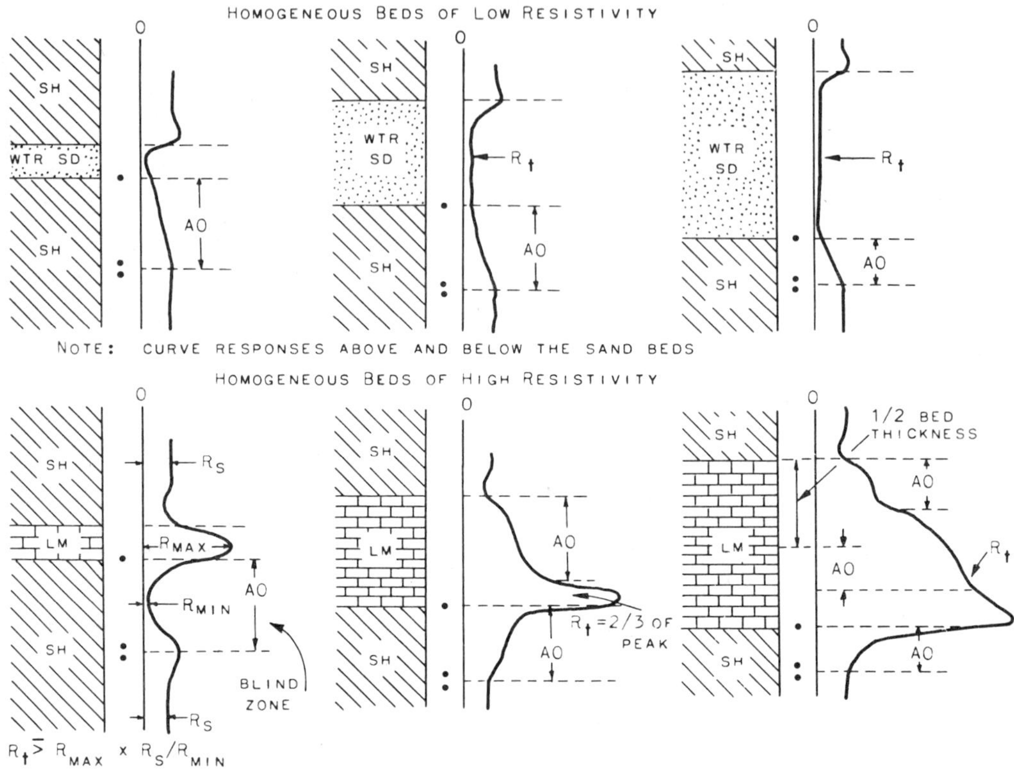

FIGURE 2.3 *Bed Thickness Effects for Lateral Devices.*

electrode to a point midway between the measuring electrodes for lateral devices (AO).

CURVE SHAPES AND MEASURING CHARACTERISTICS. Figures 2.2 and 2.3 illustrate the effects of formation resistivity and bed thickness on normal and lateral devices. Apparent values of R_{xo} and/or R_t are obtained from the short normal and long normal curves and then corrected to true values. Note the curve responses above and below the bed boundaries. A check of such responses on the log print can be an aid to spotting erroneous curves. In low- and high-resisitivity formations, R_t can be read directly from the lateral in homogeneous beds 40 ft or more in thickness. In general, the lateral reference electrodes should be well inside the bed to eliminate surrounding bed effects if R_t values are to be read directly. For high-resistivity beds less than 40 ft thick, R_t can be calculated as shown in the lower portion of figure 2.3.

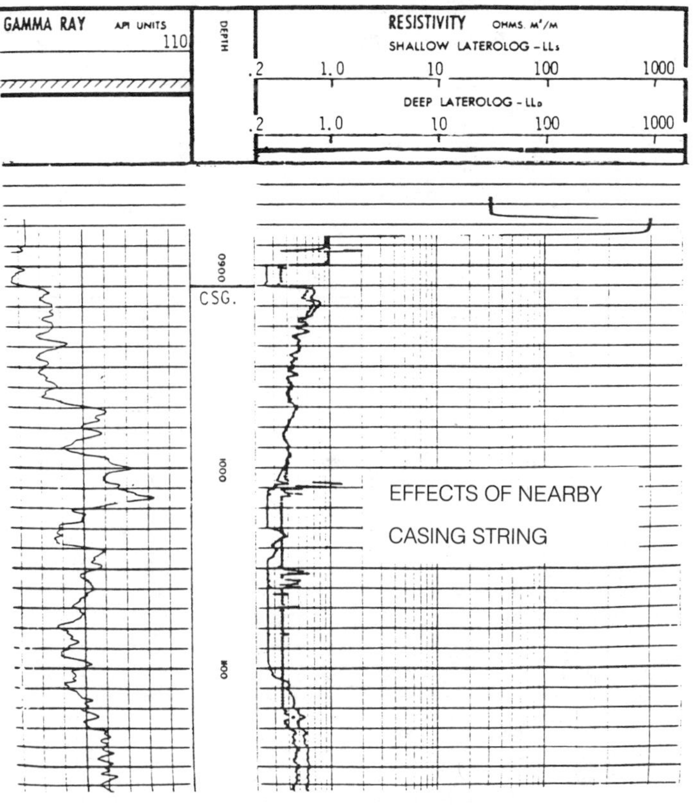

FIGURE 2.4 *Effects of Casing in Nearby Well on Resistivity Logs.*

ELECTRICAL LOG LIMITATIONS. Electrical logs have the following limitations:

1. Current paths are unfocused. The curve responses are influenced by mud resistivity, hole size, and bed thickness.
2. Usually, true formation resistivity can be obtained directly from only the lateral curve in homogeneous beds at least 40 ft thick.
3. The nonsymmetrical lateral has shadow or dead zones that restrict the evaluation of true resistivity in thin beds (see fig. 2.3).

Log Quality Control

Some conditions will affect all resistivity devices and should, therefore, be watched for regardless of the device being used. For example, the proximity of casing to the borehole will cause uncharacteristically low readings. Figure 2.4 shows a resistivity log from a directional well that passed close to a neighboring well. Another common characteristic is the appearance of a resistivity log where some gross failure in the

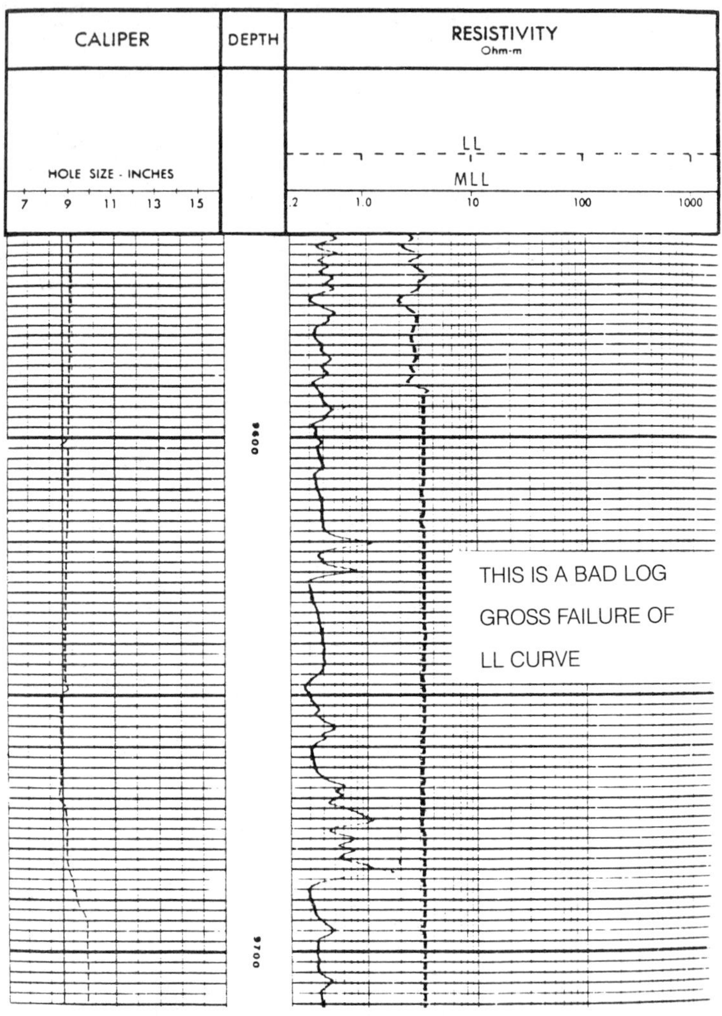

FIGURE 2.5 *Gross Failure of a Resistivity Device.*

measuring system occurs (see fig. 2.5). An even more insidious type of quality problem can arise when two log runs are spliced together for a composite. Figure 2.6 shows an example where two different types of resistivity surveys were mixed and depths from the two runs do not match. (Between 1100 and 1200 ft, the composite shows 150 ft of log.)

THE SP LOG

Spontaneous potential (SP), one of the first logging measurements ever made, was discovered by accident. It appeared in the borehole as

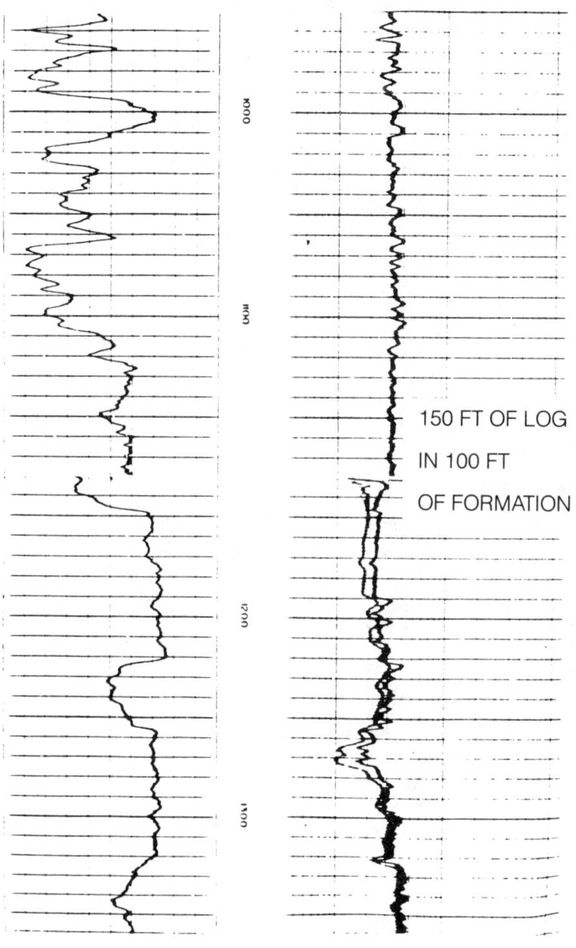

FIGURE 2.6 *Bad Composite Resistivity Log.*

a DC potential that caused perturbations of the old electric logging systems. Its usefulness was soon realized and it has remained one of the few well log measurements to have been in continuous use for over fifty years.

The SP measurement has a number of functions. It is used, for example, for correlation; as an indication of lithology; as an indication of porosity and permeability; and as a means to estimate R_w and, hence, salinity of the formation water.

Figure 2.7 shows a typical SP log. It is represented as a solid curve and shows departures to the left from a *base* line or *shale* line on the right to a *sand* line on the left in the "cleanest" nonshale zones. The scale of the log is in millivolts, abbreviated mV. Notice that there is no absolute scale in mV, only a relative scale of so many mV per division.

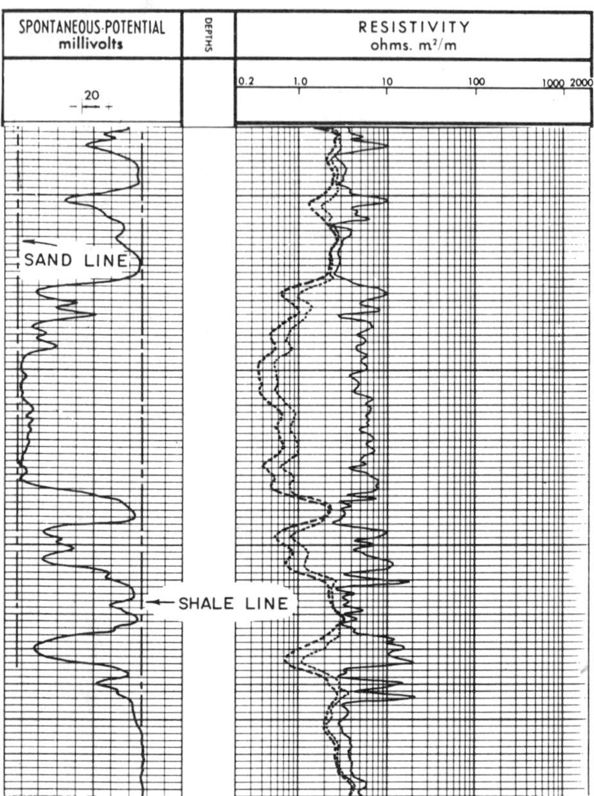

FIGURE 2.7 *Typical SP Log with Resistivity Recording. Courtesy Schlumberger Well Services.*

Recording the SP

The SP can be recorded very simply by suspending a single electrode in the borehole and measuring the voltage difference between that electrode and a ground electrode, which usually takes the form of a "fish" and makes electrical contact with the earth at the surface. A generalized illustration of the SP recording system is shown in figure 2.8. Such SP electrodes are built into nearly all logging tools. For example, the SP can be recorded using an induction log, a laterolog, a sonic log, a sidewall core gun, etc. However, the SP cannot be recorded in oil-based muds since there is no conductive path through the mud, and a conductive path is essential for generation of a spontaneous potential.

Factors Affecting the SP

Spontaneous potential is usually accurately and easily measured. There are, however, some circumstances where SP readings need careful handling:

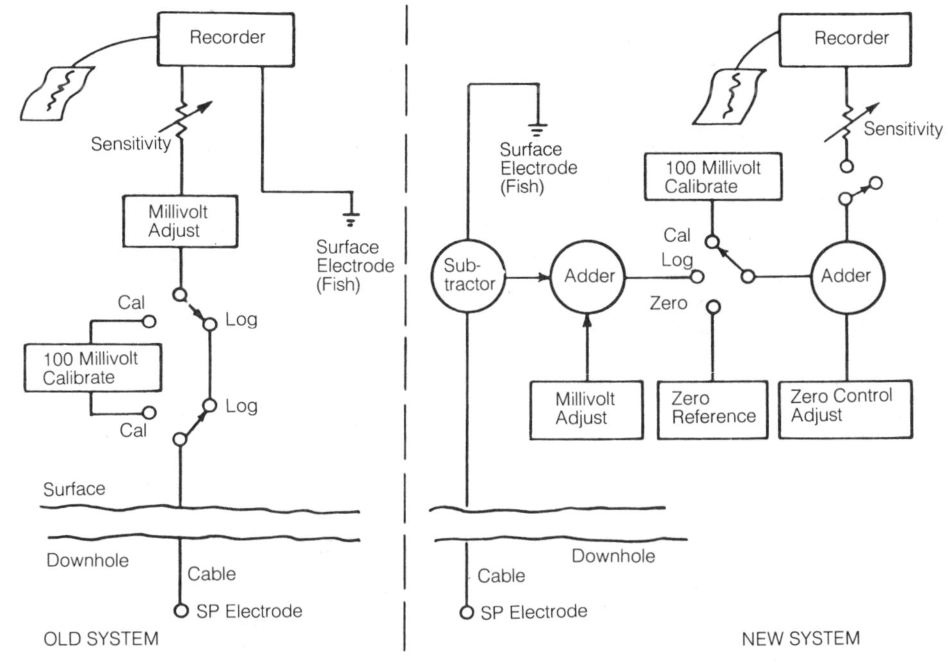

FIGURE 2.8 *Schematic Representation of the SP Measure Circuits. Courtesy Schlumberger Well Services.*

1. Oil-based muds: Due to a complete lack of an electrical path in the mud column, no SP will be generated.
2. Shaly formations will reduce the measured SP: This phenomenon permits the formation shaliness to be determined if a clean sand with the same water salinity is available for a legitimate comparison.
3. Hydrocarbon saturation will reduce SP measurements: Thus, only water-bearing sands should be selected for R_w determination from the SP.
4. Unbalanced mud columns, with differential pressure into the formation, can cause "streaming" potentials that augment the SP: This effect is noticeable in depleted reservoirs. There is no way to handle it quantitatively. (This phenomenon is called the electrokinetic SP.)
5. Bed thickness can affect the SP measurement quite dramatically: In thin beds, the SP does not fully develop. A number of correction charts are available to correct the SP in these cases.
6. The magnitude and direction of SP deflections depend on the relative salinities of the connate water and the mud filtrate: With salty connate water and fresh mud filtrate, SP deflections in porous zones will be to the left of the SP shale baseline. With fresh formation water and salt mud, the deflection may be reversed. In the case where both salinities are the same there will be no SP deflections.

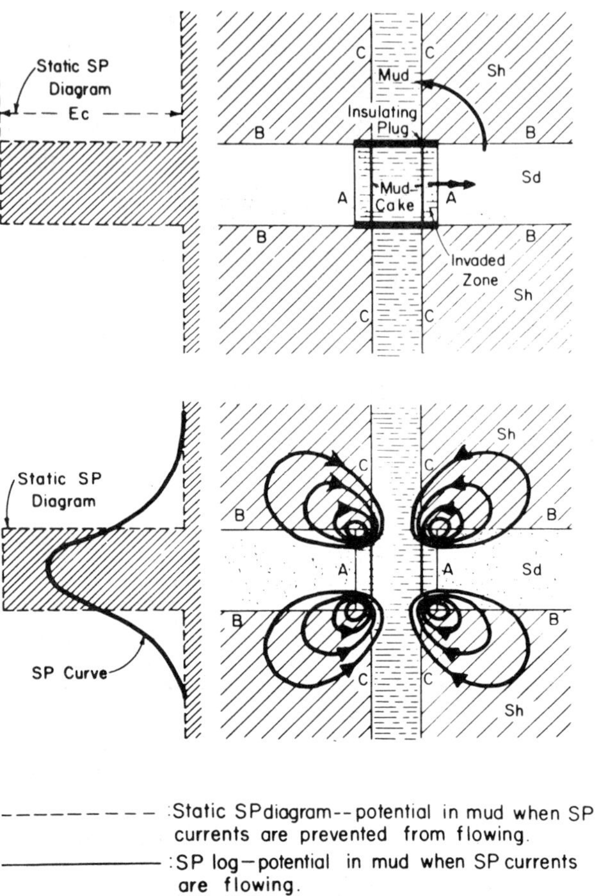

FIGURE 2.9 *SP and Current Distribution In and Around a Permeable Bed.*

Some typical SP shapes are illustrated for permeable beds (fig. 2.9), highly resistive formations (fig. 2.10), and thin beds (fig. 2.11).

SP Quality Control

SP logs are prone to errors, but they are errors that can be easily detected in the field:

1. A poor ground can cause the SP baseline to move. For example, placing the SP ground in a mud pit whose level is changing or on the ocean floor where currents can drag it back and forth can cause problems.
2. Cyclical or saw-tooth SP profiles can indicate a magnetized cable winch drum or intermittent SP circuit contacts. A log of this type is useless.

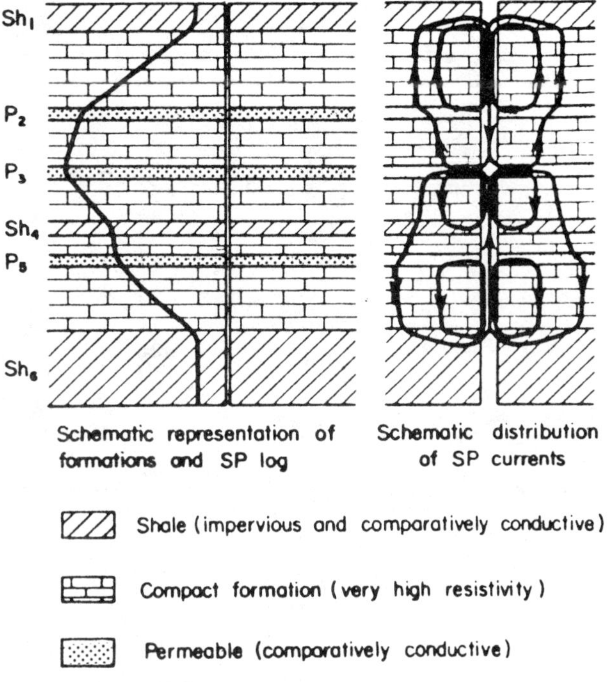

FIGURE 2.10 *SP In and Around Highly Resistive Formations.*

3. Sometimes the polarity of the SP is reversed due to incorrect galvanometer connections. If the SP moves in the wrong direction, incorrect connections could be the cause.
4. Some tools carrying SP electrodes have two dissimilar metals (steel and bronze, for example) in close proximity. Since the two metals have different electrochemical potentials, the interface via the mud column acts as a small battery. This condition is known as bimetallism and could show itself as an extraneous potential added to the SP. It can be cured by wrapping insulating tape around the offending metal parts of the sonde.
5. Baseline shifts can become quite pronounced in some geographical areas. As the tool is raised up the hole, the SP baseline will gently shift to the left by an amount that, typically, is a few millivolts per thousand feet. Eventually the SP will disappear off the left track edge and a backup trace will appear from the right. This is normal. If a downhole ground is used (cable armor), the baseline will move very rapidly near the casing shoe. Some service-company engineers will "ride" the SP millivolt box and purposely keep the baseline straight. If this technique is used, a notation should be made, preferably on the log.

Figure 2.12 illustrates many of these SP anomalies and their associated curve shapes.

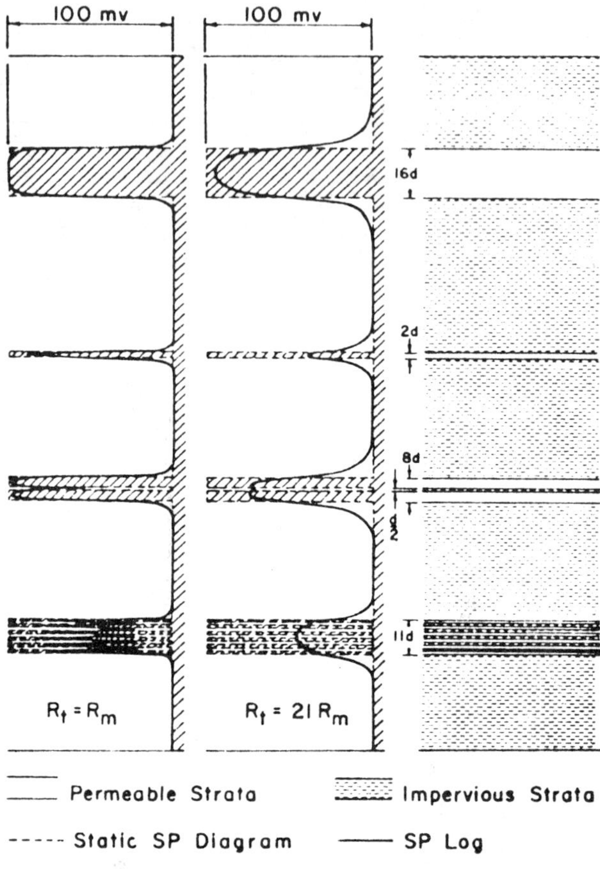

FIGURE 2.11 *Schematic SP Representations for Beds of Different Thickness and Different Contrasts of* R_t/R_m.

SP Field Examples

Cyclically repeating anomalies on an SP curve are illustrated in figure 2.13, where they appear every 12 ft: they are due to magnetism on the winch. Some of the more dramatic effects of electrical interference on an SP log are illustrated in fig. 2.14: be it lightning, arc welding, or AC interference, the SP faithfully mirrors the presence of the interference. Figure 2.15 shows a section from an induction log where the SP failed: above 1880 ft, the SP curve is either a straight line or shows large spikes. This log should be rerun.

INDUCTION LOGGING

All of the logging systems used before the introduction of induction logging depended upon the presence of an electrically conductive fluid in the borehole in order to transmit electric current to the formation. In most of the rotary-drilled wells, the drilling fluid is a water-base mud that conducts electricity. However, some wells are drilled with nonconductive fluids such as oil-emulsion muds; oil-base muds; or fresh water, air, and gas. Under such conditions, it is impossible to obtain a satisfactory electrical log using the old electric logging tools.

Induction logging does not depend upon physical contact with the walls of the wellbore to energize the formation. The induction log is similar to a transformer in that the transmitter coil is energized with alternating current, which induces into the formation a secondary current flow that is proportional to the electrical conductivity of the formation and to the cross-sectional area affected by the energizing coil. The higher the conductivity of the formation (the lower the resistivity), the larger the formation current will be. This current in turn induces a signal into a receiver coil, the intensity of which is proportional to the formation current and conductivity. The signal detected by the receiver coil is amplified, converted to a DC signal, and recorded at the surface.

Induction logging equipment provides an accurate and detailed record of the formations over a wide range of conductivity values. The accuracy is excellent for conductivity values higher than 20 mmho/m (resistivity values less than 50 $\Omega \cdot$ m) and is acceptable in lower conductivity ranges (down to 10 mmho/m). Beyond this limit, the induction log continues to respond to formation conductivity variations, but with diminished accuracy. There is a small uncertainty of about ± 1 mmho/m on the zero of the present sondes.

When to Use an Induction Log

Induction logs are recommended for use:

1. where fresh water or oil-based mud is employed,
2. where the R_{mf}/R_w ratio is greater than 3,
3. where R_t is less than 100 Ω, and
4. where bed thickness is greater than 30 ft.

The induction log is the only resistivity device that will work in oil-based mud (where oil is the continuous phase) or air-filled holes. Figure 2.16 illustrates the preferred working conditions for the induction tool. The top right portion of this chart shows the conditions in which the induction log should be run, i.e., where both R_{mf}/R_w and porosity are relatively high.

Tool Availability and Ratings

Two commonly used induction devices are the single and dual types. Each of these tools can be combined with other sensors such as the sonic tool, thereby allowing both sonic and resistivity logs to be recorded simultaneously in one trip into the well. Typical ratings for these tools are 350°F and 20,000 psi. Table 2.2 gives a summary of

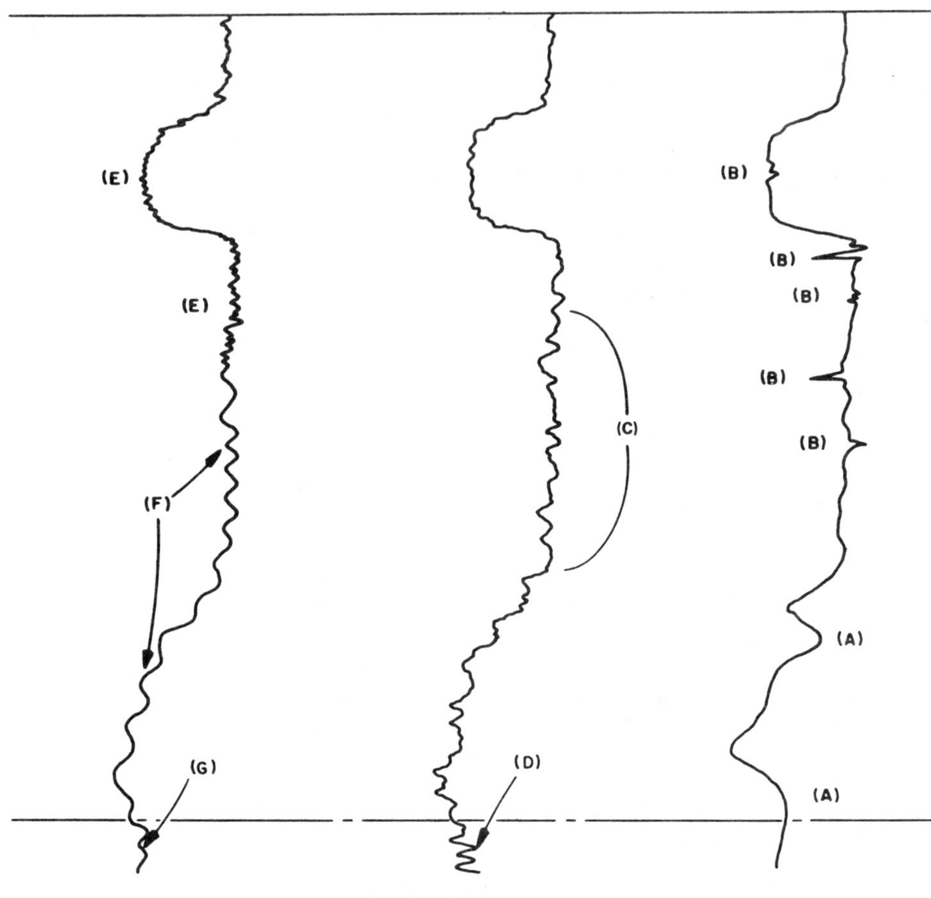

FIGURE 2.12 *SP Curve Examples: (A) Bimetallism; (B) Radio transmission, lightning, arc welding, coffee pot; (C) Different potential between casing, rig, and truck with wireline intermittently rubbing on Kelly bushing; (D) Always check on noise before pick-up; (E) Crosstalk between circuits or conductors; (F) Magnetism on the SP; (G) Pick-up noise indication due to magnetized drum or drive chain.*

currently available induction devices together with dimensions and ratings.

Presentations and Scales

Induction logs and combination induction–sonic logs can be run with a variety of scales and presentations. The primary measurement of conductivity always appears on a linear scale. However, it is not always presented. Induction resistivity can appear on a linear scale or a logarithmic scale. When the sonic curve is present as well, a split grid is usually employed. Figure 2.17 shows a combination induction–

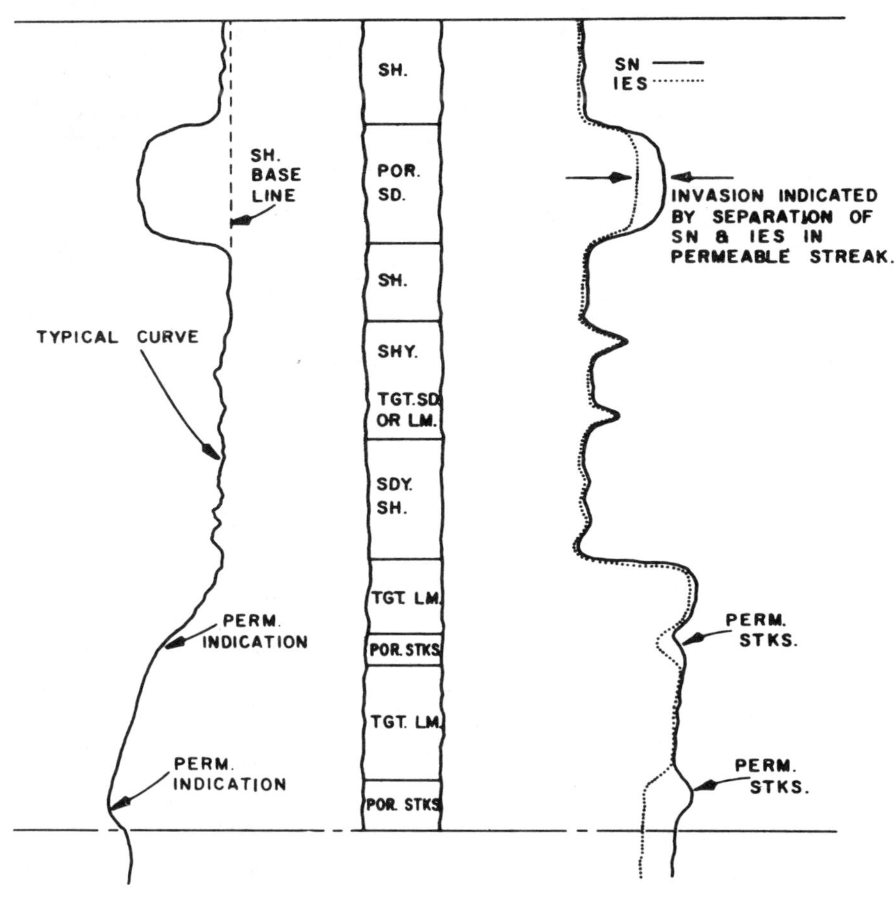

sonic, gamma ray tool string, and figure 2.18 illustrates the various possibilities available.

Theory of Induction Devices

Induction tools work like mine detectors or the airport security devices through which the passengers walk. A transmitter coil, with an alternating current passing through it, sets up an alternating magnetic field. The alternating magnetic field induces an alternating potential in the formation. Depending on the conductivity of the material in the formation, a large or small alternating current is caused to flow in the formation. The alternating magnetic field associated with this current induces an AC potential in a receiving coil, which can be measured and thus becomes an indicator of the conductivity of the formation. Figure 2.19 illustrates this concept by reference to a simple two-coil device.

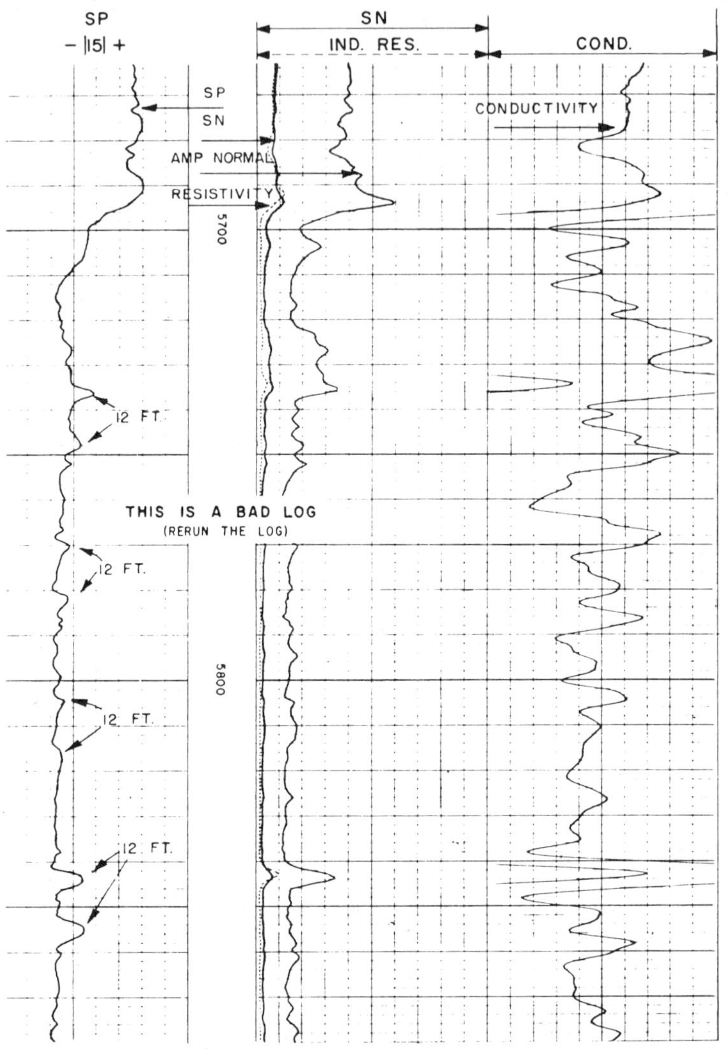

FIGURE 2.13 *Magnetism on the SP.*

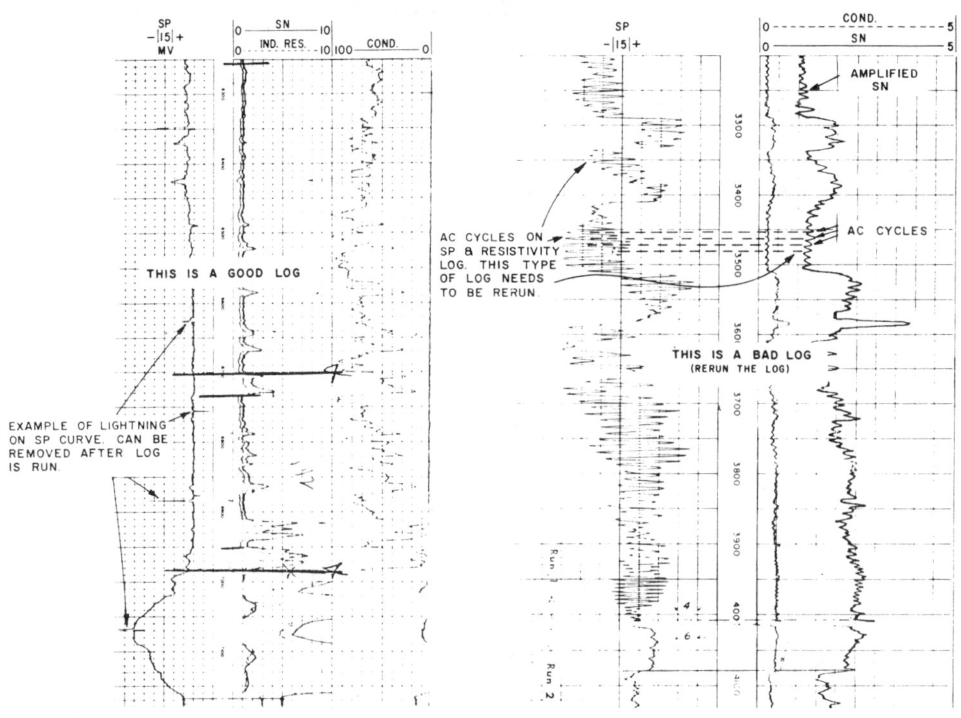

FIGURE 2.14 *Effects of Lightning and AC Interference on the SP.*

The oscillator attached to the transmitter is held at a constant 20-kHz frequency and with a constant current so that the magnetic field strength is also constant. The induced current in the formation is thus large when the resistivity is low (conductivity high) and the measured signal (the voltage appearing on the receiver coil) is also greatest in low resistivity formations. In very high resistivity formations, the measured signal becomes very small and resolution is lost. This loss of resolution in high resistivities becomes a problem at about 75 $\Omega \cdot m^2/m$. Above 100 $\Omega \cdot m^2/m$, the induction tool is not very useful for quantitative measurements.

Induction tools require no special borehole fluids and work equally well in air, oil, or water-based muds. In fact, in air-filled holes or oil-based muds (where oil is the continuous phase), the induction tool is the only resistivity tool that will work.

Calibration of the system is a two-point method. The sonde is suspended high off the ground well away from conductive materials and is adjusted by zero shift to read zero conductivity (infinite resistivity). A circular ring of metal in the form of a test loop is then placed around the sonde and a scale-sensitivity adjustment is made to the

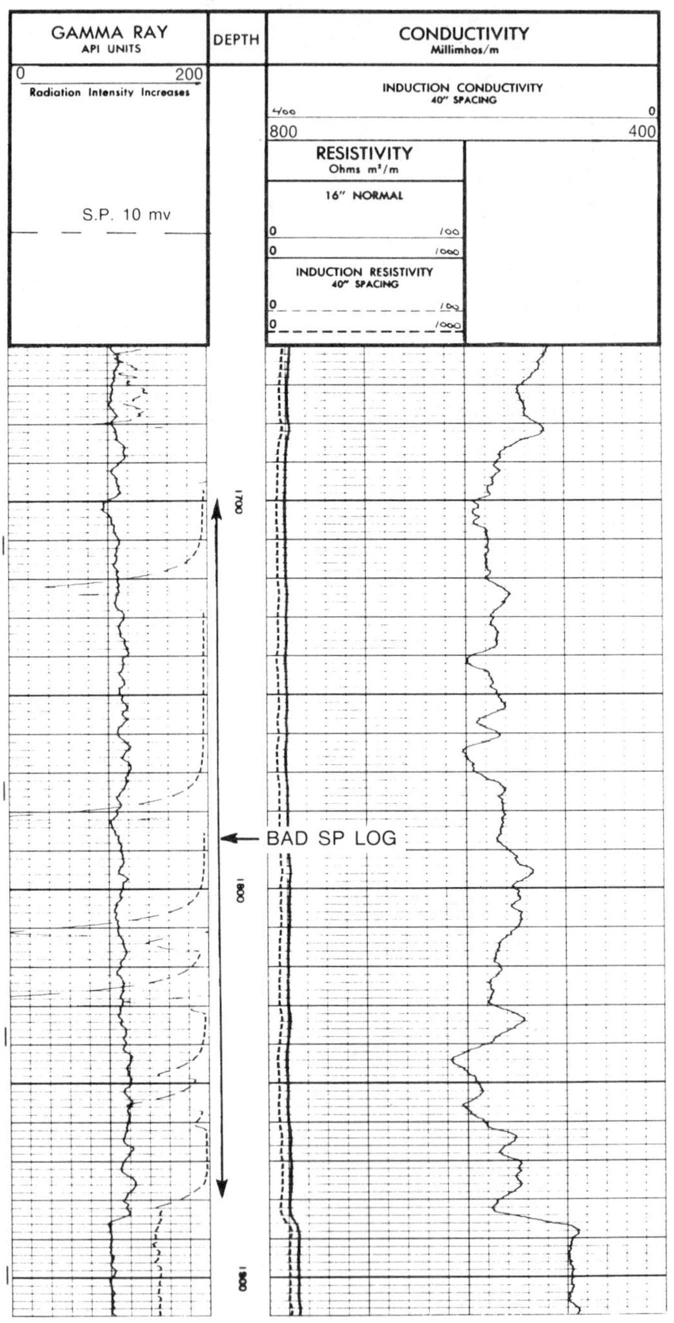

FIGURE 2.15 *Bad SP Example.*

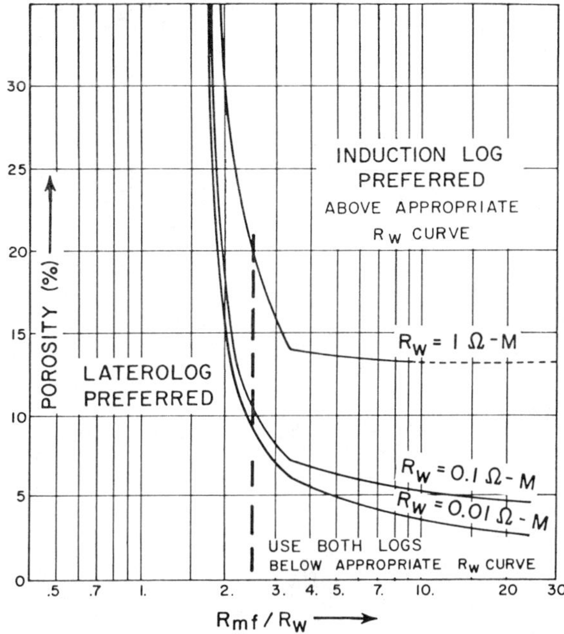

FIGURE 2.16 *Preferred Conditions for Induction Tools. Courtesy Schlumberger Well Services.*

TABLE 2.2 *Induction Tool Sizes and Ratings*

Tool Name	Service Company	Maximum Temp (°F)	Maximum Pressure (psi)	Diam. (inches)	Remarks
Dual Induction	S	340	18,000	3.375	
Spherically Focused Log	S	350	20,000	3.875	
Induction Spherically Focused Log	S	350	20,000	3.875	
Induction Electric	S	350	20,000	3.875	
Survey	S	350	20,000	2.750	Limited areas
	S	500	25,000	3.375	No 16-in. normal
Dual Induction Laterolog	G	350	20,000	4.00	
Induction Electrical Log	G	350	20,000	3.88	
Dual Induction	DA	350	18,000	3.625	
Focused Log	DA	400	20,000	3.625	
Induction Electrolog	DA	350	18,000	3.625	
		350	20,000	2.688	
		400	25,000	3.375	
Dual Induction Log	W	300	10,000	3.375	Limited areas
	W	325	15,000	3.625	Standard tool
	W	350	20,000	4.000	Limited areas
Induction Electrical Log	W	300	20,000	3.375	

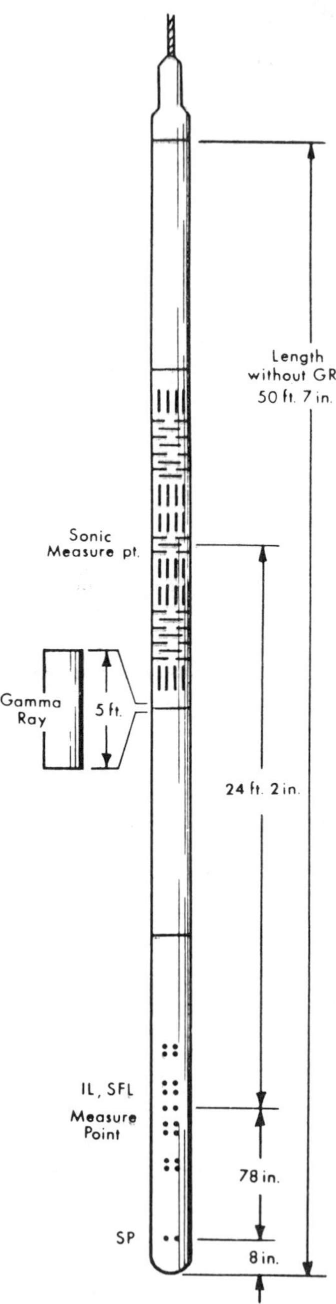

FIGURE 2.17 *Induction–Sonic Tool Combination. Courtesy Schlumberger Well Services.*

known value of the test loop. Depending on the tool, this will be in the order of the equivalent of a 1- or 2-Ω formation.

When the tool is at the bottom of a 10,000-ft well, there is no way a test loop can be placed around the sonde; therefore, an internal calibrator is included in the tool. This will have a nominal value of 1 or 2 Ω and its precise value is determined monthly by reference to the test loop. These internal calibrators shift with aging but behave reasonably well under normal use. A check of the zero-conductivity point when the tool is in the hole is accomplished by simply opening the receiving coil. Any extraneous signal is cancelled out by a zero adjustment.

Skin Effects

To fully understand the induction response, several factors must be considered. In addition to the emf induced by the magnetic field of the transmitter coil, other emfs are induced in the ground loops and affect the amplitude and phase of the eddy currents. These additional emfs are produced by linkage of each ground loop with its own magnetic field (a ground loop has self-inductance), and with the magnetic fields of the other nearby ground loops (there is mutual inductance between different ground loops). In this multiple cross-coupled system, the resultant eddy currents will not be independent of one another.

It can be predicted that, with increasing distance from the source (i.e., from the transmitter coil), there will be attenuation in the amount of power transmitted because:

1. the dissipation of energy by the flow of eddy currents in the region near the source decreases the energy available for transmission to regions farther out; and
2. regions at a distance from the source are shielded from the magnetic field of the transmitter coil by the annulling effect of magnetic fields of opposite sign from the eddy currents in the conductive medium closer to the transmitter. (In a sense, the shielding of the outer regions is equivalent to a reflection of the energy back toward the source.)

As a consequence of these interactions, there is a reduction of the receiver-coil signal; i.e., a reduction of high conductivity. This reduction in the apparent conductivity reading is commonly termed *skin effect.* If σ_g is the conductivity reading that would be observed in a given configuration of media without skin effect and σ_a is the conductivity actually observed, then the difference-conductivity, σ_s, is the skin effect, that is,

$$\sigma_s = \sigma_g - \sigma_a.$$

An amount σ_s is added to the observed reading by means of a skin-effect compensating network. It is not linear and can best be illustrated by figure 2.20. In practical terms, the tool will read a resistivity

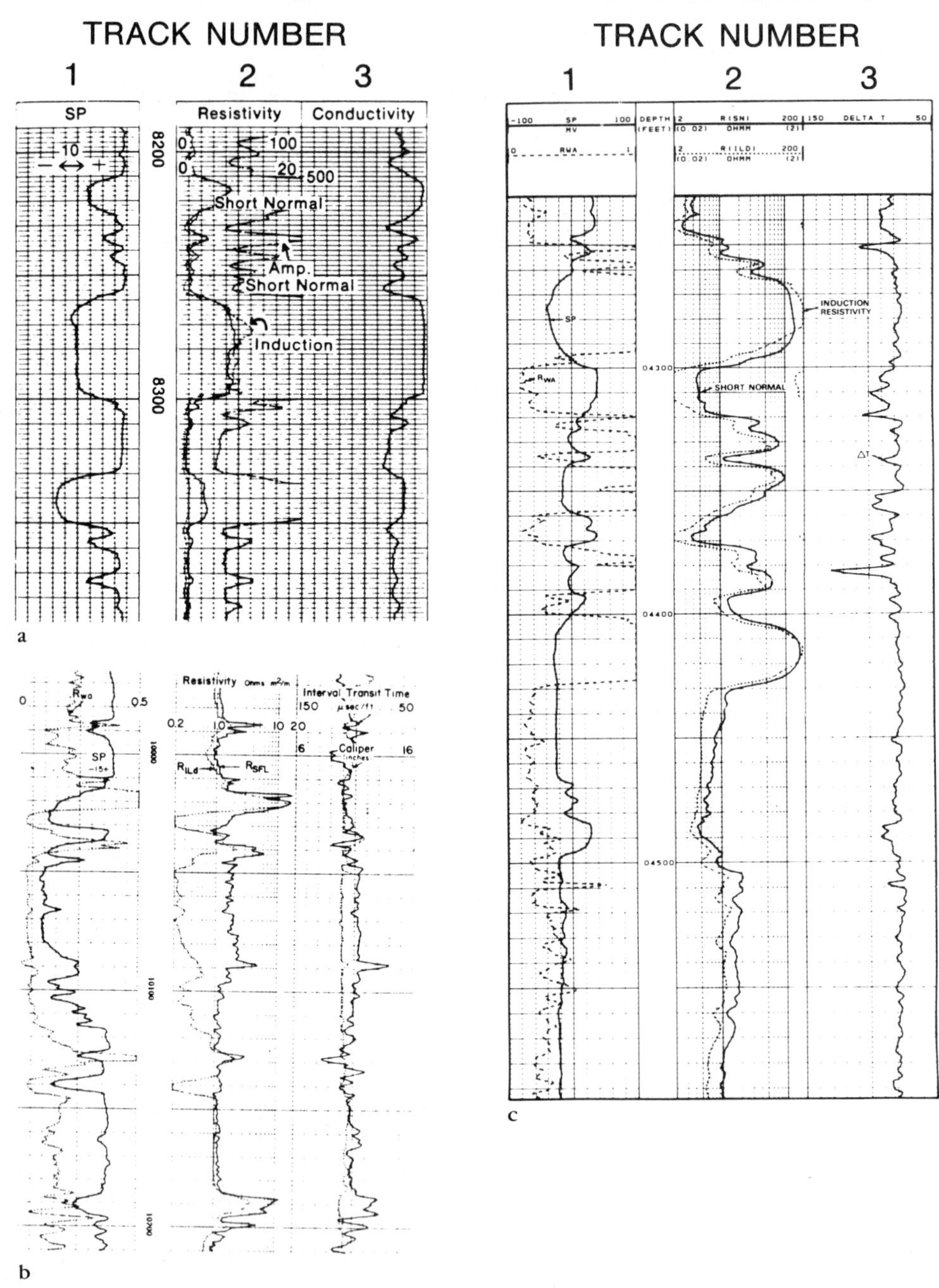

FIGURE 2.18 *Various Presentations of the Induction Log: (a) IES Linear, (b) ISF Logrithmic (courtesy Schlumberger Well Services), (c) IEL Logrithmic (courtesy Gearhart Industries, Inc.), (d) IEL Linear, (e) IEL Logrithmic (courtesy Dresser Atlas), and (f) IEL Linear (courtesy Welex, a Halliburton Company).*

TRACK NUMBER

1 2 3

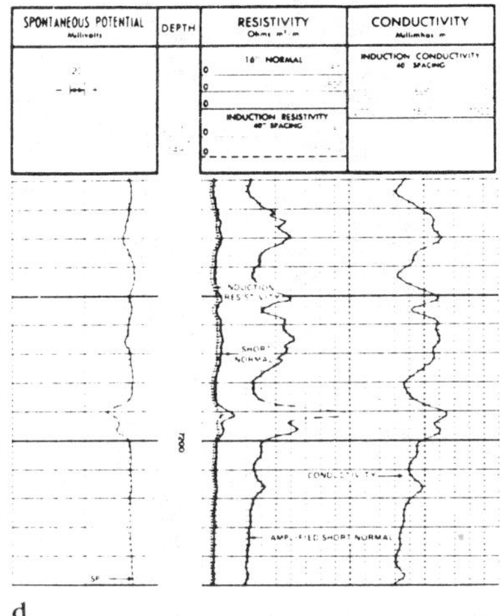

d

SPONTANEOUS POTENTIAL
Millivolts
DEPTH
RESISTIVITY
CONDUCTIVITY
16" NORMAL
INDUCTION CONDUCTIVITY
40" SPACING
INDUCTION RESISTIVITY
40" SPACING
SHORT NORMAL
CONDUCTIVITY
SP

e

TRACK NUMBER

1 2 3

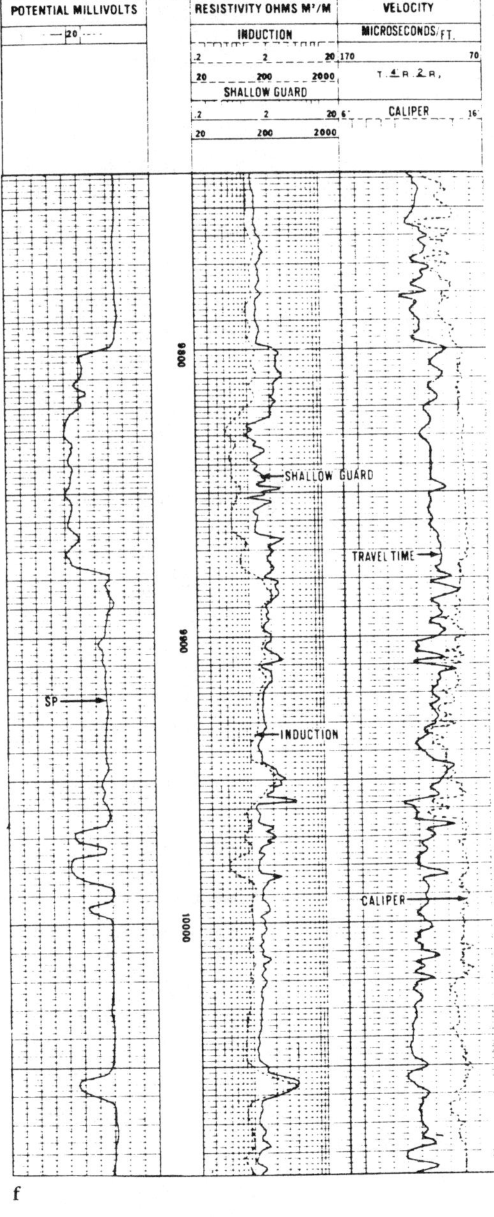

f

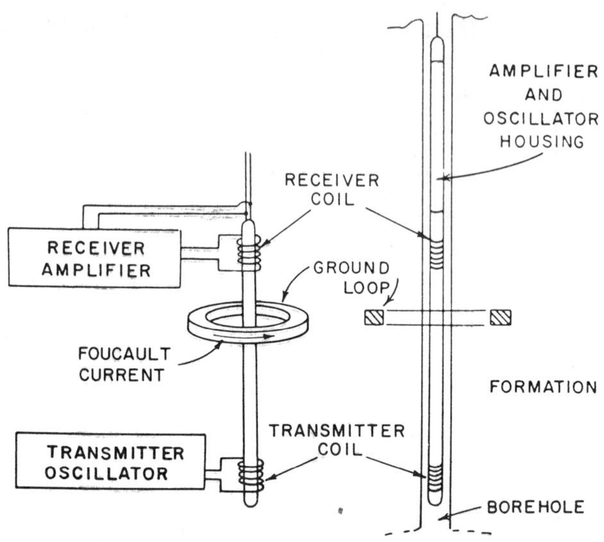

FIGURE 2.19 ***Basic Two-Coil Induction Log System. Courtesy Schlumberger Well Services.***

that is too high unless the skin-effect compensation is working properly. A failure to switch in the compensation can cause an erroneous log.

Tool Calibration

Induction-log calibration can be performed on land at any time.The sonde is placed in a zero-conductivity environment; this is normally done by raising the sonde up in the air well away from metallic objects. This defines a zero point. A calibration loop is then placed around the sonde to give a known conductivity signal, usually 500

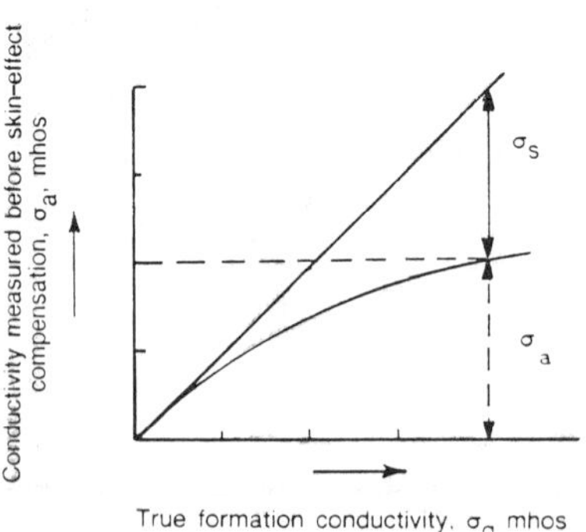

FIGURE 2.20 ***Correction of Formation Conductivity for Skin Effect.***

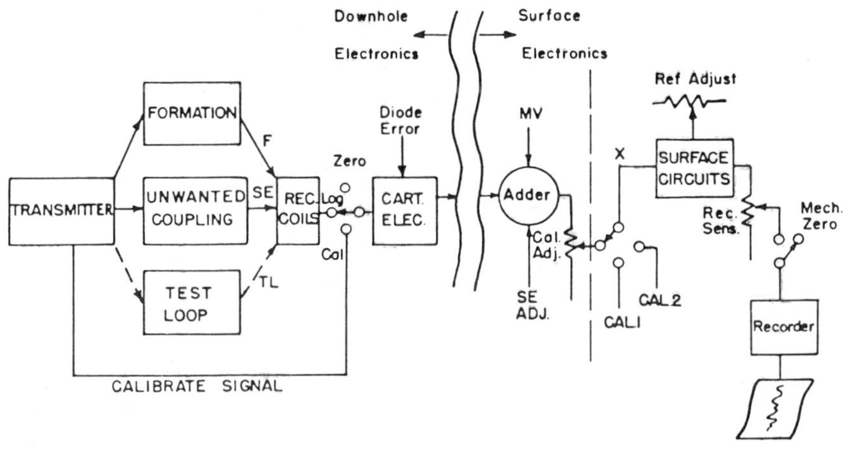

FIGURE 2.21 *Typical Induction Calibration System. Courtesy Schlumberger Well Services.*

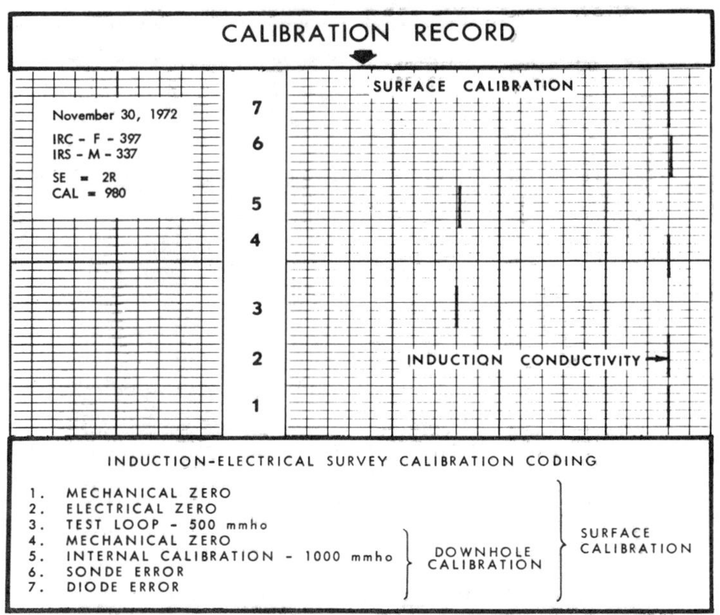

FIGURE 2.22 *Induction Surface Calibration System. Courtesy Schlumberger Well Services.*

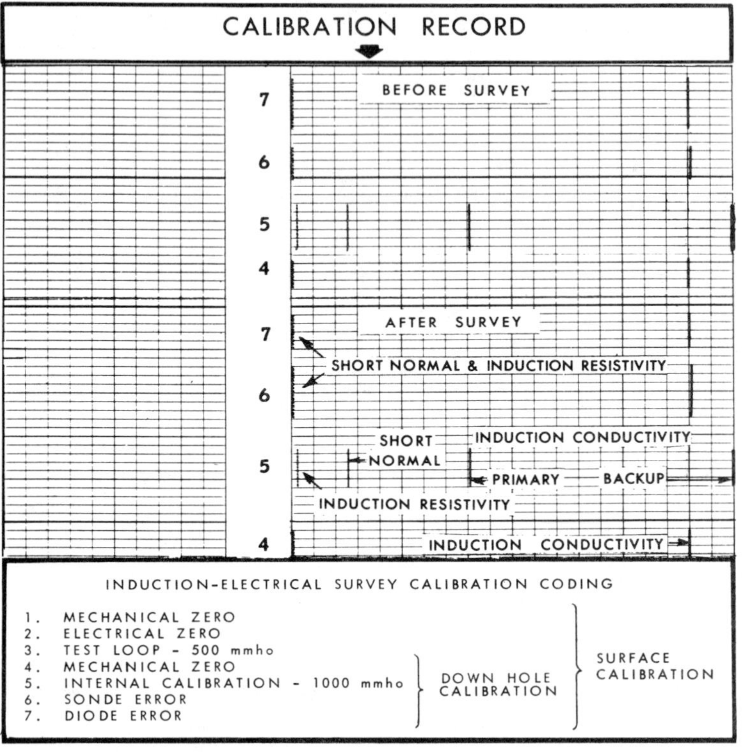

FIGURE 2.23 ***Before and After Survey Induction Calibration Records. Courtesy Schlumberger Well Services.***

mmho. This calibration is performed monthly. It is almost impossible to perform on an offshore rig due to the surrounding metal structure. The sonde and its associated electronic cartridge form a matched set and should always be used together.

The overall calibration system for induction devices is illustrated in figure 2.21. Here, and in other chapters, details of analog calibrations will be given even though few are currently used. Computer calibration tails are documented in Appendix J. Logs recorded in analog form will exist in files for years to come. By including analog calibration tails in this text a means is offered both to check on old log accuracy and to follow more clearly the methodology of log calibration in general.

SURFACE CALIBRATION.* The surface calibration of induction tools is performed through a memorizer panel. The calibration steps, illustrated in figure 2.22, are as follows:

*The calibration information that follows has been provided, with permission, by Schlumberger Well Services.

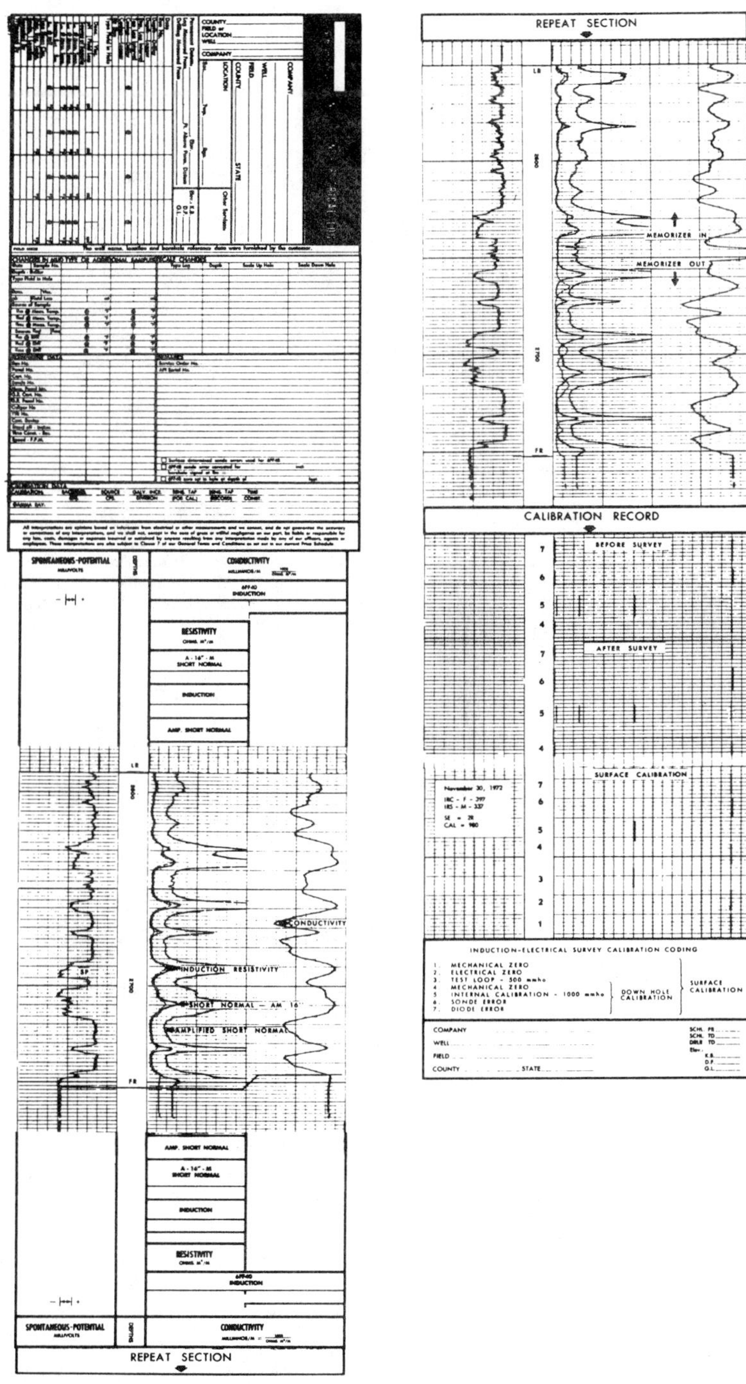

FIGURE 2.24 *A Complete Induction Survey. Courtesy Schlumberger Well Services.*

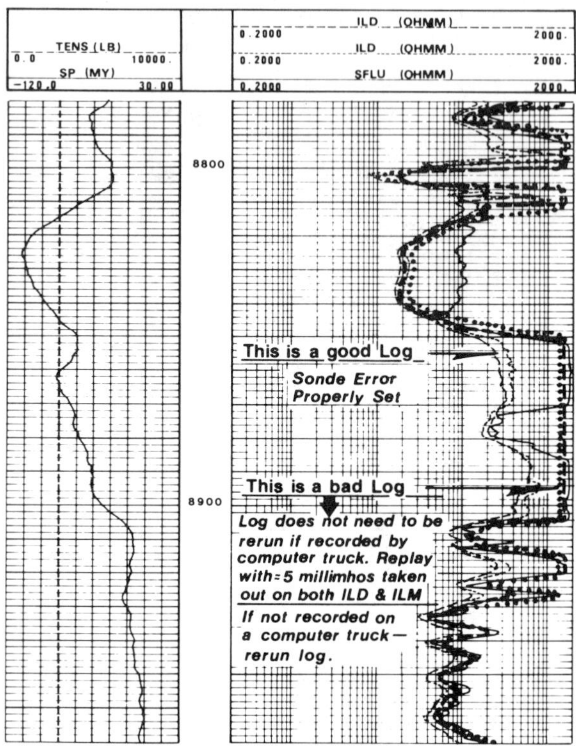

FIGURE 2.25 *Correct and Incorrect Sonde Error Settings.*

1. Mechanical zero: Galvanometer circuits are open. The conductivity galvanometer may be set at 19 divisions (right edge), 9 divisions, or 8 divisions (as in fig. 2.22). If set at 9 or 8 divisions to permit recording negative sonde error (Step 6) values, the galvanometer is later reset at 10 divisions for logging.
2. Electrical zero: Tool is in the "measure" condition, with the sonde suspended in a zero-conductivity area. Galvanometer reads same as Step 1.
3. Test loop: Conductivity scale is now on 500. The automatic skin-effect correction is applied. The test loop is in position. Galvanometer reads 10 divisions above zero (500 mmho).
4. Mechanical zero: Same as Step 1.
5. Internal calibrate: Tool is in the "measure" condition, on the 1000-mmho/track scale. Sensitivity has previously been set by reference to the test loop response. Galvanometer reads 1000 mmho, plus or minus up to 50.
6. Sonde error: The undesirable signal generated by the slight imperfection of the sonde coils has previously been cancelled by an equal and opposite voltage from the panel. Now the sonde is disconnected by

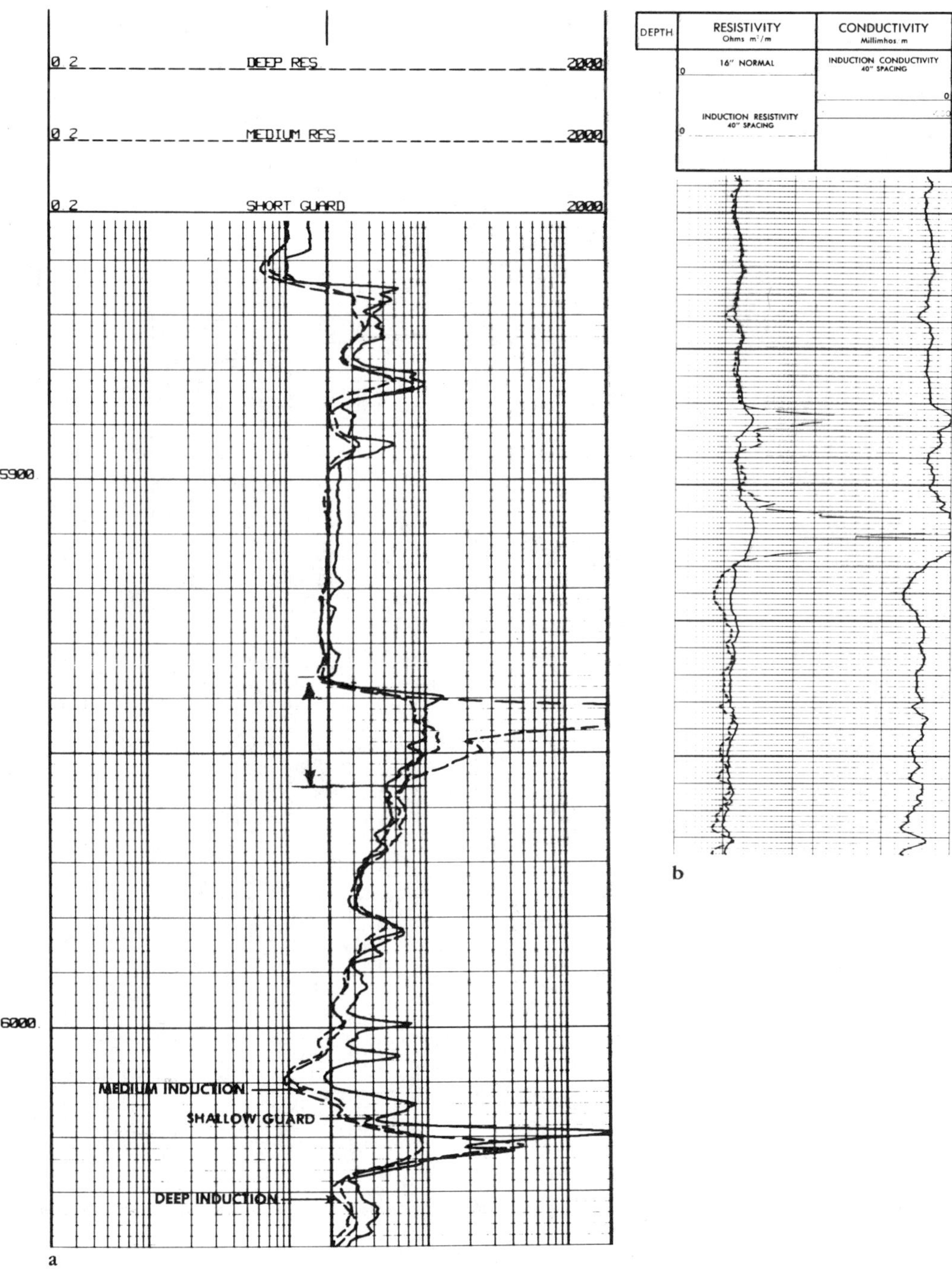

FIGURE 2.26 *Sonde Error Examples: (a) Dual Induction with Bad Deep Induction and (b) Induction Electrolog with Bad Deep Induction.*

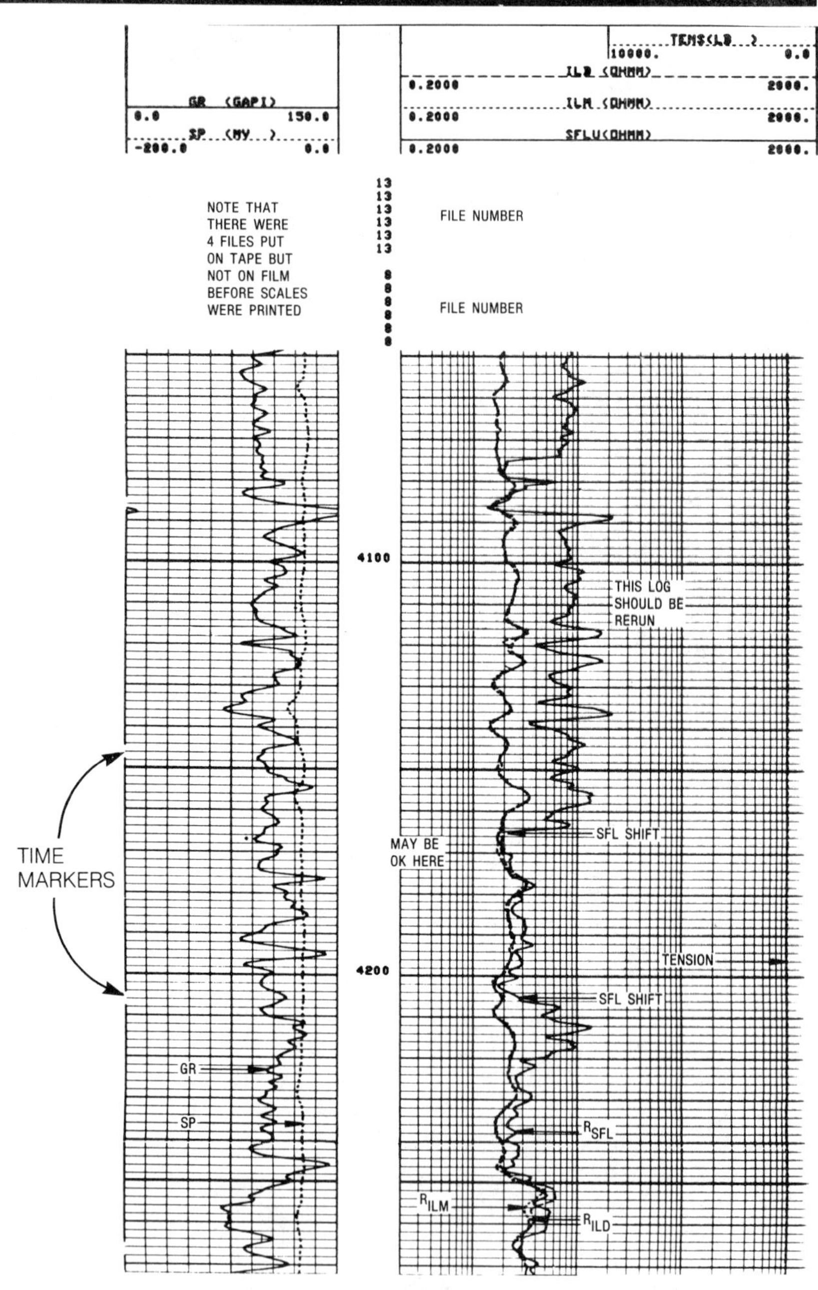

FIGURE 2.27 *Example of Bad Induction SFL Log.*

the zero relay, and the magnitude of this "negative sonde error" is recorded. Conductivity scale of 100 is used. Galvanometer reads a few millimho (normally no more than 1 division) left or right of zero.

7. Diode error: Same settings as Step 6, but the voltage source for the "negative sonde error" is switched off. Now the only signal is that produced by the measure-circuit electronics. In present equipment, this is very near zero. Galvanometer reads within 0.1 or 0.2 division of zero.

This surface test film is dated to insure currency and is identified by tool serial numbers.

BEFORE SURVEY CALIBRATION: Performed downhole prior to each logging operation, the before survey calibration of induction tools is done through a memorizer panel. The steps involved (see fig. 2.23) are as follows:

4. Mechanical zero: Same as Step 1 surface calibration. Conductivity galvanometer is at 8 divisions (see Step 1) of Track 3; short normal galvanometer is at zero, Track 2 (left edge).
5. Internal calibration, IL, and SN: Resistivity scale on desired logging scale, calibrate relay on. The calibrate resistors introduce signals equivalent to 1000 mmho (approximately) and 10.5 $\Omega \cdot$ m. The conductivity galvanometer reads the same as in Step 5, Surface Test Film (10 divisions $\pm$ 0.5). The short normal galvanometer reads 2.62 divisions, Track 2 (logging scale is 40 $\Omega \cdot$ m).
6. Sonde error: Zero relay on, conductivity scale 100. Conductivity galvanometer is set at the sonde error reading established in Step 6 of the surface calibration. Short normal galvanometers should read zero, Track 2.

6a. Downhole sonde error: Alternatively, the sonde error can be established downhole, providing there is a thick, in-gauge, highly resistive formation. If this technique is used, the fact will be entered on the log heading.

7. Diode error: Same as Step 7, surface calibration. Conductivity galvanometer reads zero $\pm$ 0.2 divisions.

AFTER SURVEY CALIBRATION. The purpose of after survey calibration is to verify that the calibrations have not drifted significantly from their proper settings. Induction tool after survey calibrations are performed downhole after each logging operation:

Steps 4–7 are the same as for before logging; recording is made without changing zero or sensitivity settings.

Quality Control

A complete induction survey including all calibrations and a repeat section is illustrated in figure 2.24. Log quality checks should include the following:

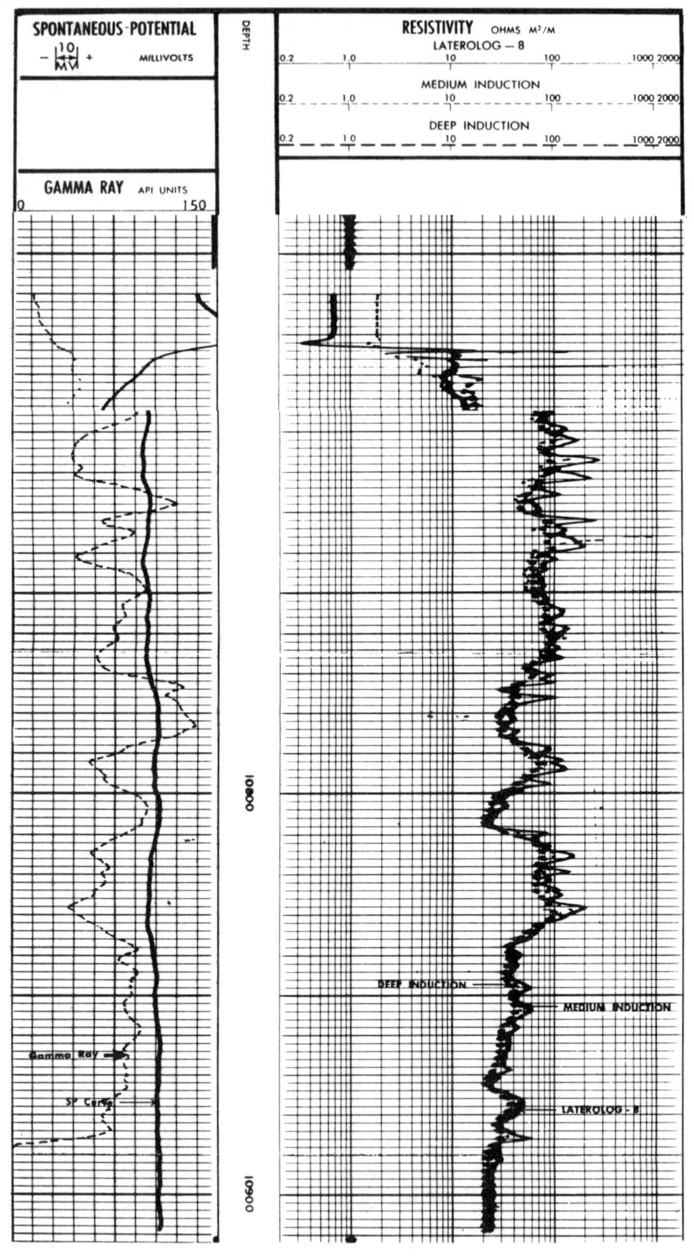

FIGURE 2.28 *Comparison of Dual-Induction Log with Dual Laterolog on a Subsequent Run.*

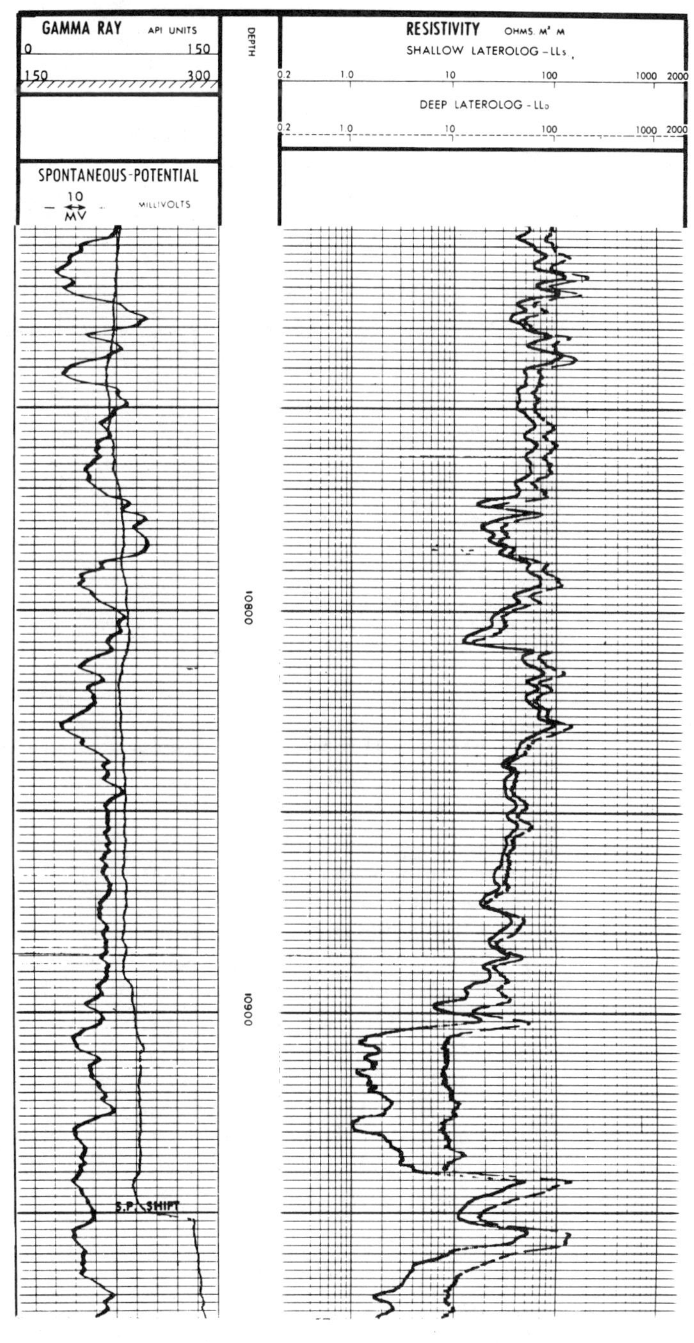
GAMMA RAY API UNITS
0 150
150 300
SPONTANEOUS-POTENTIAL
10 MV
MILLIVOLTS
DEPTH
RESISTIVITY OHMS. M² M
SHALLOW LATEROLOG - LLs
0.2 1.0 10 100 1000 2000
DEEP LATEROLOG - LLD
0.2 1.0 10 100 1000 2000
10800
10900
S.P. SHIFT

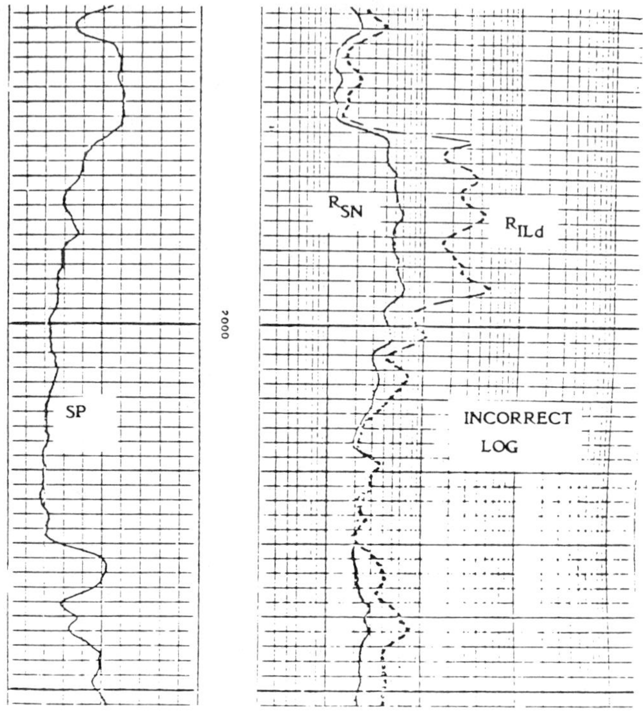

FIGURE 2.29 *Misleading Induction Log (No Skin-Effect Correction).*

1. Recording speed (maximum 6000 ft/hr).
2. Correctness of readings:
   a. Separation between R_{IL} and R_{SN} or R_{SFL} can be explained by curve response charts, borehole effects, and/or invasion consequences.
   b. Short normal does not read less than zero.
   c. Induction reciprocated resistivity is correct for all conductivity readings.
   d. Conductivity curve does not read less than zero, except momentarily in zones of extreme contrast.
   e. Skin-effect correction is recorded at the bottom of the survey.
3. Accuracy and stability of calibrations:
   a. Sonde and diode error indicate no change between before and after readings.
   b. Calibration signal (approximately 1000 mmho) should not change more than 3%.
4. Test films to be attached:
   a. 200 ft of repeat with memorizer out and in,
   b. downhole calibration before survey,
   c. downhole calibration after survey, and
   d. master surface calibration.

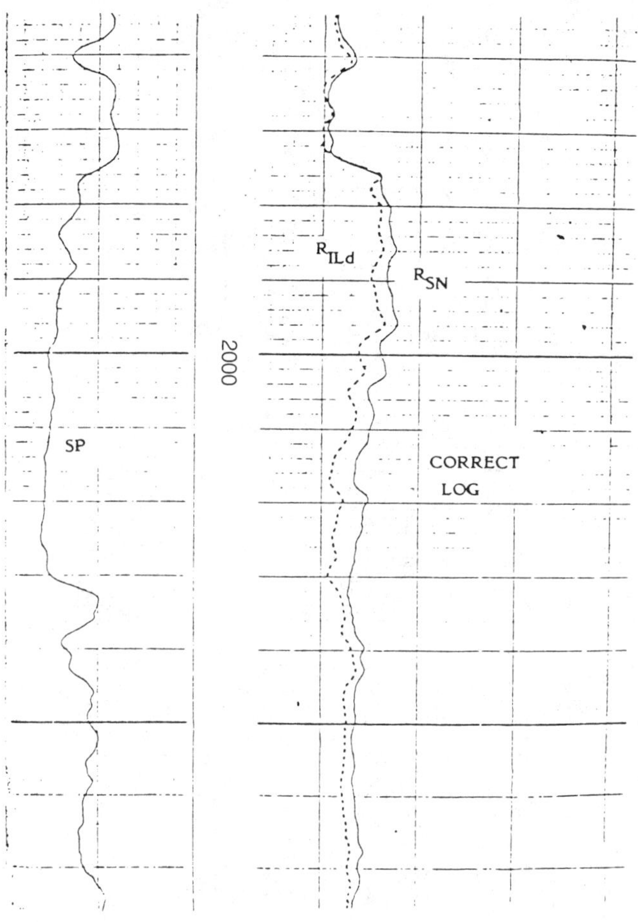

FIGURE 2.30 *Correct Induction Log (with Skin-Effect Correction).*

Field Examples

SONDE ERRORS. An incorrectly set sonde error can be detected by the flat-topped portions on the induction curves above 1000 ohms. Figure 2.25 illustrates two induction log runs, one with the sonde error correctly set and one with the sonde error incorrectly set. Figure 2.26 shows two bad induction logs both suffering from incorrect calibration of the sonde error. These logs should be rerun.

INTERMITTENT SFL. Figure 2.27 illustrates a bad induction SFL log (spherically focused log). Note the intermittent shifts of the SFL curve.

INDUCTION/LATEROLOG COMPARISON. Figure 2.28 shows a comparison between an initial dual-induction log and a subsequent dual laterolog

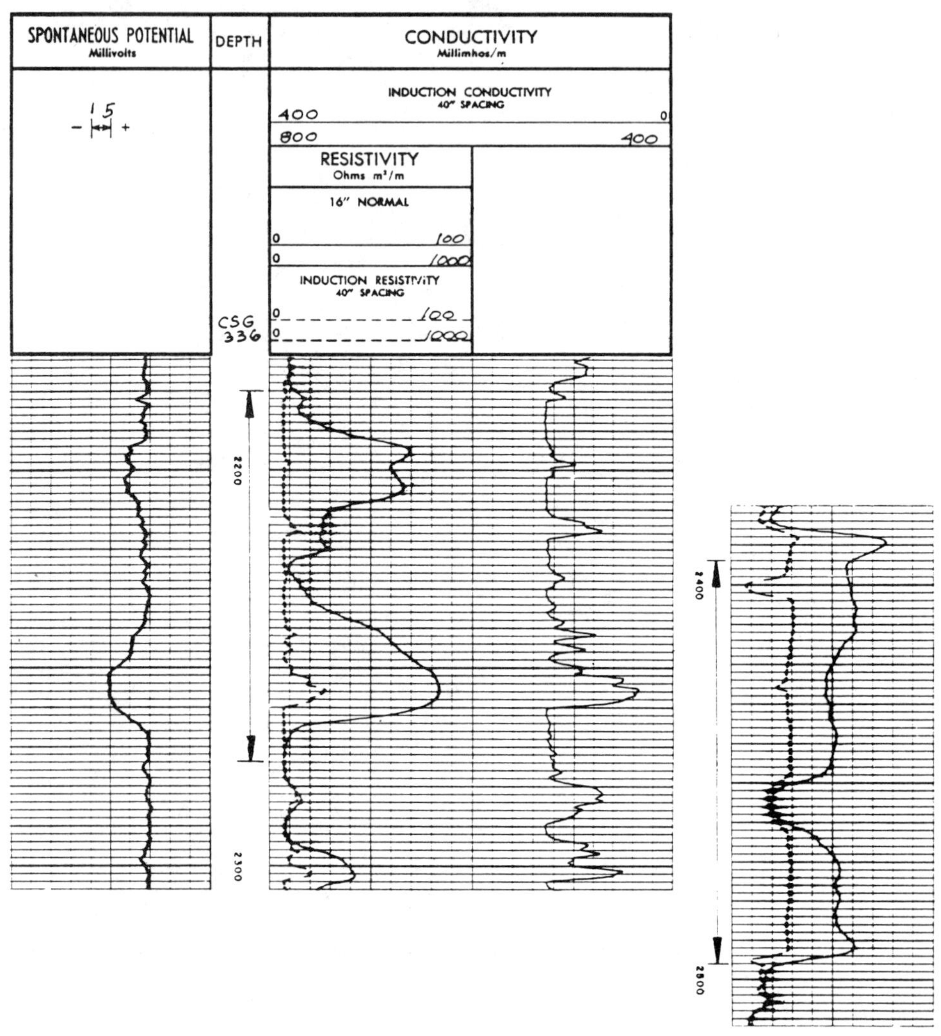

FIGURE 2.31 *Intermittently Bad Induction.*

on a later run. Both logs are good. Any differences can be explained by invasion effects and hole-size corrections.

SKIN-EFFECT PROBLEMS. Figure 2.29 shows an induction log that appears to indicate an oil/water contact at 1997 ft. This proved not to be the case. The skin-effect correction had not been applied. Figure 2.30 shows the same interval as relogged with the skin-effect correction applied. Note that in figure 2.29, the induction reads higher than the short normal in the shales—a clue to the lack of skin-effect correction.

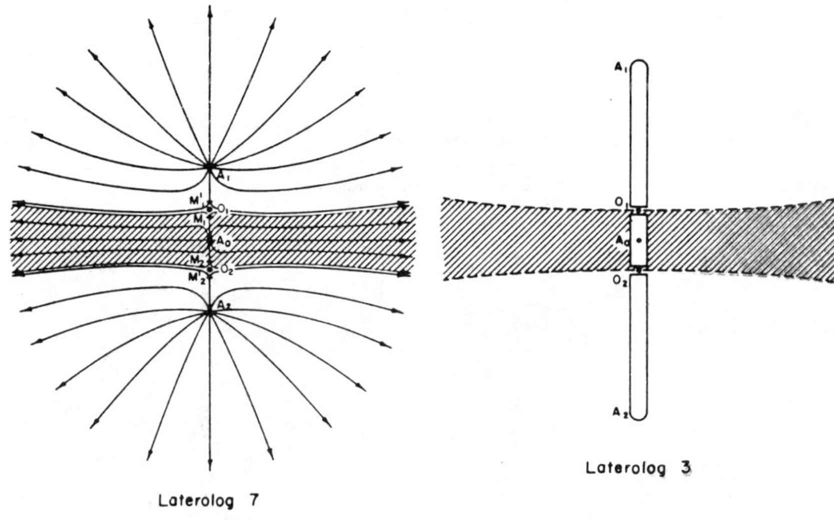

FIGURE 2.32 *Schematic Diagrams of Early Laterolog. Courtesy Schlumberger Well Services.*

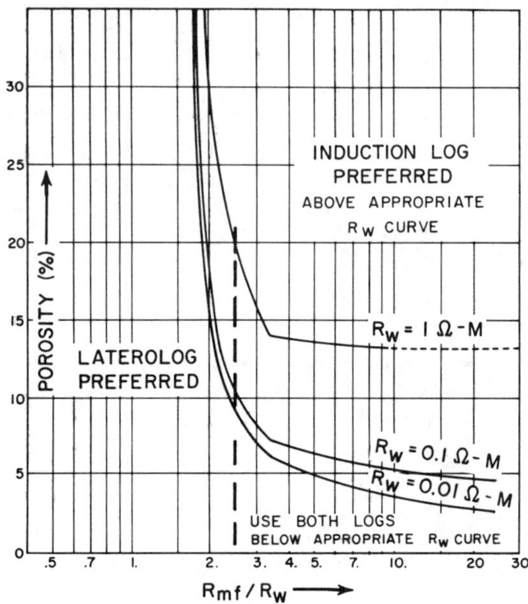

FIGURE 2.33 *Preferred Ranges of Application of Induction Logs and Laterologs. Courtesy Schlumberger Well Services.*

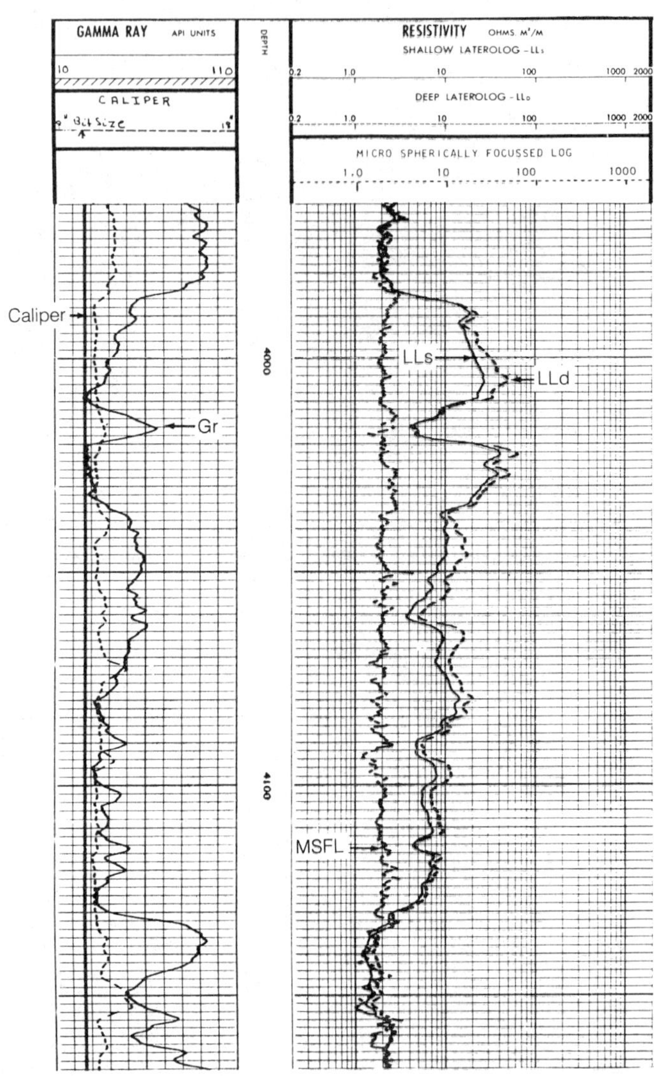

FIGURE 2.34 *Typical Dual Laterolog Presentation. Courtesy Schlumberger Well Services.*

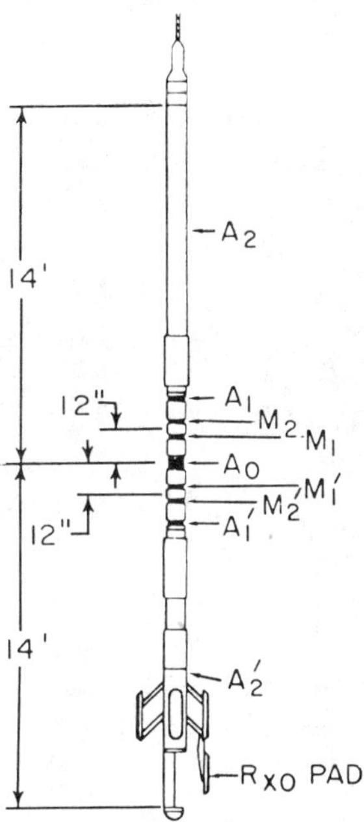

FIGURE 2.35 *Schematic Diagram of the Dual Laterolog–R_{xo} Tool. Courtesy Schlumberger Well Services.*

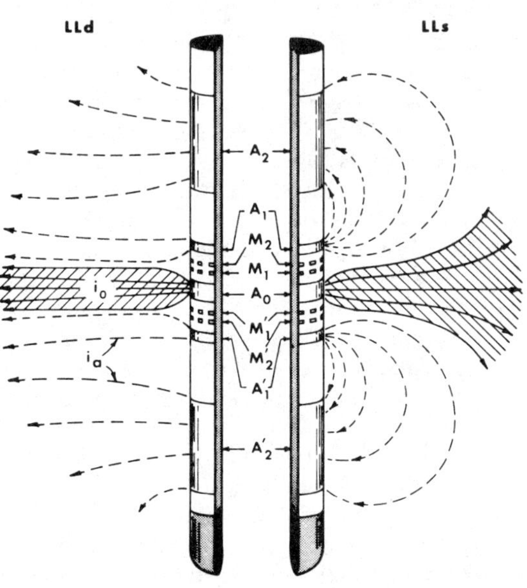

FIGURE 2.36 *Representative Current Patterns for the Deep and Shallow Laterologs. Courtesy Schlumberger Well Services.*

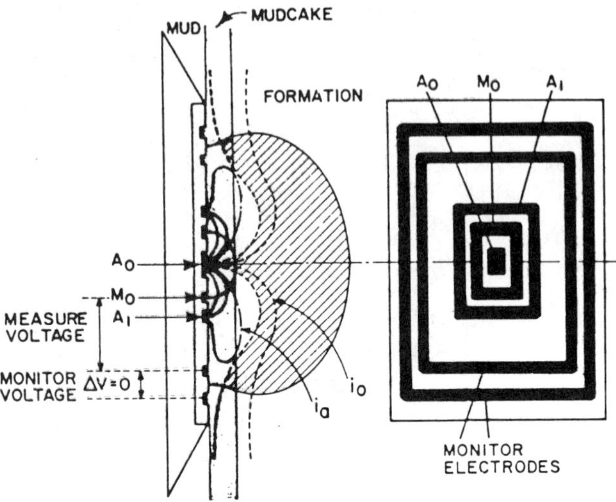

FIGURE 2.37 *Electrode Arrangement of MSFL (*right*) and Current Distribution (*left*). Courtesy Schlumberger Well Services.*

INTERMITTENT PROBLEMS. Figure 2.31 shows an induction log where the readings remain constant over several intervals. This log is bad and should be rerun with correctly operating equipment.

THE LATEROLOG

In the 1920s, Conrad Schlumberger put forward the idea of a "guarded electrode" in an attempt to improve on the then-current electrical logs that had undesirable borehole effects. His idea was not put into practice until H. G. Doll designed a working guard-electrode system in 1949. From this starting point, laterologs evolved in a number of ways. The Laterolog 7, which used small guard electrodes, was later joined by the Laterolog 3, which used long guard electrodes (fig. 2.32). Both operated on the same principle: a constant survey current, i_0, was "forced," or focused, into the formation by bucking currents from the guard electrodes. By monitoring the voltage required to maintain the fixed current, i_0, the formation resistivity was measured. From these tools evolved the Conductivity Laterolog, which maintained a constant voltage on the measure electrode while current variations monitored the formation conductivity.

Today, the laterolog tool most commonly used is the 28-ft Simultaneous Dual Laterolog. It is neither a conductivity nor a resistivity laterolog, but rather a hybrid using a constant product of current and voltage (perhaps it should be called a *joule* laterolog). The design of this tool solves many of the problems associated with the earlier laterologs. It is now the standard, basic resistivity tool for logging salt-mud environments.

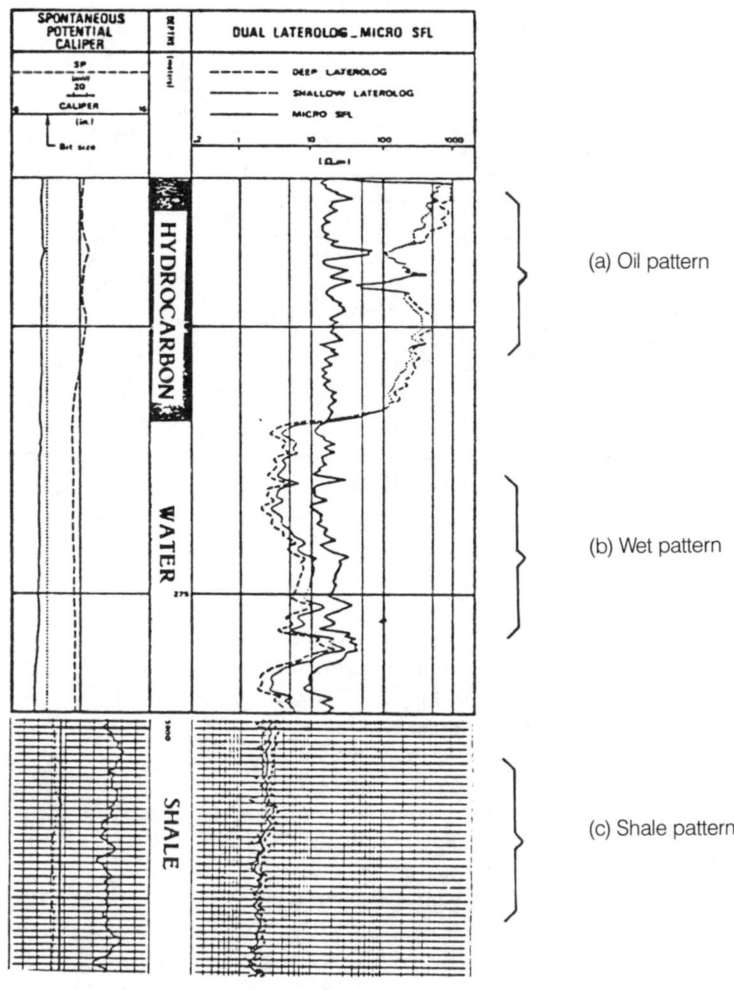

FIGURE 2.38 *Dual Laterolog Patterns.*

When to Use a Laterolog

Laterologs should be used when the following conditions exist:

1. Seawater or brine mud is in the hole.
2. The R_{mf}/R_w ratio is less than 3.
3. Hole size is less than 16 in.

The laterolog will be superior to the induction log when R_t exceeds 100 Ω. It will also give a better estimate of R_t than will the induction log when bed thickness is less than 10 ft.

Figure 2.33 should be referred to when there is doubt as to whether

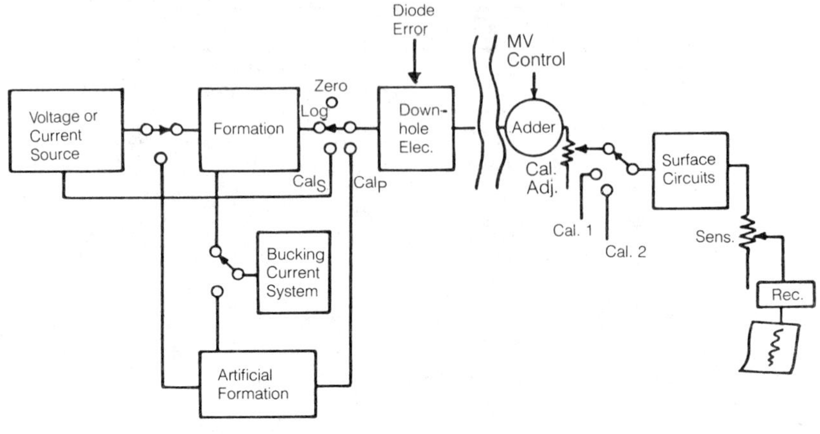

FIGURE 2.39 ***Calibration Schematic for Focused-Current Tools. Courtesy Schlumberger Well Services.***

the laterolog should be run. This figure shows a plot of the R_{mf}/R_w ratio versus porosity. The laterolog is preferred for use when the crossplot of R_{mf}/R_w versus ϕ falls in the left side of the chart.

The Dual Laterolog Tool

The Schlumberger Dual Laterolog tool makes three resistivity measurements. The laterolog deep (LLd), the laterolog shallow (LLs), and a MicroSpherically Focused Log (MSFL). In addition to these resistivity measurements, auxiliary curves such as a caliper, gamma ray, and spontaneous potential may be recorded. The resistivity curves are presented on a standard four-decade logarithmic scale (fig. 2.34).

The Dual Laterolog tool is known as a 28-ft Simultaneous Dual Laterolog and is abbreviated in a number of ways (e.g., DLT, DLL, DST). The part of the tool carrying the caliper and MSFL devices is optional. When it is attached, the tool has a minimum OD of 5¼ in.; when it is not attached, the tool has an OD of 3⅝ in. Temperature and pressure ratings are a standard 350°F and 20,000 psi. Figure 2.35 is an illustration of the tool with its associated measure electrodes.

As this figure shows, the tool is a pad-type device, i.e., once the tool is at the bottom of the well, arms with contact pads on them are extended to fit against the sides of the borehole wall. The mechanics of measuring both a deep and shallow laterolog from a single set of electrodes are handled by circuitry inside the tool. The LLd uses long-focusing electrodes and a distant return electrode while the LLs uses short-focusing electrodes and a near return electrode. The current paths for these devices are shown in figure 2.36. The current paths for the MSFL, which has five rectangular electrodes mounted on a pad carried on one of the caliper arms, are shown in figure 2.37.

Dual Laterolog "Fingerprints"

The characteristic behavior of the DLL tool in zones with movable hydrocarbons makes quick look interpretation very simple. The golden rule is that the pattern $R_{LLd} > R_{LLs} > R_{MSFL}$ (fig. 2.38(a)) is a good indication that hydrocarbons are present. Conversely, the pattern $R_{MSFL} > R_{LLs} > R_{LLd}$ (fig. 2.38(b)) is a good indication that the zone is wet (100% water saturation). Any relative ordering of the curves other than these two cases suggests little or no invasion and indicates that the zone is of low permeability (fig. 2.38(c)).

Calibration

Focused-current systems such as the Dual Laterolog are calibrated by means of a downhole *artificial formation* (see fig. 2.39), which, in most cases, is provided by a precision resistor network. For one type of dual laterolog, the artificial formation is a signal provided by a 31.62-Ω precision resistor in the cartridge. This signal is then divided by exactly 31.62 in the surface panel resulting in a 1-Ω · m calibration signal. The 1000-Ω · m signal is derived similarly by multiplying the downhole signal by 31.62. The final calibration check confirms the validity of the adjustments by reading directly the 31.62-Ω · m output of the cartridge.

BEFORE SURVEY CALIBRATION.* A typical analog unit calibration for the laterolog entails the following steps (refer to fig. 2.40):

1. Mechanical zero: Galvanometer circuits are open. Laterolog deep (LLd) and laterolog shallow (LLs) are set at 1 Ω · m; SP at 0 division of Track 1 and gamma ray (GR) at 10 divisions of Track 1. (Note: the example in fig. 2.40 does not include an SP recording.)
2. Record set: The panel injects a signal into the SP circuit to allow full-scale deflection of any SP scale. The SP galvanometer sensitivity is then adjusted to 10 divisions of Track 1. The LLd and LLs read mechanical zero on this step.
3. Test-1 (electrical zero): The downhole cartridge calibrate relays are actuated, thus switching to a 31.62-Ω calibrate resistor. Panel circuits divide the 31.62 Ω · m signal by 31.62 yielding 1 Ω · m. Laterolog galvanometer circuits are then electrically set to match mechanical zero. SP circuits are shorted and SP reads mechanical zero.
4. Test-2: The cartridge calibrate relay is still actuated, but the surface panel now multiplies the 31.62 Ω · m laterolog signals from the cartridge by 31.62, then adjusted to 1000 Ω · m. SP reads mechanical zero.
5. Tool cal: The cartridge calibrate relay is still actuated but the surface panel is now in the measure position, thus the LLd and LLs read 31.62 Ω · m. This checks the offset of phase sensitive detectors and also the linearity of the surface equipment. SP reads mechanical zero.

*The calibration information that follows has been provided, with permission, by Schlumberger Well Services.

SIMULTANEOUS DUAL LATEROLOG

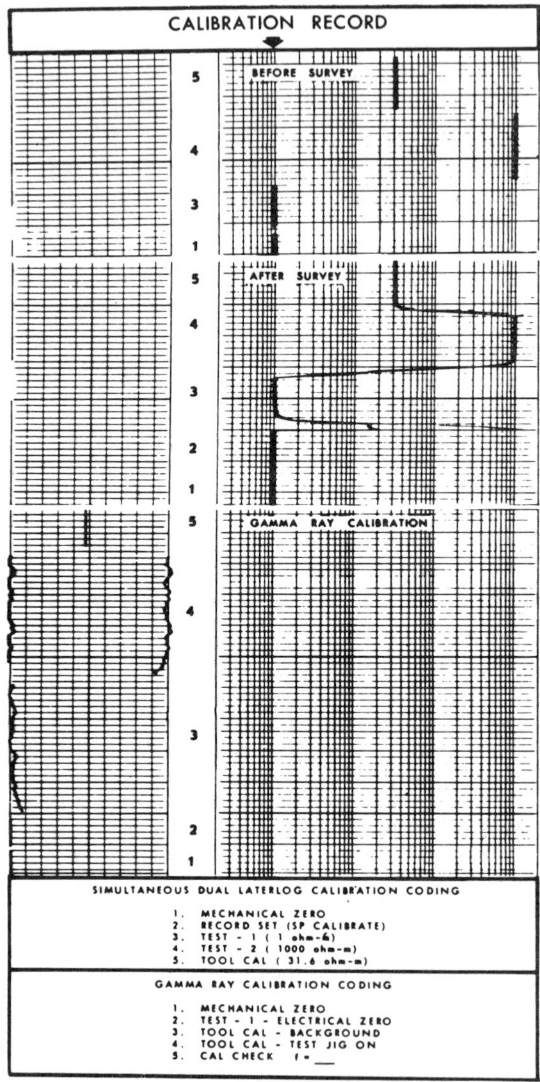

FIGURE 2.40 *Analog Unit Laterolog Calibration Tail. Courtesy Schlumberger Well Services.*

```
                BEFORE SURVEY CALIBRATION SUMMARY

 PERFORMED:      78/10/10
 PROGRAM FILE:   MILL     (VERSION     10.2       78/ 6/27)

DLT                 ELECTRONICS CALIBRATION SUMMARY

              MEASURED                         CALIBRATED
           ①ZERO   ②PLUS                    ZERO        PLUS        UNITS
  LLD       0.0     34.5                     0.0        31.6        OHMM
  LLS       0.0     33.9                     0.0        31.6        OHMM

MSFL                ELECTRONICS CALIBRATION SUMMARY

              MEASURED                        ⑤CALIBRATED
           ③ZERO   ④PLUS                    ZERO        PLUS        UNITS
 MSFL       0.0    1003.                     0.0        999.9       MMHO
   I1       5.8    206.0                     0.0        199.9       MMHO

SGTE                DETECTOR CALIBRATION SUMMARY

              MEASURED
            BKGD      JIG          CALIBRATED            UNITS
   GR        43       195              164               GAPI

MSFL                CALIPER CALIBRATION SUMMARY

              MEASURED                          CALIBRATED
            SMALL    LARGE                    SMALL       LARGE       UNITS
  CALI       9.4      13.6                     8.0         12.0         IN
                 AFTER SURVEY TOOL CHECK SUMMARY

 PERFORMED:      78/10/10
 PROGRAM FILE:   MILL     (VERSION     10.2       78/ 6/27)

DLT                         TOOL   CHECK

                      ZERO                        PLUS
             BEFORE          AFTER          BEFORE       AFTER      UNITS
  LLD          0.0            0.0            31.6         31.6      OHMM
  LLS          0.0            0.0            31.6         31.6      OHMM

MSFL                        TOOL   CHECK

                      ZERO                        PLUS
             BEFORE          AFTER          BEFORE       AFTER      UNITS
 MSFL          0.0           -0.1           999.9        1001.      MMHO
   I1          0.0            0.0           199.9        205.3      MMHO
```

TOLERANCE

Curve	Zero	Plus	Units
LLs	±.1	±2%	ohmm
LLd	±.1	±2%	ohmm
MSFL	±2	±2%	mmho/m

FIGURE 2.41 *Computer-Unit Laterolog Calibration. Courtesy Schlumberger Well Services.*

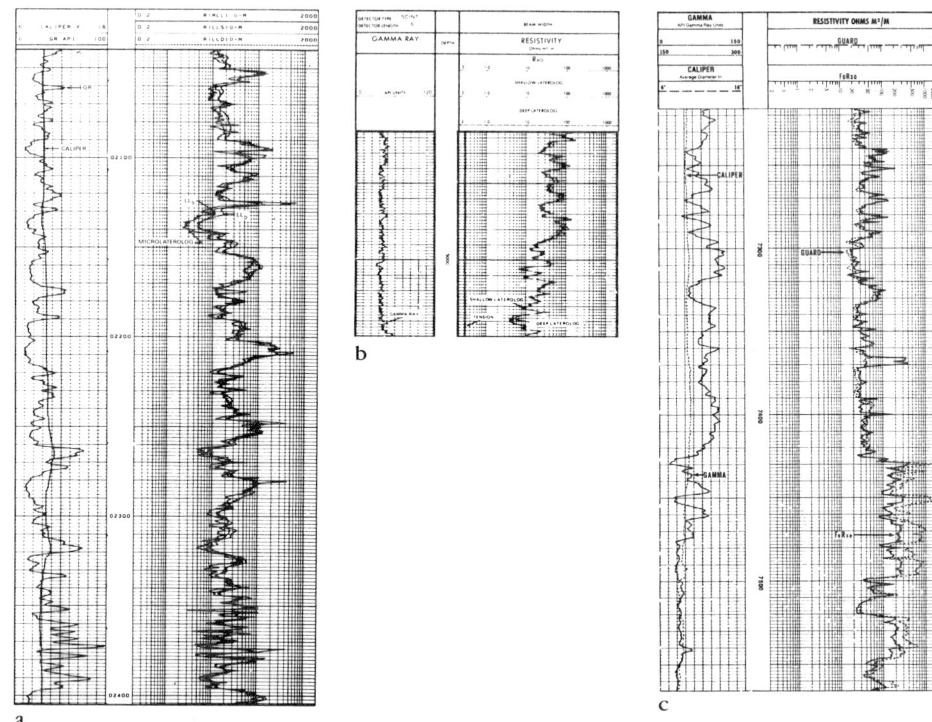

FIGURE 2.42 *Service Company Laterolog Presentations: (*a*) DLL-Microlaterolog (courtesy Gearhart Industries, Inc.), (*b*) Dual Laterolog (courtesy Dresser Atlas), and (*c*) Guard-F_oR_{xo} (courtesy Welex, a Halliburton Company).*

AFTER SURVEY CALIBRATION. Steps 1 through 5 are the same as for the before survey, but are recorded without changing zero or sensitivity settings. The purpose is to verify that the calibrations have not drifted significantly from their proper settings.

COMPUTER-UNIT CALIBRATION. A typical computer-unit calibration for the laterolog is illustrated in figure 2.41. (Refer to the section on Calibration Systems under Log Quality Control in chapter one for an explanation of computer-unit calibrations and to Appendix J for a more complete treatment.) Standard presentation of the log is as illustrated in figure 2.42, which shows logs from a number of different service companies, and figure 2.43, which shows a complete laterolog presentation.

Log Quality Control

The LLd, LLs, and MSFL curves should all be on-depth and track with each other in impermeable or uninvaded formations. Separation is expected in permeable, invaded formations. High readings may be

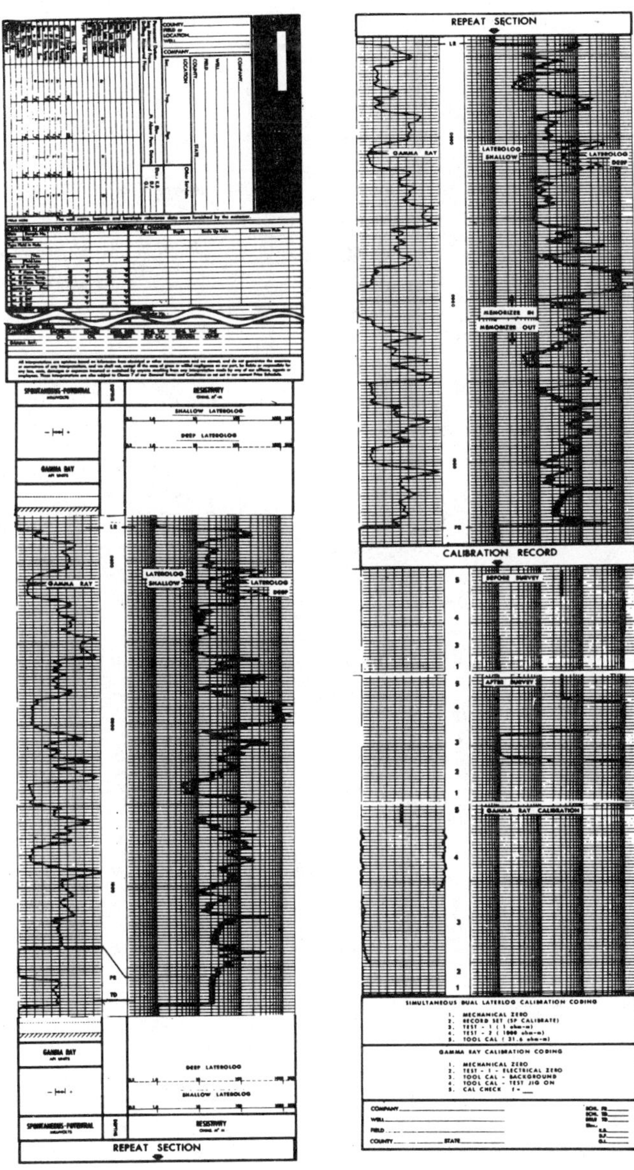

FIGURE 2.43 *Complete Laterolog Presentation. Courtesy Schlumberger Well Services.*

expected in salt and anhydrite formations. Log quality checks should include the following:

1. Recording speed (maximum 5000 ft/hr).
2. Correctness of readings:
   a. Separation between laterolog deep (LLd) and laterolog shallow (LLs) can be confirmed by curve response charts, borehole effect, and/or invasion consequences.
   b. Centralizers are run on wells where hole conditions permit. The type of centralizer is noted on the heading.
3. Test films indicate no change between before and after survey.
4. Test films to be attached:
   a. 200 ft of repeat,
   b. downhole calibration before survey,
   c. downhole calibration after survey, and
   d. gamma ray calibration if GR was recorded.

Field Examples

Figure 2-44 illustrates a bad log. Points to note are (a) serious cyclic noise spikes on the Deep Laterolog, and (b) no time marks.

Figure 2.45 illustrates a repeat section and a main run of a Dual Laterolog. The shallow laterolog is defective on the main run. This log should be rerun.

MICRORESISTIVITY

Microresistivity tools offer a means to perform the following important functions and parameter determinations:

correlation
flushed-zone saturations (S_{xo})
residual oil saturation *(ROS)*
hydrocarbon movability
hydrocarbon density (ρ_{hy})
invasion diameter *(d_i)*
invasion corrections to deep resistivity devices
thin-bed identification

A variety of tools, old and new, are available. Each tool has its own special characteristics. This section will present their names, uses, and idiosyncracies.

Microresistivity Tools

The following list covers the majority of the present-day microresistivity and shallow-focused tools.

16-in. SN	Short Normal Log
LL8	Laterolog 8
SFL	Spherically Focused Log

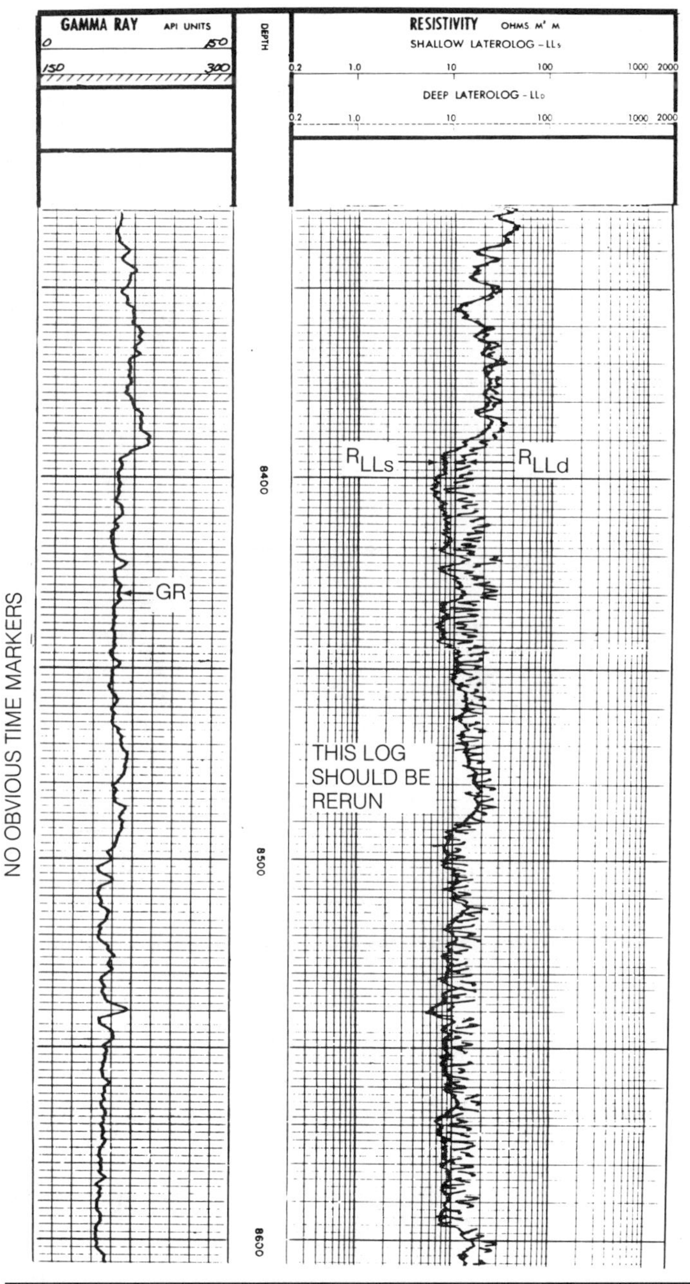

FIGURE 2.44 *Example of Bad Deep Laterolog.*

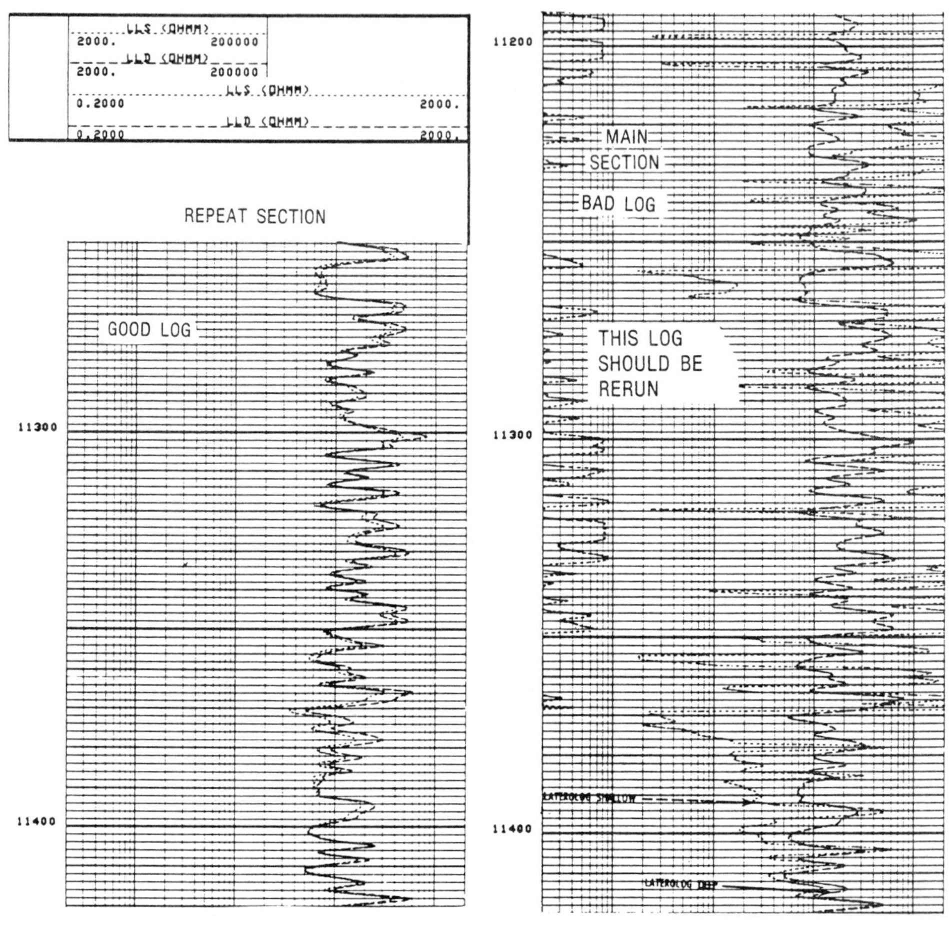

FIGURE 2.45 *Example of Bad Shallow Laterolog.*

MLL	Microlaterolog
PL	Proximity Log
MSFL	Microspherically Focused Log
ML	Microlog

These tools can be divided into two main groups: the *mandrel tools* and the *pad-contact tools.* The mandrel tools have electrodes placed on a cylindrical mandrel that is run into the hole. They do not require physical contact with the formation. The pad-contact tools have their electrodes embedded in an insulating pad that is carried on a caliper arm, which is forced against the borehole wall. The sizes and ratings of the microresistivity tools are summarized in table 2.3. These devices are usually run in combination with some specific deep-resistivity device. For example, the following are usually run together:

TABLE 2.3 *Microresistivity Tool Types and Ratings*

Name	OD (inches)	Type
16-in. SN	2¾ to 3⅞	Mandrel
LL8	3⅜ & 3⅞	Mandrel
SFL	3½	Mandrel
MLL	5⅟₁₆	Pad
PL	5½	Pad
MSFL	5¼	Pad
ML	5⅟₁₆	Pad

MSFL and dual laterolog
LL8 and dual induction
SFL and induction or dual induction log
SN and induction log
MLL and microlog
PL and microlog

The microlog is worthy of special mention since it is an underrated device that should be run more frequently than it is. It was one of the first microresistivity devices on the market and has had a spectacular career. Originally, it was used as a pseudoporosity device. When that function was improved upon with modern porosity devices, the micrsame was relegated to the pile of has-beens by many people in the logging industry. But it is still a valuable tool because it offers a superb visual identification of porous and permeable zones. Figure 2.46 shows a micrology and proximity log presentation. The presence of permeability is indicated wherever the micronormal curve reads higher than the microinverse curve, and the microinverse curve reads close to R_{mc}. That is, the microlog, recording two resistivity curves at shallow depths, looks for a resistivity contrast between the mudcake and the flushed zone. If no porosity or permeability is present in the formation, there is no filtrate invasion, and thus no mudcake buildup; hence, there will be no separation between the resistivity curves.

Depth of Investigation

Each microresistivity tool has its characteristic depth of investigation. It is important to know what this is for each of the tools in order to select the right one for the job. If invasion is shallow, then a tool with a shallow depth of investigation is needed if the tool is to read R_{xo} without undue influence from R_t. Conversely, in situations where deep invasion exists, a deep-investigation tool will assure a reading of R_{xo} free from any effects of R_{mc}.

As with other tools, no single value for the depth of investigation can be used. Rather, a pseudogeometric factor must be used. This factor indicates how much of the total tool signal is received from an

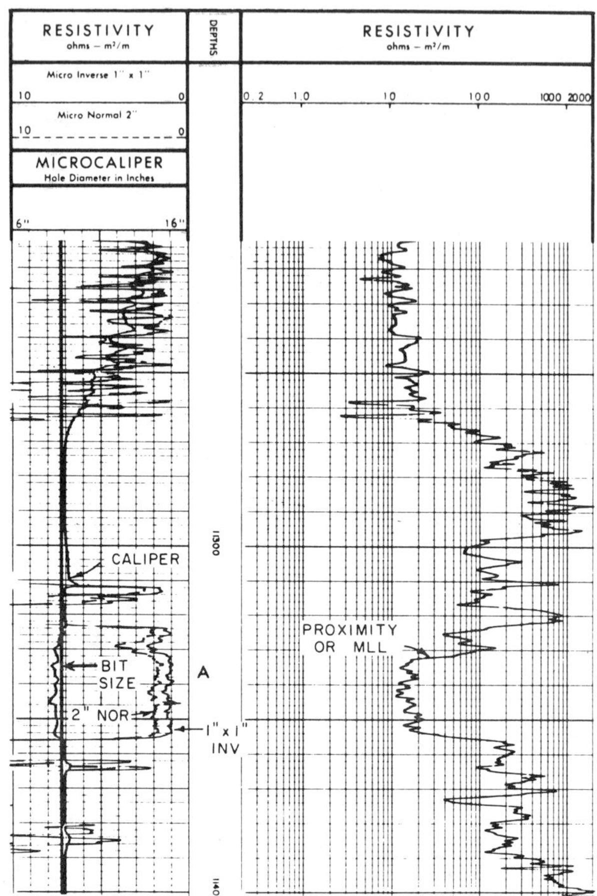

FIGURE 2.46 *Presentation of Proximity Log–MicrYolog. Courtesy Schlumberger Well Services.*

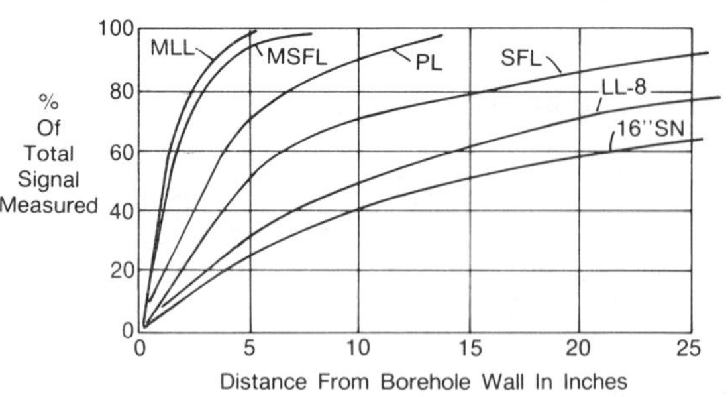

FIGURE 2.47 *Depth of Investigation of Microresistivity and Shallow-Focused Tools.*

TABLE 2.4 *Microresistivity and Shallow Focused Tool Depths of Investigation*

Rank	Tool	90% Response at (inches)
1. ML	Microlog	1
2. MLL	Microlaterolog	4
3. MSFL	Microspherically Focused	4½
4. PL	Proximity Log	10
5. SFL	Spherically Focused Log	24
6. LL8	Laterolog-8	49
7. SN	16-in. Short Normal	70+

annular formation volume represented by distances (expressed in inches) from the borehole wall (see fig. 2.47 and table 2.4).

Bed Resolution

Just as each of the microresistivity tools has its characteristic depth of investigation, so too does each tool have its own characteristic bed resolution; that is, some tools are better than others at distinguishing thin beds. Tools with large-bed resolution values are "blind" to thin shale and/or sandstone layers. For example, 3-in. shale streaks will not be "seen" by a short normal log. The bed resolution of each microresistivity device is shown in table 2.5.

Environmental Corrections

Microresistivity devices of the mandrel type are subject to aberrations due to the size of the wellbore. These effects can be quite severe. The pad-contact tools, however, are only affected by excessive mudcake thickness. Provided that pad contact with the formation is maintained, the pad-contact tools are unaffected by the size of the wellbore. Refer to service-company chart books for details on these corrections.

TABLE 2.5 *Bed Resolution of Microresistivity Tools*

Rank	Tool	Minimum Bed Resolution (inches)
1. MLL	Microlaterolog	1½
2. ML	Microlog	3
3. MSFL	Microspherically Focused	12
4. PL	Proximity Log	12
5. SFL	Spherically Focused Log	12
6. LL8	Laterolog-8	12
7. SN	16-in. Short Normal	18

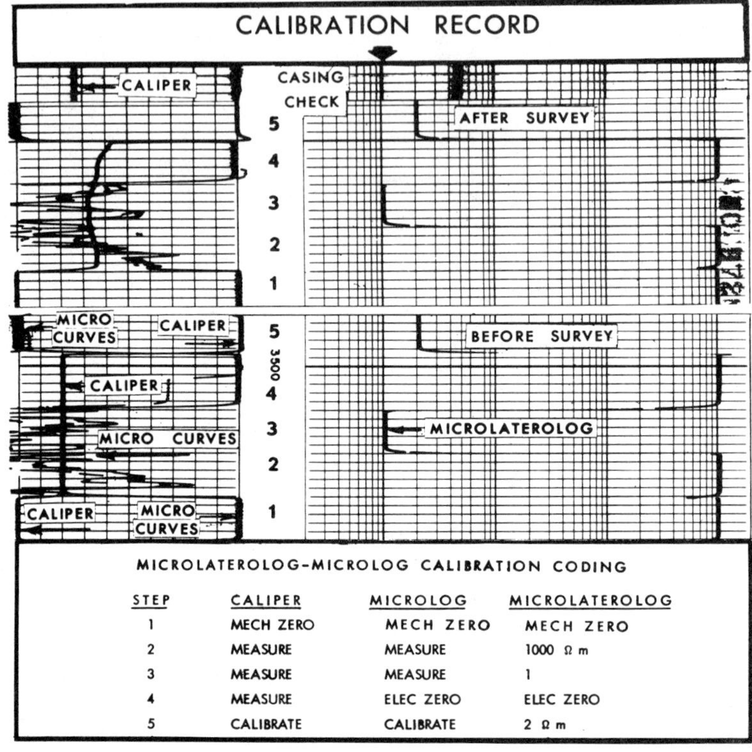

MICROLATEROLOG-MICROLOG CALIBRATION CODING

STEP	CALIPER	MICROLOG	MICROLATEROLOG
1	MECH ZERO	MECH ZERO	MECH ZERO
2	MEASURE	MEASURE	1000 Ω m
3	MEASURE	MEASURE	1
4	MEASURE	ELEC ZERO	ELEC ZERO
5	CALIBRATE	CALIBRATE	2 Ω m

FIGURE 2.48 *Calibration Record for the Microlaterolog–Microlog.*

*Calibration of the Microlaterolog–Microlog and the Proximity–Microlog**

PRINCIPLES OF CALIBRATION. The accuracies of the Microlaterolog, Proximity-Log, Microlog, and Caliper measurements rely upon automatically regulated downhole circuits. These produce known cable signals for each set of conditions over the various ranges of response. A small diode-error signal on the Microlaterolog curve is cancelled during calibration.

The surface circuits and galvanometers are first adjusted by means of precisely regulated calibration signals from the panel. The caliper circuit is adjusted at the surface by opening the arms in a gauge ring and setting the measure circuit so as to read the correct diameter. The log circuits are then adjusted with the tool in the hole, by means of regulated calibration signals generated in the electronic cartridge.

*The following calibration and log quality-check information has been provided, with permission, by Schlumberger Well Services.

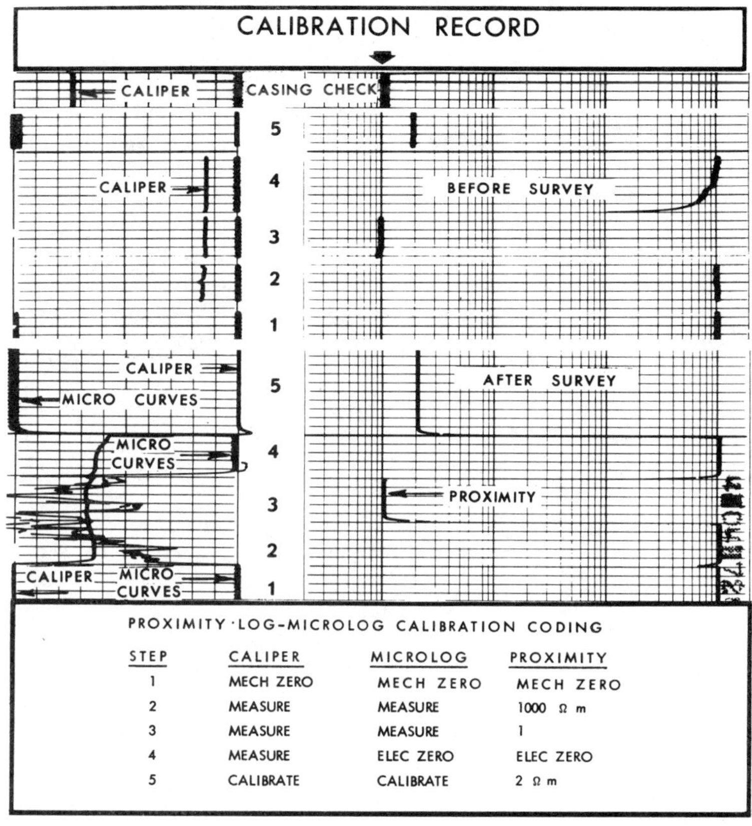

STEP	CALIPER	MICROLOG	PROXIMITY
1	MECH ZERO	MECH ZERO	MECH ZERO
2	MEASURE	MEASURE	1000 Ω m
3	MEASURE	MEASURE	1
4	MEASURE	ELEC ZERO	ELEC ZERO
5	CALIBRATE	CALIBRATE	2 Ω m

FIGURE 2.49 *Calibration Record for the Proximity Log–Microlog.*

Cartridge circuitry is the same in the Microlaterolog–Microlog and the Proximity–Microlog combination tools.

Sample calibration records for the Microlaterolog–Microlog and the Proximity–Microlog are shown in figures 2.48 and 2.49.

BEFORE SURVEY CALIBRATION

1. Mechanical zero: Galvanometer circuits are open; Microlaterolog—or Proximity-Log—galvanometer at 1000 Ω · m; microlog galvanometers at zero of Track 1 (right edge of track); caliper galvanometer at left edge of Track 1.
2. "1000" position:
   a. Microlog and caliper are in the measure position (caliper galvanometer may be off-screen if tool is closed. Microlog signals may respond to formation).
   b. The downhole signal is removed from the Microlaterolog—or

Proximity-Log—circuit and a 1-mmho (1000-Ω · m) formation signal is injected. Galvanometer reads 1000 Ω · m.

3. "1" position:
   a. Caliper and Microlog remain on measure.
   b. The downhole signal is removed from the Microlaterolog—or Proximity-Log—circuit and a panel signal equivalent to a 1000-mmho (1-Ω · m) formation signal is injected. Galvanometer reads 1 Ω · m.
4. "1000-Ω · m" position (electrical zero):
   a. Caliper remains on measure.
   b. The Microlaterolog—or Proximity-Log—and Microlog zero relays are actuated in the cartridge.
   c. A 1-mmho (1000-Ω · m) signal is injected from the panel to the Microlaterolog—or the Proximity-Log—galvanometer, which is then adjusted to 1000 Ω · m, thus cancelling any diode error.
   d. The microlog galvanometers are adjusted to read zero.
5. "2-Ω · m" position:
   a. The Microlog and Microlaterolog—or Proximity-Log—calibrate relays are actuated, removing any sonde signals.
   b. A caliper signal equivalent to 10 in. is injected by the cartridge; since the caliper zero is normally at 6 in., the galvanometer will read 16 in. (right edge of Track 1).
   c. The cartridge puts in a 5-Ω · m signal to the microlog channel and the panel is automatically switched to the 5-Ω · m scale. The galvanometers will read full scale (left edge of Track 1).
   d. The cartridge puts in a 500-mmho signal to the Microlaterolog—or the Proximity-Log—channel. The galvanometer reads 1 Ω · m.

AFTER SURVEY CALIBRATION. The before survey sequence is repeated, except that no controls are changed from their logging positions. Any circuit drifts will show up as differences from the before survey settings.

Quality Control

SPECIFIC LOG QUALITY CHECKS*

1. Recording speed: 2000 to 2500 ft/hr.
2. Correctness of readings:
   a. No negative readings.
   b. All ratios derived from comparing R_{MLL}—or R_{PL}—to resistivity of other devices can be explained by invasion and R_{xo}/R_t values.
   c. No large positive separation on ML resistivity curves with the ratio $R_{2''}/R_{1''}$ greater than 2. (Except oil zone with high ROS, or irregular holes.)
   d. Caliper checks. (Caliper checks casing ID. It may be necessary to record several feet up into casing due to the presence of a cement sheath inside the casing.)

*This quality check information was provided by Schlumberger Well Services.

3. Test films indicate no change between before and after survey.
4. Test films to be attached:
   a. 200 ft of repeat,
   b. mud log (choose largest part of the hole),
   c. downhole calibration before survey,
   d. downhole calibration after survey, and
   e. caliper check in casing (recorded on each logging run).

IN GENERAL. Quality control for microresistivity devices can be summarized by the following maxims.

Beware of washed out holes because (a) pad-contact tools lose contact with the formation and "float" in the mud column and (b) mandrel tools give inaccurate readings.

Beware of thick mudcakes because pad-contact tools require large corrections.

If hole conditions are bad, forget about trying to measure R_{xo} because either the tool will stick or the pad will tear up. Either way, no usable log reading will be obtained.

Field Examples

BAD SHORT NORMAL. Figure 2.50 illustrates a section of log where the Short Normal failed somewhere above 700 ft. Proper evaluation of the bed at 510 to 540 ft is not possible.

INTERMITTENT SFL. Figure 2.51 shows an SFL log with intermittent spikes appearing every 13 ft on the log. Apart from the spikes, the actual values recorded are acceptable. It would probably be unnecessary to rerun this log.

DIELECTRIC LOGGING

Tools Available

There are two logging devices that make use of the differing dielectric constants of formations: the Dielectric Constant Log (DCL) marketed by Gearhart, and the Electromagnetic Propagation Tool (EPT) marketed by Schlumberger.

The dielectric constant of water differs by an order of magnitude from that of oil, gas, or rock matrix (see table 2.6). The dielectric constant of a formation is greatly dependent on the formation water saturation; moreover, it has been shown to be essentially independent of water salinities in relatively freshwater formations. Dielectric-constant tools have proved their usefulness in places where formation waters are so fresh that resistivity tools cannot differentiate between hydrocarbons and water. They will undoubtedly be helpful also in areas of low but variable water salinity, and possibly elsewhere.

Empirical equations relating porosity and water saturation to the dielectric-constant measurements have been developed for both the

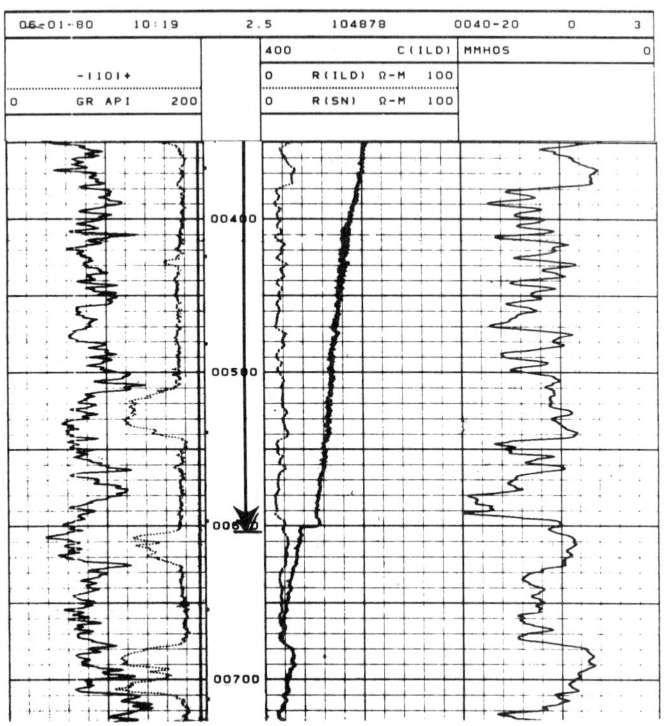

FIGURE 2.50 *Short Normal Example.*

DCL and EPT logs. However, more development and experience are required to demonstrate that reliable quantitative data can be derived from the measurements.

Overview of Theory

Electromagnetic waves propagated through a medium are affected by three parameters: (1) conductivity, (2) dielectric constant or dielectric permittivity, and (3) magnetic susceptibility. At low frequencies (i.e., in the kilohertz range), conductivity has the largest effect on electromagnetic wave propagation. As the frequency is increased, the influence of dielectric permittivity (dielectric constant) increases.

A dielectric is a material that has a low electrical conductivity compared to metals. It is characterized by its *dielectric constant* and *dielectric loss,* both functions of frequency and temperature. The dielectric constant is an independent, well-defined, basic electrical property of matter.

Water is one of the very few materials found abundantly in nature that has permanent electric dipoles and thus high dielectric constants. Air is taken as the normalizing medium and is given a dielectric constant of 1. In comparison, water has a dielectric constant of approx-

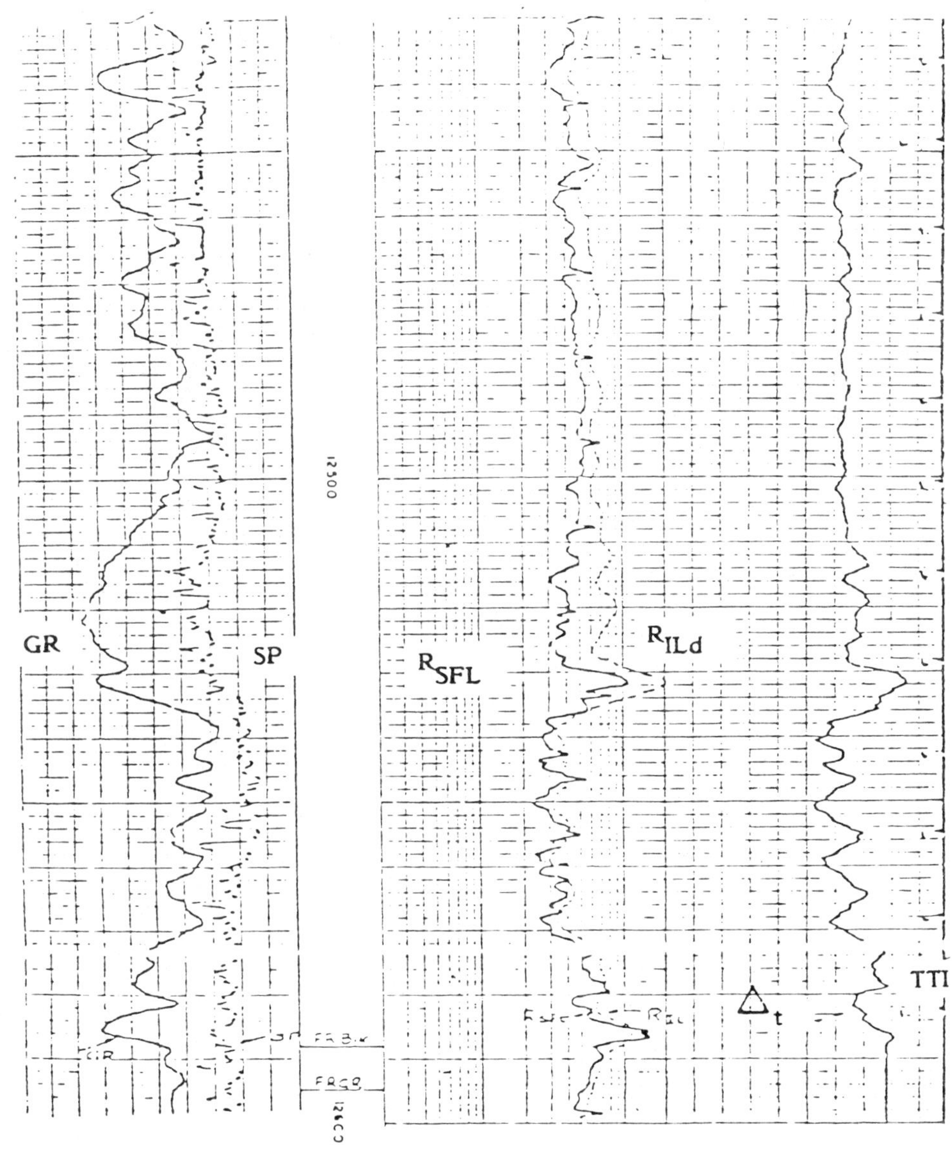

FIGURE 2.51 *Intermittent SFL.*

TABLE 2.6 *Dielectric Constants of Sundry Materials*

Material	Relative Dielectric Permittivity
Gas or air	1.0
Oil	2.2
Water	56–80
Quartz	4.7
Limestone	7.5
Dolomite	6.9
Anhydrite	6.5

imately 80, while crude oil, gas, and rock matrix have dielectric constants in the order of 2 to 11 (table 2.6). Therefore, the dielectric-constant logs should be able to differentiate between oil-bearing and water-bearing formations when resistivity measurements cannot.

Magnetic susceptibility has negligible effects on electromagnetic wave propagation at the high frequencies used by the dielectric-constant tools. However, it may become a factor if much ferrous material is encountered.

Neither the DCL or the EPT measures the dielectric constant of the formation directly. These tools measure the phase shift and attenuation of a microwave-frequency electric field propagated through the formation. These two parameters are directly related to the dielectric constant of the formation, which is largely governed by its water content. The theory is that measurements made in the frequency range where the dipole polarization of water dominates should lead to a derivation of water content that is practically independent of salinity except at high concentrations. This theory is supported by considerable field experience.

Tool Configurations and Limitations

The DCL and EPT tools differ significantly in design, and have different depths of investigation and different limiting conditions.

DCL TOOL. The Dielectric Constant Log marketed by Gearhart is a mandrel-type tool developed by Texaco Research at their laboratory in Bellaire, Texas. It operates at a frequency of 30 MHz. A schematic of the tool is shown in figure 2.52. Its depth of investigation is probably several feet. In low-resistivity muds ($R_m < 1\ \Omega$), results are improved by running long, soft-rubber fins that break up borehole eddy currents.

This tool works best in clean, freshwater formations with $R_t \approx 10$ or $15\ \Omega \cdot \text{m}$. Best results are obtained when the tool is centralized. Deep invasion effects are very significant, especially when filtrate resistivities differ greatly from formation-water resistivities. Repeatability is poor when logging in high-salinity muds. Resistivity from the DCL tool seldom agrees with resistivities measured by other logging tools; how-

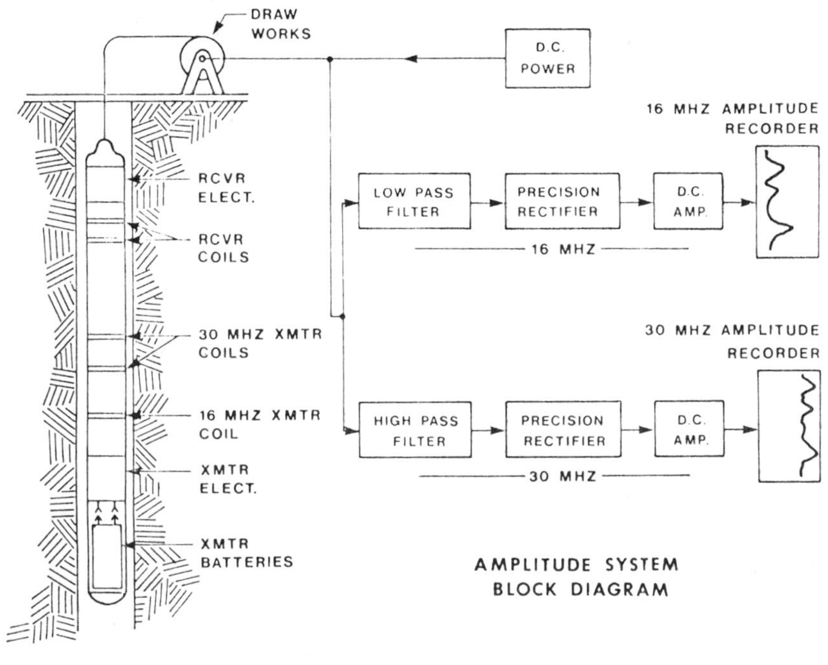

FIGURE 2.52 *DCL Tool Schematic. Courtesy Gearhart Industries, Inc.*

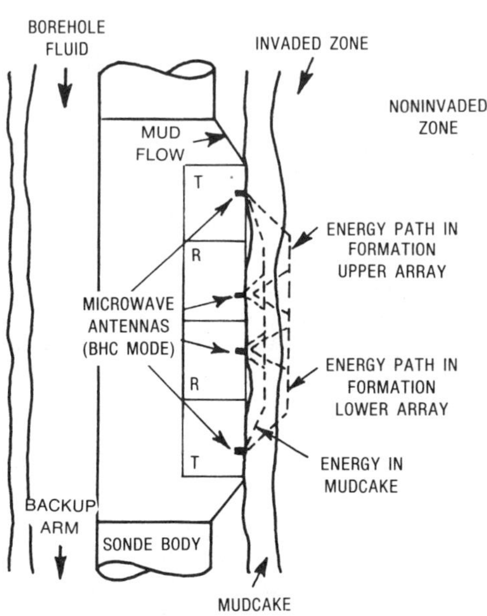

FIGURE 2.53 *EPT Tool Schematic. Courtesy Schlumberger Well Services.*

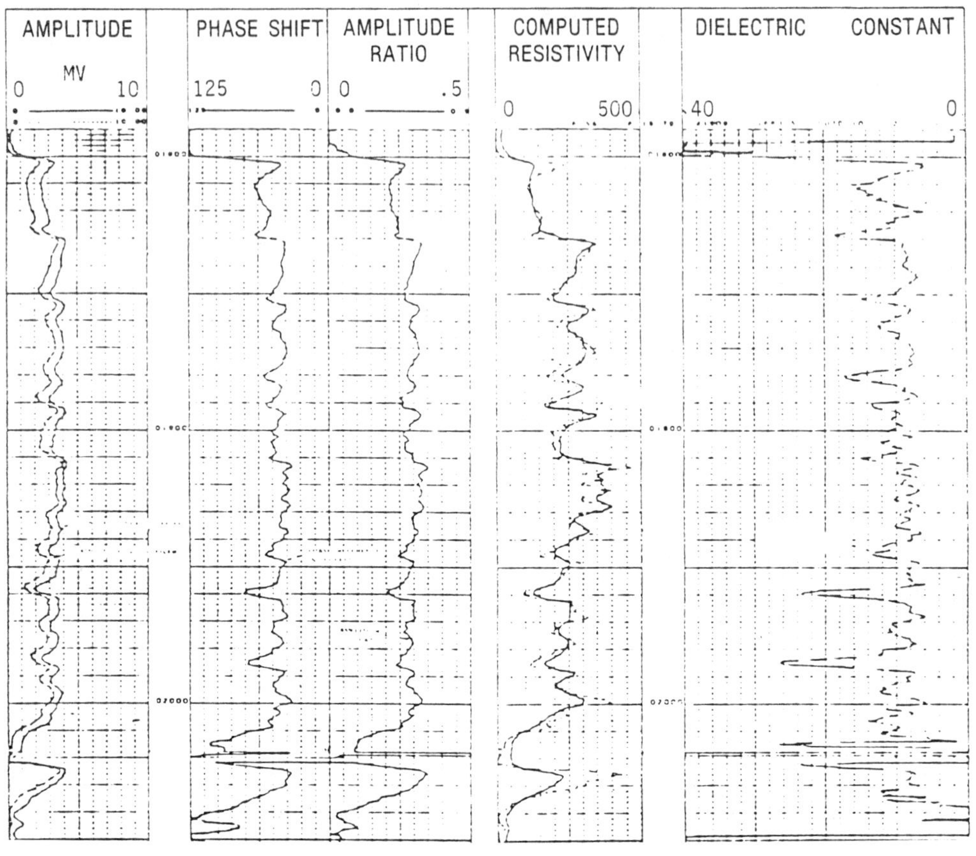

FIGURE 2.54 *DCL Log Format. Courtesy Gearhart Industries, Inc.*

ever, the curves have similar character and can be correlated. The difference in resistivity measurements arises largely from the differences in frequency and depth of investigation. DCL resistivity may be more accurate than induction log resistivities when formation resistivities are greater than 100 Ω · m. The DCL can be run inside fiberglass casing.

EPT TOOL. The Electromagnetic Propagation Tool marketed by Schlumberger consists of two transmitting and two receiving antennas arranged in a vertically symmetric configuration, T-R-R-T, on a metallic pad, which is pressed against the borehole wall (see fig. 2.53). The spacing between the transmitting and the closest receiving antenna is 8 cm. The distance between the receiving antennas is 4 cm. The tool operates at a frequency of 1.1 GHz. Its depth of investigation is approximately 4 cm; thus the EPT samples the invaded or flushed zone of the formation. Use of the two-transmitter–two-receiver system re-

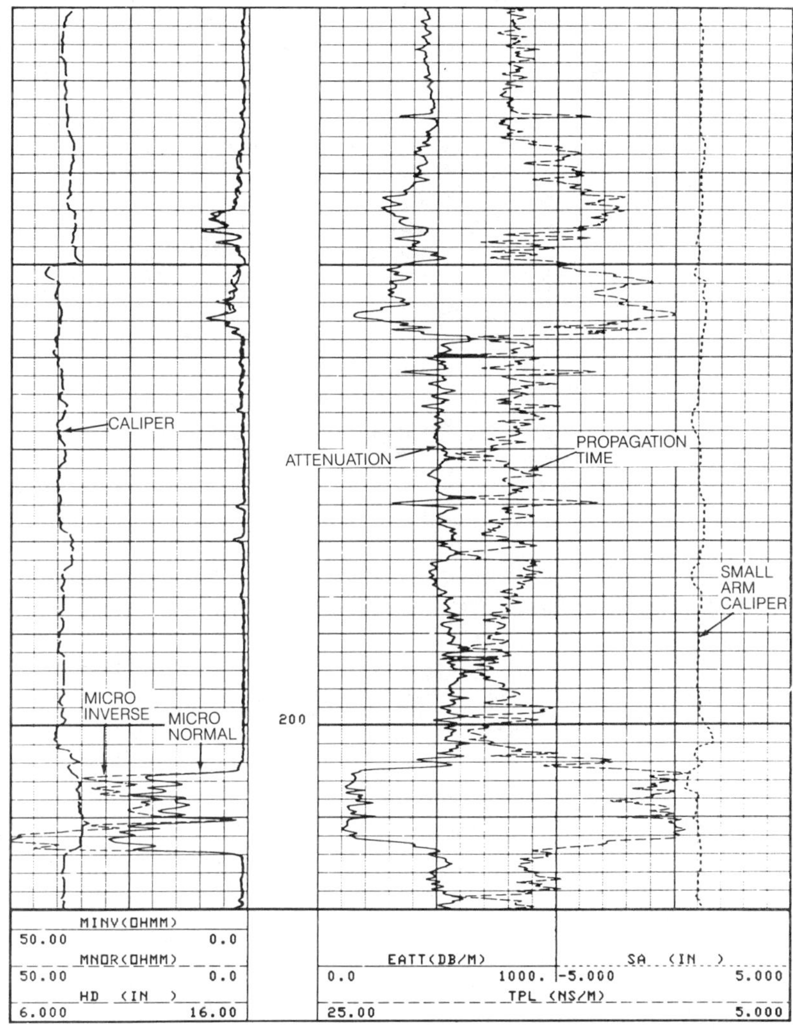

FIGURE 2.55 *DPT Log Format. Courtesy Schlumberger Well Services.*

sults in cancellations or reduction of the effect caused by mudcakes, irregularities, or pad tilt.

The EPT tool is designed for fresh-mud logging applications in 6¼- to 11-in. boreholes. It must be run in an open hole. The tool investigates only the flushed or invaded portion of the formation. The signals are unreliable in oil- or gas-filled boreholes, and may be unreliable in salt mud. Good pad-to-wall contact is required. Rugose boreholes may result in unreliable data. Reliable data can be obtained up to a maximum mudcake thickness of 1 cm (⅜ in.). The maximum temperature rating of the tool is 210°F. Signal levels become too low for reliable interpretation when formation resistivities are less than 2 or 3 Ω · m. Experience indicates that if the power level at the far receiver drops below −50 dB/m, the signal is too weak for reliable detection, and a flag on the log shows when this condition exists.

Log Formats

A sample of the DCL format is shown in figure 2.54. In Track 1, two amplitude curves are presented. They are used for quality control. Generally, when the far-receiver amplitude drops below 0.1 V the data should not be used. In Track 2, the phase-shift curve is presented. An amplitude-ratio curve is presented in Track 3. The computed relative-dielectric-constant curve and resistivity curve shown in Tracks 4 and 5 are optional.

An example of an EPT log format is shown in figure 2.55. Track 1 presents a caliper together with micronormal and microinverse curves. Tracks 2 and 3 present the travel time (TPL) and the attenuation (EATT). An additional small arm caliper (SA) is shown on Track 3 to monitor borehole rugosity.

Calibration

A single-point measurement in air is used to calibrate the amplitude ratio and phase difference for the DCL tool. The EPT reference signals in the sonde circuitry introduce phase-shift and attenuation signals of known value, which are used to calibrate the tool. A further validation check is made by a tool-in-air measurement prior to logging.

CHAPTER THREE

SONIC (ACOUSTIC) LOGGING

Sonic tools measure the speed of elastic waves in subsurface formations. These measurements are useful for a number of reasons and are used by many professional disciplines. Specifically, the measurements may be used to calculate, measure, detect, and/or estimate the following:

porosity
lithology
integrated travel time
true time scales
fractures
synthetic seismograms
bond between casing, cement, and formation

Curves recorded on a typical sonic log (see fig. 3.1) are the interval transit time, Δt, in microseconds per foot (the reciprocal of speed); integrated travel time; caliper, gamma ray; and/or SP.

THE BHC AND LSS SONIC LOGS

Tools Available

Tools available for sonic measurement include the borehole compensated (BHC) tool, a slim-tool version that can be run through tubing, and the long-spacing sonic (LSS) tool. Typical ratings for these tools are shown in table 3.1.

Operating Principle

Sonic tools are comprised of transmitter transducers that convert electrical energy into mechanical energy and receiver transducers that do the reverse. In its simplest form, the measurement is made in an uncompensated mode (fig. 3.2). At time 0 (T_0), the transmitter emits a small shock wave, which travels through the mud to the borehole wall where it is refracted through the formation. Part of the energy traveling through the formation in turn is refracted back into the mud column and finds its way to the first receiver at time T_1, and to the second receiver at time T_2. The difference in the two times is referred to as Δt and represents the time taken for a compressive wave to travel through the formation a distance equal to the spacing between the two receivers. If this distance is one foot, then the formation travel

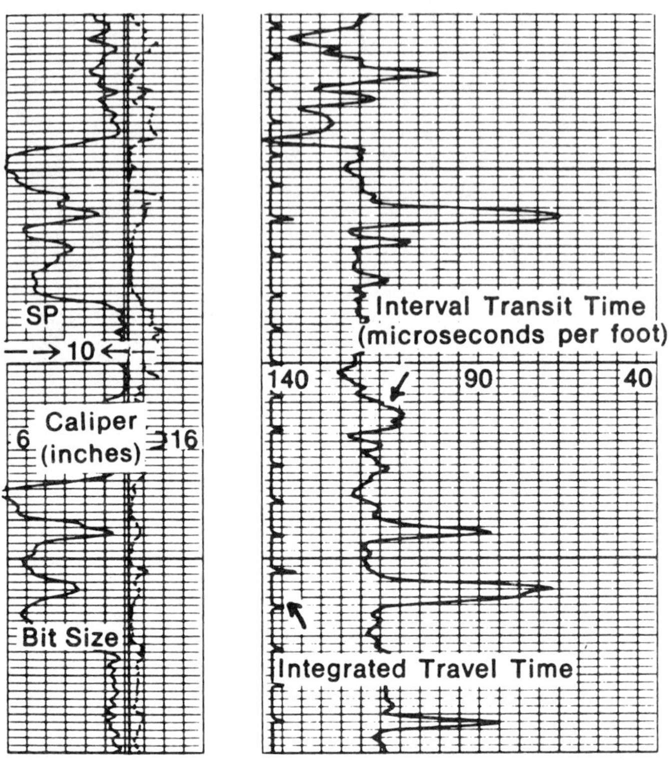

FIGURE 3.1 *Presentation of Sonic Log. Courtesy Schlumberger Well Services.*

time, Δt, is expressed in microseconds per foot. See values for some common media in table 3.2.

Early sonic tools assumed that the travel paths through the mud to the two receivers were equal. This was true in the case of a smooth borehole of unchanging size, but was not true if the borehole was of varying size or if the sonde tilted with respect to the axis of the borehole. These difficulties were overcome by the introduction of the BHC (borehole compensated) sonic tool.

Figure 3.3 illustrates the principle of the BHC sonic tool. Two transmitters and four receivers are employed and two values of Δt are measured and averaged. The net result of this system is the elimina-

TABLE 3.1 *Sonic Tools Available*

Tool	OD (inches)	Max. Psi	Max. °F
Borehole compensated	3⅜ or 3⅝	20,000	350
Slim-hole	1 11/16	16,500	300
Long-spacing	3⅝	20,000	350

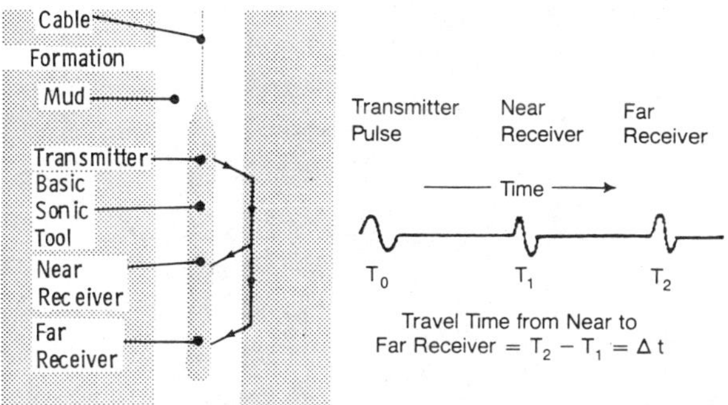

FIGURE 3.2 *Basic Sonic Device.*

tion of errors in Δt due to sonde tilt and hole-size variation. Even so, there are practical limits to the working range of the tool (e.g., in large holes).

The long-spacing sonic (LSS) tool was introduced in an attempt to overcome environmental problems. For example, when a shale formation is drilled, the shales exposed to the mud frequently change their properties by absorption of water from the drilling mud. The travel time for elastic waves therefore changes too. In order to read the travel time in the undisturbed formation further from the borehole, a longer transmitter–receiver spacing is required. Typically, a long-spacing sonic tool will have a transmitter–receiver spacing of 8, 10, or 12 ft.

TABLE 3.2 Δt_{ma} *for Common Minerals and Fluids*

Material	Travel Time Δt (μs/ft)	Velocity (ft/s)	Density g/cc
Dolomite	43.5	23,000	2.87
Limestone	47.5	21,000	2.71
Sandstone	55.6	18,000	2.65
Anhydrite	50.0	20,000	2.97
Gypsum	52.5	19,000	2.35
Salt	67.0	15,000	2.03
Water (fresh)	200	5,000	1.00
Water (100,000 ppm NaCl)	189	5,300	1.06
Water (200,000 ppm NaCl)	176	5,700	1.14
Oil	232	4,300	
Air	919	1,088	
Casing	57	17,000	

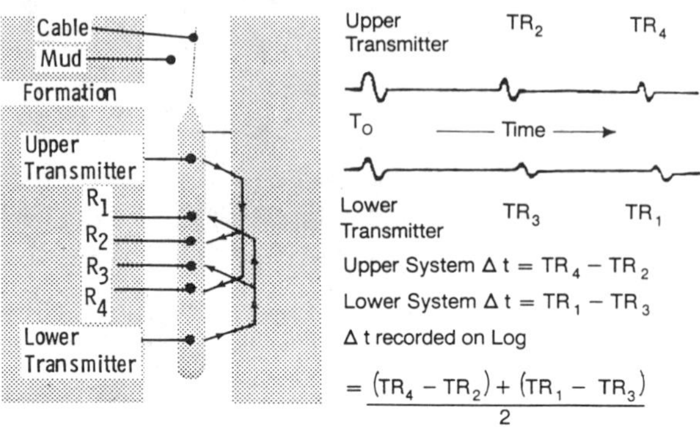

FIGURE 3.3 *Borehole Compensated Sonic Tool.*

Lengthening the spacing on a sonic device achieves two ends:

1. A valid sonic log may be recorded in a bigger hole with a long-spacing device than with a conventionally spaced tool.
2. The zone investigated by the tool is deeper into the formation with a long-spacing device.

Figure 3.4 illustrates maximum detectable travel time as a function of spacing. Figure 3.5 illustrates depth of investigation of a sonic device as a function of spacing.

The long-spacing tools make their measurements in a *depth-derived* mode. That is, the tool achieves borehole compensation by memorizing travel times measured when the tool is at one depth and combining those with travel times recorded at a shallower depth when an alternate combination of transmitters and receivers is activated.

Figure 3.6 shows a comparison of a conventional borehole compensated (BHC) sonic log with a long-spacing sonic (LSS) log.

Figure 3.7 shows a typical sonic wave train as "seen" at the receiver. Compressional waves, taking different paths from the transmitter to the receiver, take different times to arrive. The actual travel-time measurement is determined at the first arrival peak, E_1. However, the tool's internal trigger mechanism for detecting this peak is subject to some errors. Figure 3.8 illustrates two common problems: (1) The bias level is set too high and the travel time is triggered on the E_3 peak, causing an erroneously long time to be measured (this is known as *cycle skipping*). And (2) the bias is set too low and the travel time is triggered by noise, causing an erroneously short travel time (this problem is known as *noise*). It is not always possible to distinguish the difference between cycle skipping and noise in the BHC, since two measurements are effectively averaged by the tool. Figure 3.9 illustrates a bad sonic log, which had to be rerun due to cycle skipping and noise.

All current sonic equipment provides a CRT monitor on which

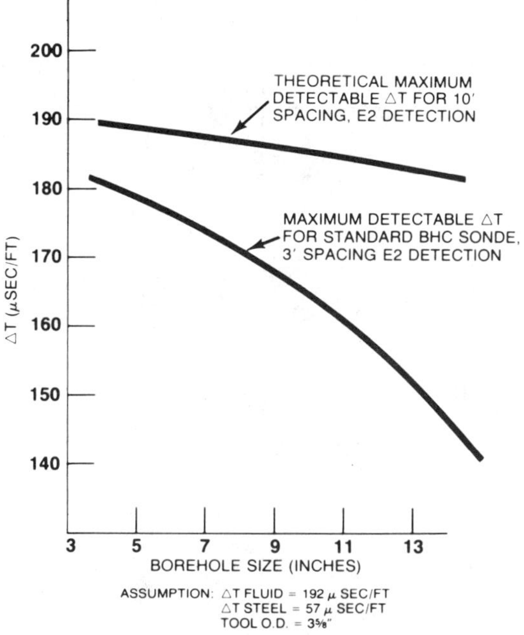

FIGURE 3.4 ***Maximum Detectable Δt for 3- and 10-ft Spacings.***

received waveforms can be observed. Equipment varies, but the correct scope picture is shown in company literature. With the logging engineer's help learn what to look for, and be sure that the scope picture is used throughout the log. This is the most important single factor in assuring correct sonic logs.

Seismic Applications of Sonic Logs

Sonic logs began as an adjunct to seismic exploration, and continue as an important source of calibration data. The integrated-time log is available as part of a sonic log. Time ticks representing 1 ms of travel time appear on one side of the depth track (fig. 3.10). Totaling these ticks over any log interval gives the acoustic travel time for that interval of formation. (Computer units now do this job automatically.) The accuracy of the integrated-time ticks can be checked by picking a section with a constant Δt. There, the depth interval for a 1-ms travel time should equal $1000/\Delta t$ ft. In casing, for example, with $\Delta t = 56$ μs/ft, the millisecond time ticks should appear every 17 or 18 ft.

Check-shot surveys are a form of sonic logging used to calibrate a seismic survey over long intervals of formation. A geophone is lowered to a predetermined depth, which is chosen according to the seismic record. An airgun generates an acoustic pulse at the surface, and a time-indexed record is made of the geophone's response.

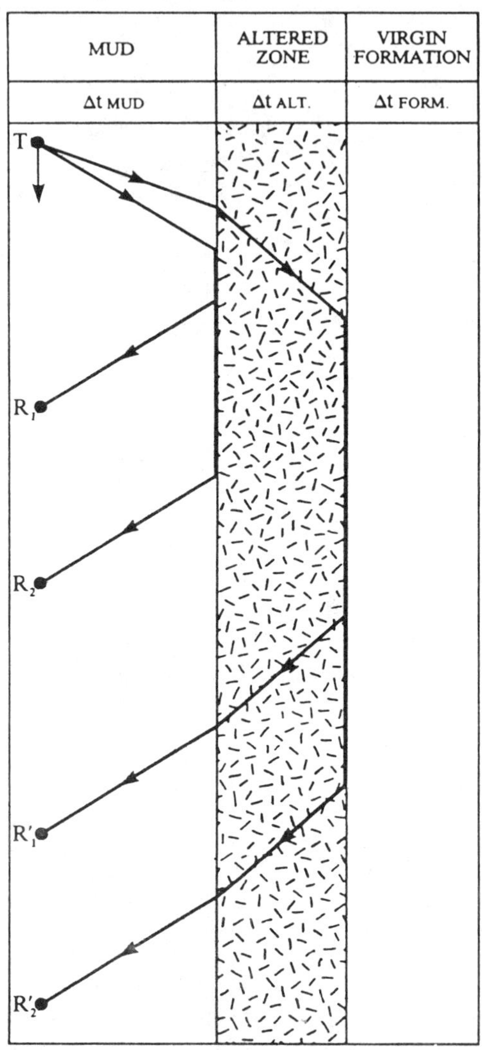

FIGURE 3.5 *Depth of Investigation of Sonic Devices. Courtesy Schlumberger Well Services.*

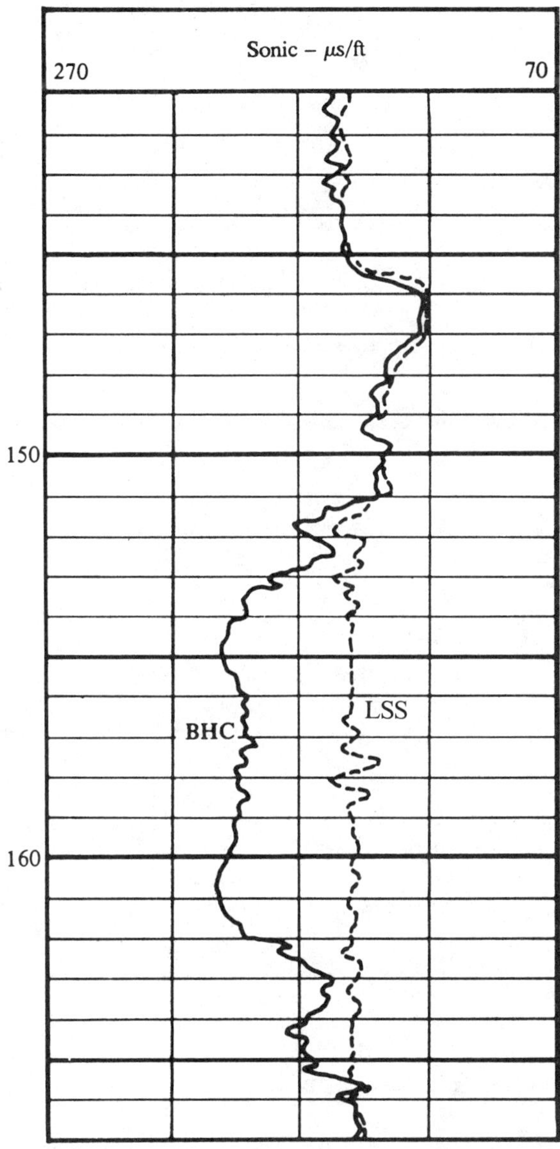

FIGURE 3.6 *Comparison of the BHC Sonic Log and LSS Sonic Log Through Interval with Shale Alteration. Courtesy Schlumberger Well Services.*

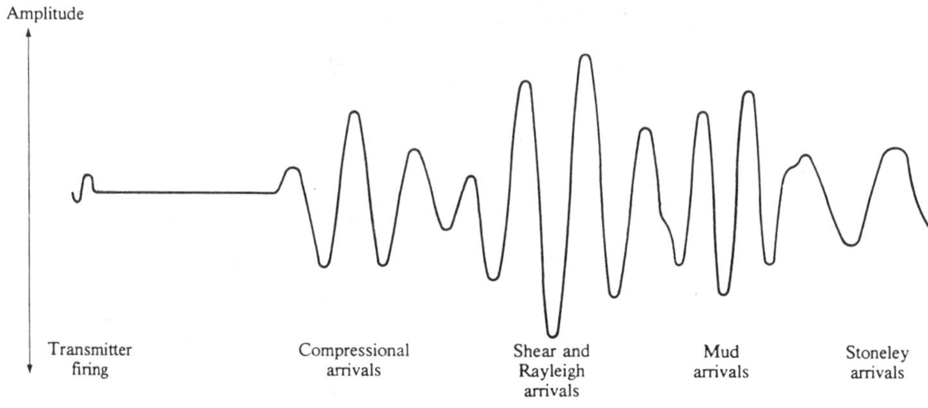

FIGURE 3.7 *Acoustic Wave Train Seen at Receiver.*

Sonic logs are used in conjunction with density logs to generate "synthetic seismograms" by computer. This technique is helpful to geophysicists.

*Calibration**

Since the measurement is of time, calibration of the BHC tool consists of relating the tool signal to an accurate clock. Such clocks are easily provided by electronic circuits and a crystal-controlled oscillator. For example, the signal received at the sensing circuit of one type of BHC tool consists of timed bursts of oscillations. Calibration consists of adjusting the galvanometer response to correspond to the known time values of two calibration signals, and cross-checking the readings at intermediate values to verify linearity. In a computer unit, the same process is carried out by the software.

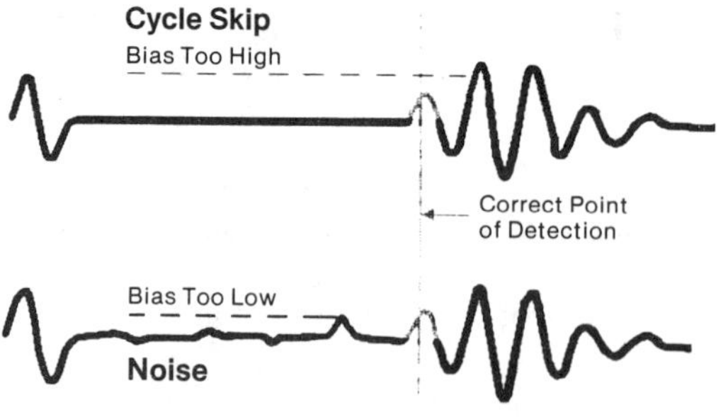

FIGURE 3.8 *Cycle Skipping and Noise.*

*The before and after survey calibration information that follows has been provided, with permission, by Schlumberger Well Services.

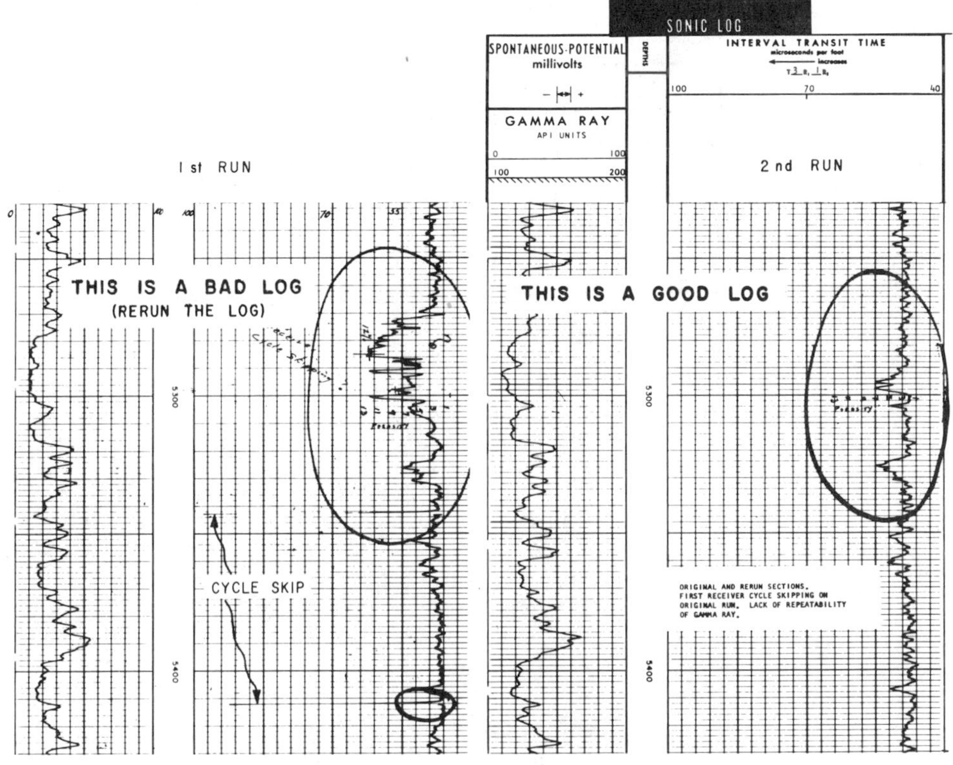

FIGURE 3.9 *Cycle Skipping.*

It should be noted that calibration is made using only the surface equipment. The sonde, with transmitters and receivers, is assumed to be in good physical condition. A bent sonde, which alters the transmitter–receiver spacing, will not be detected by normal calibration procedures yet will produce erroneous logs. Figure 3.11 schematically illustrates the calibration system of a sonic logging tool. A typical BHC calibration tail is shown in figure 3.12.

BEFORE SURVEY CALIBRATION (BHC)

Note: Sonic calibrations are performed through a memorizer panel.

1. Mechanical zero: Galvanometer circuits are open. The Δt is mechanically set at an arbitrary position, depending upon the scale to be run and local custom. In areas using a 50-100-150 scale, Δt is zeroed at 100 μs/ft, or 0 division Track 3. In most other areas, especially where changes in Δt scale are common during logging, Δt is zeroed at 40 μs/ft, or 10 divisions Track 3. The example (fig. 3.12) shows the latter case for a 40-60-80 scale.

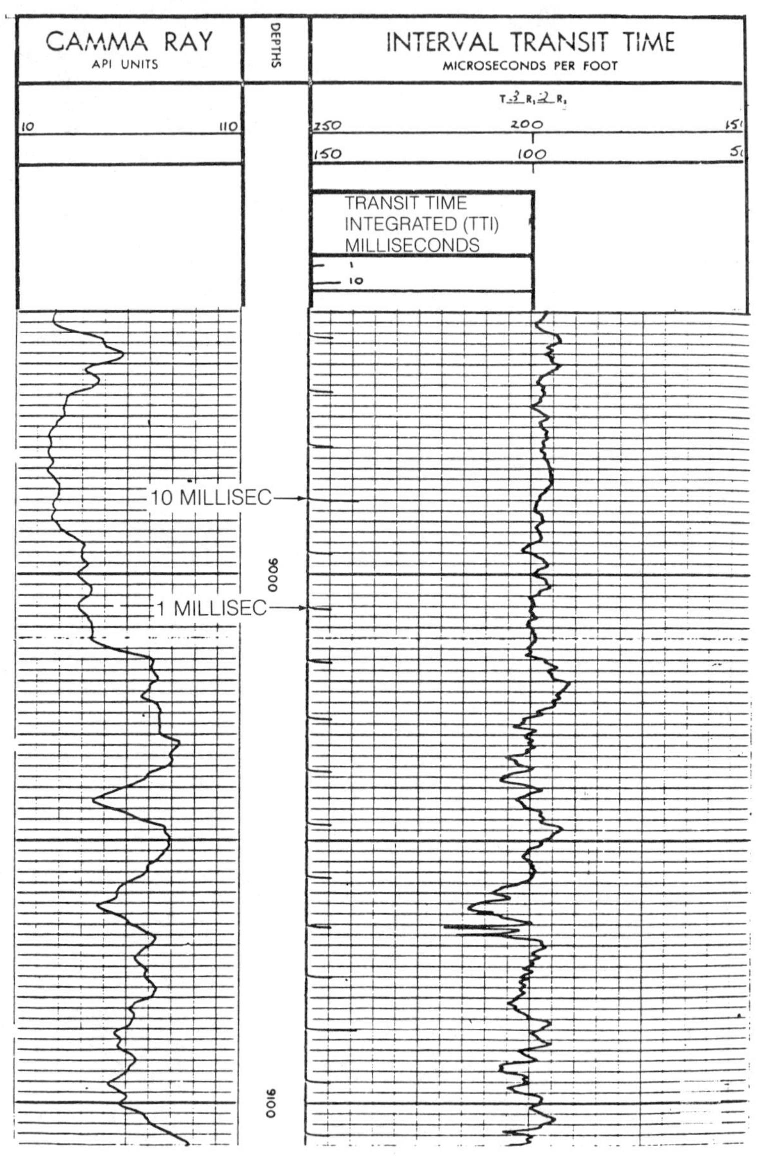

FIGURE 3.10 *Transit-Time Integration.*

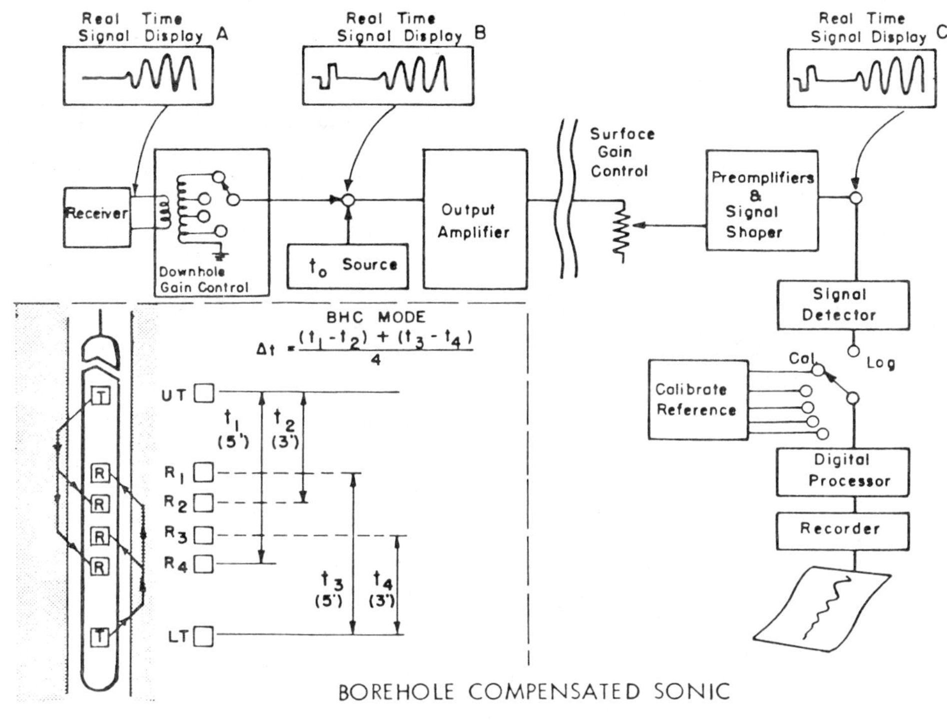

FIGURE 3.11 *Sonic Calibration Schematic. Courtesy Schlumberger Well Services.*

2. 40-μs/ft Δt reads 10 divisions, Track 3: This setting may not appear on the chosen log scale. For example, if a 50-100-150 scale is to be run, 40 μs/ft will be off-scale to the right.
3. 60-μs/ft Δt reads 0 division, Track 3.
4. 80-μs/ft Δt reads 0 division, Track 2.
5. 100-μs/ft Δt reads 0 division, Track 3 (backup galvanometer).
6. 140-μs/ft: This step does not appear on the example log (fig. 3.12) because the scale is 40-50-80 (primary) and 80-100-120 (backup); 140-μs/ft will be off-scale to the left of Track 2.

AFTER SURVEY CALIBRATION (BHC). Steps 1–6 are recorded as above, with sensitivity controls unchanged from their logging position. In some cases, the Δt scale has been changed during logging. In the example (fig. 3.12), the scale was changed at some point to 40-90-140. Therefore, the following calibration steps are for this scale:

1. Mechanical zero: Same as before survey calibration.
2. 40-μs/ft Δt reads 10 divisions, Track 3.
3. 60-μs/ft Δt reads 6 divisions, Track 3.
4. 80-μs/ft Δt reads 2 divisions, Track 3.

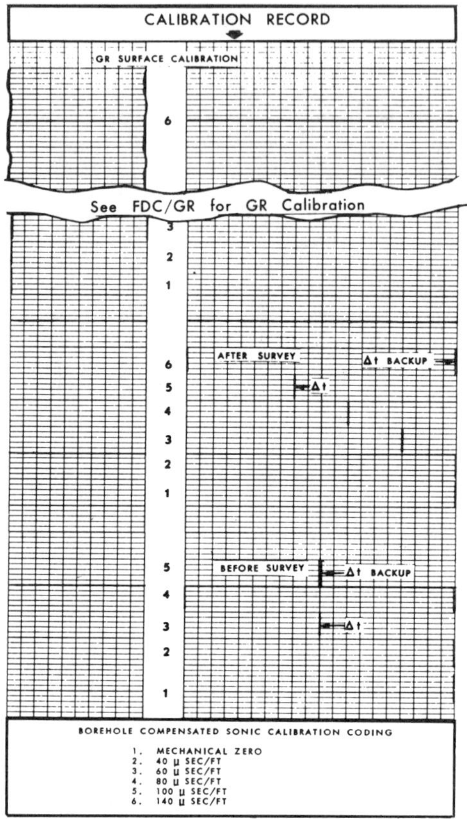

FIGURE 3.12 *Sonic Calibration Film. Courtesy Schlumberger Well Services.*

5. 100-μs/ft Δt reads 8 divisions, Track 2.
6. 140-μs/ft Δt reads 0 divisions, Track 2.

Log Quality Control

Sonic logs routinely include an SP and/or gamma ray log and a three-arm-caliper curve. The sonic curve should be smooth with no spikes or interruptions. Sometimes this is hard to achieve—notably in very soft formations and gas-bearing zones with little invasion. In these conditions, normal BHC equipment tends to "cycle skip," that is, trigger on different wave peaks at the different receivers. This is usually seen on the log as erratic spikes toward longer Δts, or sometimes as a sustained plateau of Δt readings that are about 33 μs/ft too high.

Δts in marker beds such as zero-porosity limestone, anhydrite, salt, etc., should be very near the known values. Readings in clean, compacted sands will reflect porosities with good accuracy. In shales or shaly beds, the sonic log will reflect porosities that are too high.

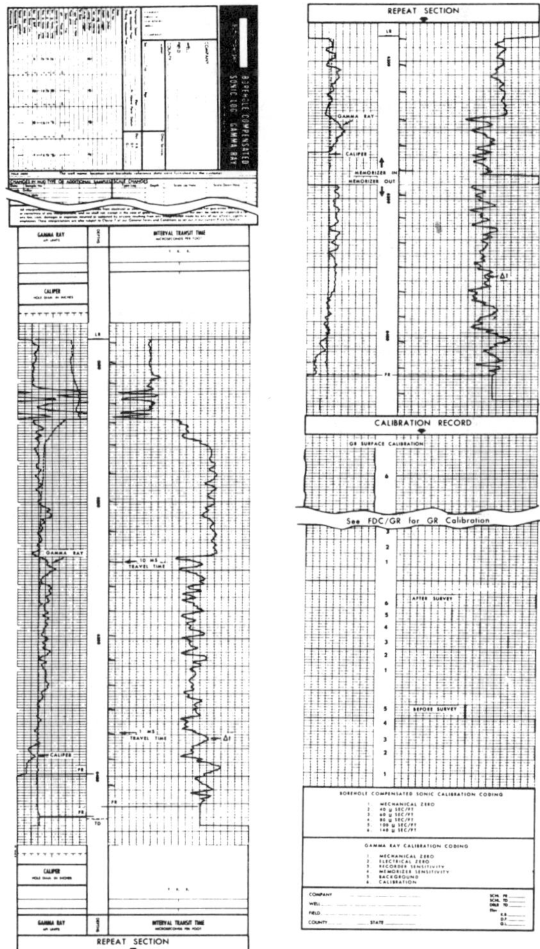

FIGURE 3.13 *Complete Sonic Log. Courtesy Schlumberger Well Services.*

BHC sonic tools are less susceptible to borehole washouts or irregularities than are other porosity tools. Where such conditions occur in zones of interest it becomes especially important to insure a good sonic log. When log analysis programs force a rejection of density and neutron data in bad hole conditions, it becomes necessary to turn to sonic data as a basis for porosity computations. Thus, a good sonic log can make all the difference. Quality control efforts should go into watching the scope while the log is being run.

Caliper defects may show up as stair-steps, straight sections, or very hashy curves, reflecting damaged or sticking arms. Normally, such logs will have to be rerun.

A complete log is shown in figure 3.13. The values shown in table

3.2 may be used to verify tool response. Specific log quality checks for a sonic log include the following:

1. Recording speed: 1800 to 4000 ft/hr maximum.
2. Correctness of reading:
   a. Any section with noise spikes or cycle skips should be repeated at least once in an attempt to improve the data.
   b. Values recorded in formations of known lithology agree with the known transit times.
   c. Porosity values derived from transit time are confirmed by other porosity logs when shale, hydrocarbons, matrix, and secondary porosity effects are considered.
   d. No transit times less than 40 μs/foot are recorded.
   e. Porosity values, when recorded, are computed correctly for any given transit time.
   f. If porosity is recorded, the matrix and fluid velocities and compaction corrections are noted on the heading.
   g. Value of transit time recorded in unbonded sections of casing is 56 ± 1 μs/ft.
   h. Caliper checks:
      (1) Casing ID is correctly recorded with sonde moving in the hole.
      (2) Caliper does not "flat top" at a hole diameter less than the limits of the tool.
3. Downhole calibrations before and after survey: No differences between these calibrations except those due to scale changes while logging.
4. Test films:
   a. 200 ft of repeat with memory out and in,
   b. calibration before survey, and
   c. calibration after survey.

Field Examples

A sonic log with noise and cycle skips due to an enlarged hole is shown in figure 3.14. The log is useless. With more careful choice of bias level and a slower logging speed a good log might have been obtained over this section.

Figure 3.15 shows a long-spacing sonic log for a shallow well (modified depths). The 10- to 12-ft curve is compared to the 8- to 10-ft curve; both are run in depth derived BHC mode. In the upper zone, there is very deep formation alteration—the deeper-spacing Δt reads shorter by almost 10 μs/ft (140 vs. 150). (The standard BHC—not shown—was mostly reading the mud.) Further down, the 10- to 12-ft curve still occasionally reads shorter by a few microseconds, usually in shalier zones, until well-compacted formations, below 227 ft, are encountered—at which point, the two long-spacing curves agree perfectly.

An erroneous, stair-steps sonic log is shown in figure 3.16. The log is useless and should be rerun. Contributing causes of this bad log

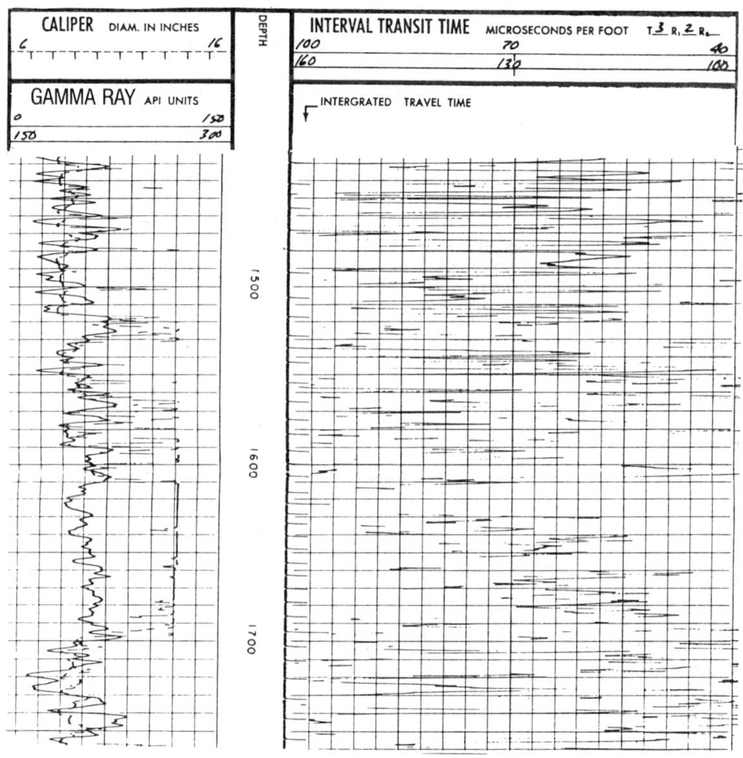

FIGURE 3.14 *A Bad Sonic Log with Excessive Noise and Cycle Skipping.*

were a malfunctioning surface panel and the running of the log from a floating vessel without a wave-motion compensator.

Figure 3.17 shows a sonic log run with a computer unit in a low-porosity carbonate. An accompanying neutron/density log showed porosities in the 1390-to-1420 m interval to be 5% on average. The sonic log revealed much higher apparent porosities in the order of 14%. The error was traced to the calibration data used in the computer unit. The log was rescaled from the raw-data tape and played back as shown in figure 3.18.

CEMENT BOND LOGGING (CBL)

Cement bond logging (CBL) is a variant of sonic logging that makes use of the observation that when sonic logs are run inside casing the signal amplitude inside well-cemented casing is much reduced, while in unsupported casing the signal is strong. The CBL format may include a gamma ray and a casing-collar log for depth control, a transit-time curve, and an amplitude measurement for evaluation of bonding.

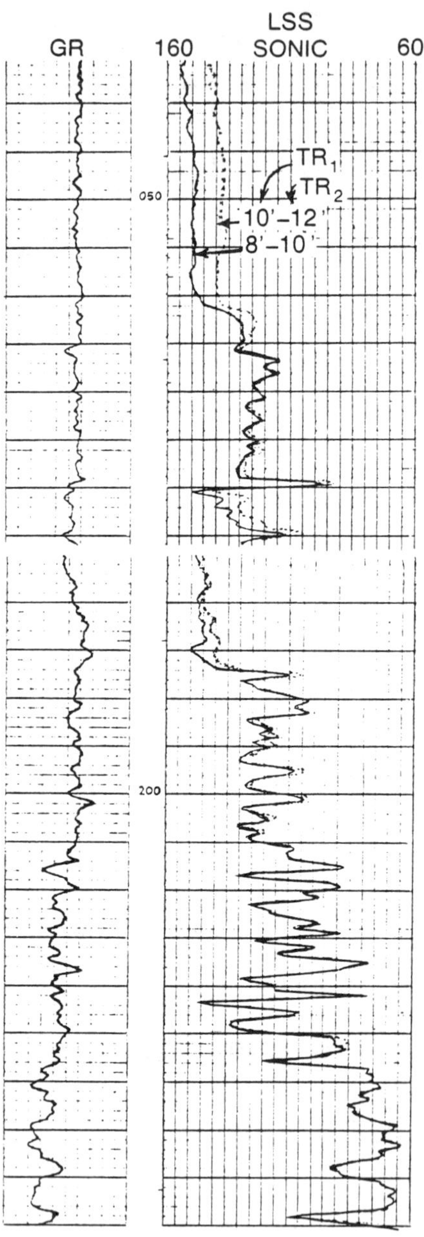

FIGURE 3.15 *Long-Spacing Sonic Logs in Altered Shales.*

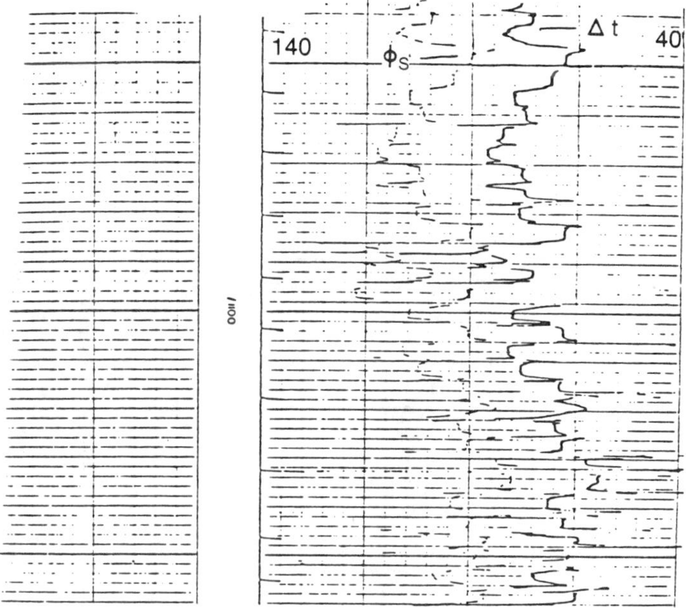

FIGURE 3.16 *Bad Sonic Log.*

There may also be a *signature* or a *variable-density* display of the actual waveforms. These displays have been found helpful in both quality control and log evaluation. Figure 3.19 shows a typical CBL presentation. On this example, gamma ray and casing-collar logs are omitted.

Measurement Principle

Predictably, a cement sheath bonded to the casing attenuates sound propagation in the pipe. CBL tools are able to differentiate between "no cement" and "solid cement." The chief problem with CBL tools is that the casing-signal attenuation is not directly related to the degree of hydraulic sealing provided by the annular cement. Hence, no matter how accurately the attenuation is measured, answers are still in terms of probabilities between the end conditions of perfect or no bonding. Thus, in the in-between range these tools are not yet able to provide unambiguous answers to the question: will the cement job prevent high-pressure fluid flows in the annulus? Even so, the CBL tool is a valuable and much-used adjunct to the inventory of logging devices. The interplay of cement presence, bonding, signature, variable-density display, and amplitude is illustrated in figure 3.20.

Cement bond logs began as auxiliaries to the sonic log, run with tools designed for delta-*T*-type logging. As the technique made a place for itself, it became important enough to motivate development of special CBL tools, which now do the majority of the bond logging.

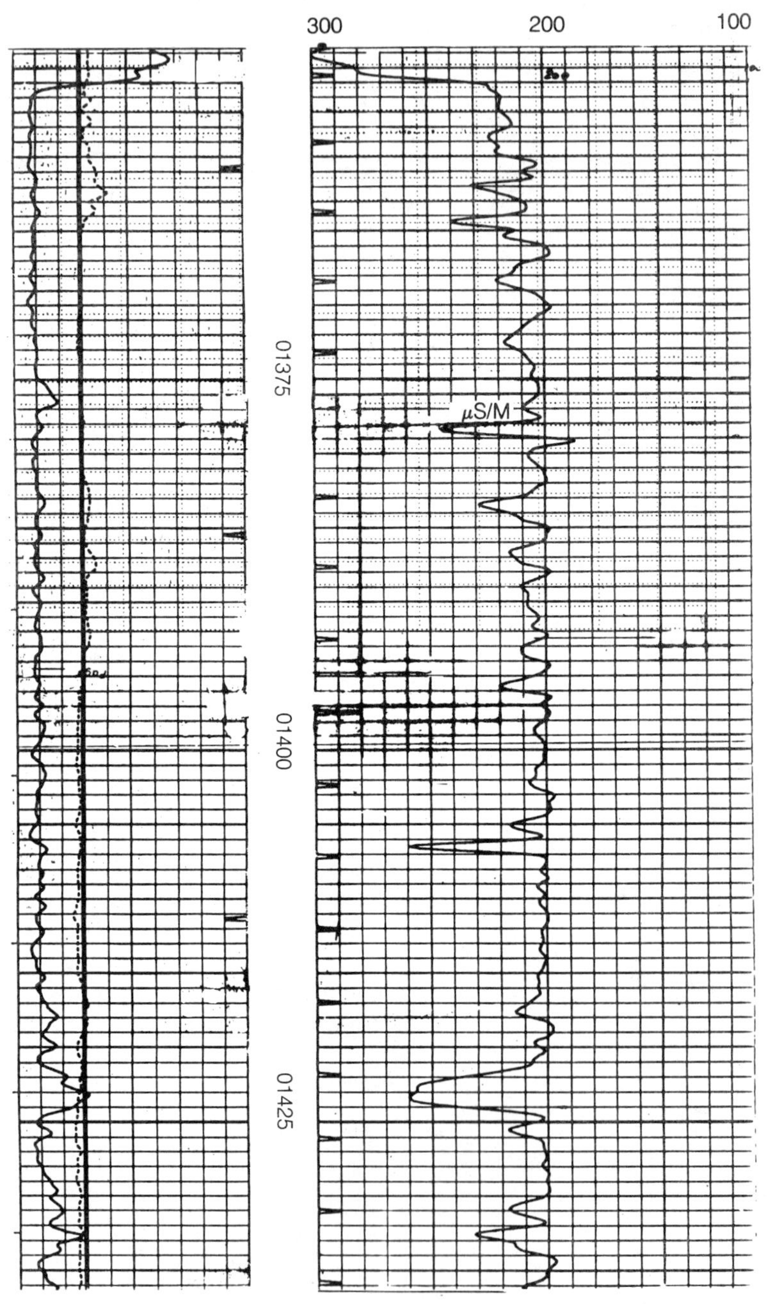

FIGURE 3.17 *Wrongly Scaled Sonic Log.*

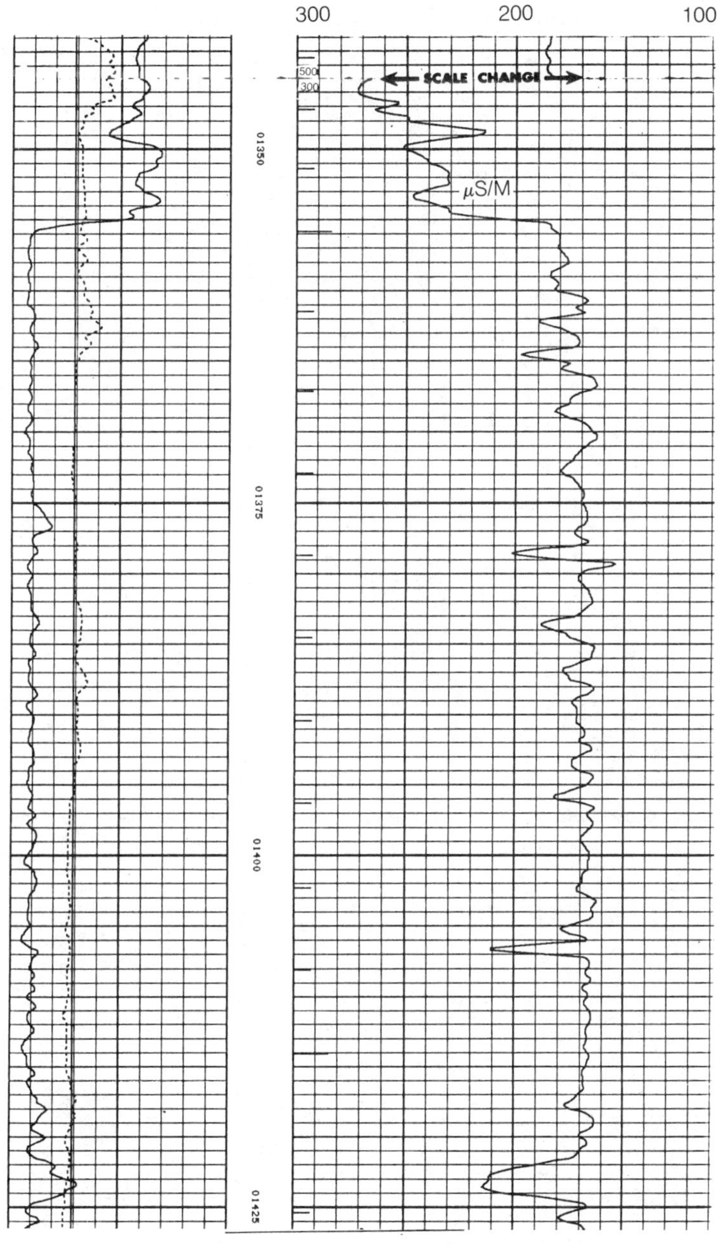

FIGURE 3.18 *Rescaled Version of Sonic Log in Figure 3.17.*

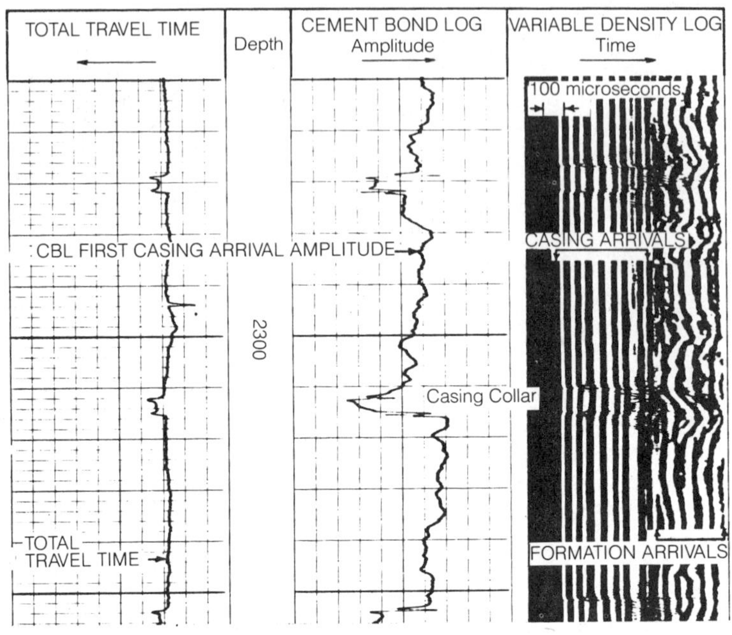

FIGURE 3.19 *Typical CBL Presentation. Courtesy Schlumberger Well Services.*

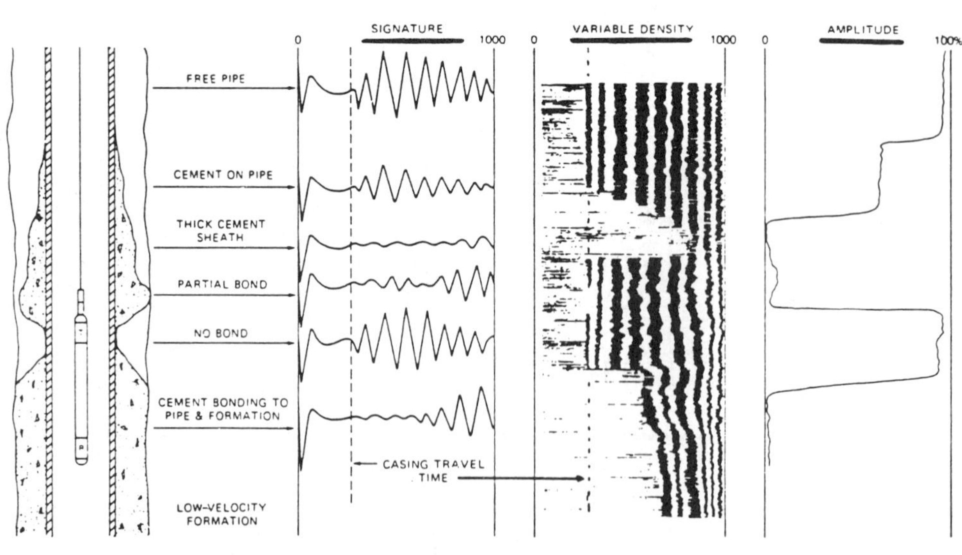

FIGURE 3.20 *Cement Bond Log Schematic. Courtesy Dresser Atlas.*

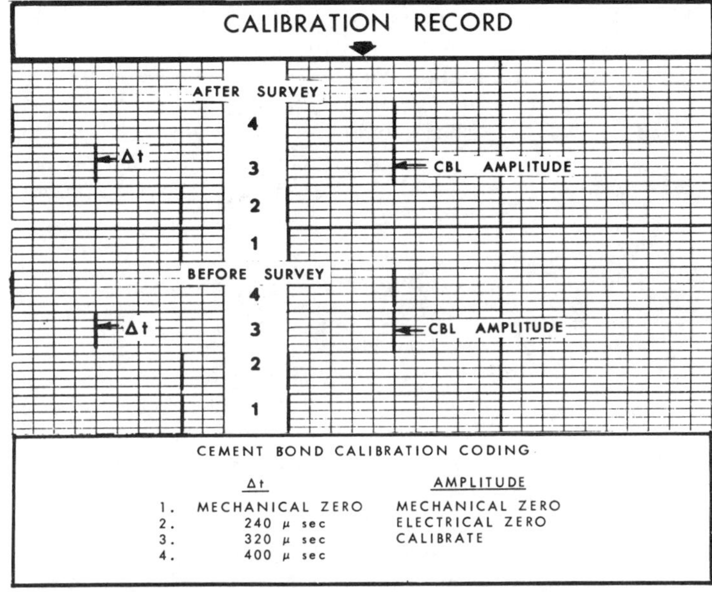

FIGURE 3.21 *CBL Calibration Record. Courtesy Schlumberger Well Services.*

Calibration

Though not universal, a common method of calibrating the received-signal amplitude in the CBL is to compare it with the amplitude of the transmitter pulse. All values are measured in millivolts. However, the precise millivolt values of the CBL curve have little objective significance. A typical service-company CBL calibration record is shown in figure 3.21.

PRINCIPLES OF CALIBRATION*

CBL Amplitude. The CBL (Cement Bond Log) amplitude calibration is based upon the level of a signal in the downhole tool. This signal represents 1 mV of receiver signal. It is amplified in the downhole cartridge and transmitted to the surface. In the surface panel the level of the signal is "normalized" to 100 mV. The amplitude circuitry responds to this voltage and produces a deflection on the amplitude galvanometer. The deflection is adjusted to 5 divisions in Track 2. The deflection is independent of the logging scale.

Single-Receiver Transit Time. The sonic tool used for the CBL transmits real-time signals representing arrivals of acoustic energy at two receivers. The receivers are spaced 3 and 5 ft from the transmitter. The signal from the 5-ft receiver is applied to the variable density

*The calibration information that follows has been provided, with permission, by Schlumberger Well Services.

recorder and the signal from the 3-ft receiver is processed by the travel-time circuitry.

The output of the travel-time circuitry is an analog signal representing the total travel time from the transmitter to the 3-ft receiver. The calibration for this signal involves applying known transit-time intervals from an oscillator to a digital-analog converter. The intervals are 240, 320, and 400 μs/ft. The galvanometer is adjusted to the proper deflections for these intervals.

BEFORE SURVEY CALIBRATION—ΔT AND AMPLITUDE

1. Mechanical zero: Galvanometer circuits are open. Δt reads 240 μs (2 divisions Track 1). Amplitude galvanometer reads 0 division Track 2.
2. Electrical zero: 240-μs signal is applied to Δt galvanometer, which is adjusted to read 240 μs. Measuring circuits are connected to amplitude galvanometer. Amplitude galvanometer is adjusted to read 0 division Track 2.
3. Calibration: Δt galvanometer reads 320 μs (6 divisions Track 1). Amplitude galvanometer is adjusted for 5 divisions deflection in Track 2. (This completes the amplitude calibration.)
4. Calibration: 400-μs signal is applied to Δt galvanometer which is adjusted to read 400 μs (10 divisions Track 1).

AFTER SURVEY CALIBRATION The four steps of before survey calibration are recorded with no adjustments made from the logging settings.

CBL Log Quality Control

GENERAL. In addition to the calibrations, repeat sections, and main log, a CBL should always include a section above the presumed cement top, where the pipe will be completely unbonded. This gives one endpoint for the log; the curve should never read higher than this. The other endpoint is given by the zero point on the log scale. The curve will never read zero, but will come close (2 to 3 mV) in well-bonded pipe.

The paradox of acoustic-amplitude-type CBL logging is that the signal of most interest is zero or near it, but the equipment triggers on a finite signal in normal operating mode (T_x detection); as the signal approaches zero, it gets harder and harder to fine-tune the system to pick up the right signal. To avoid this, the more sophisticated tools allow logging with T_0 detection; that is, a detection window set at a selected time interval after the first pulse. This time will normally be close to the casing transit time.

As with normal delta-T-sonic logging, good quality control with the CBL requires the use of an oscilloscope picture. With most equipment, this is the only way to be sure that the amplitude measurement is made on the first-arriving half-cycle of acoustic energy, essential for meaningful interpretation (see fig. 3.22).

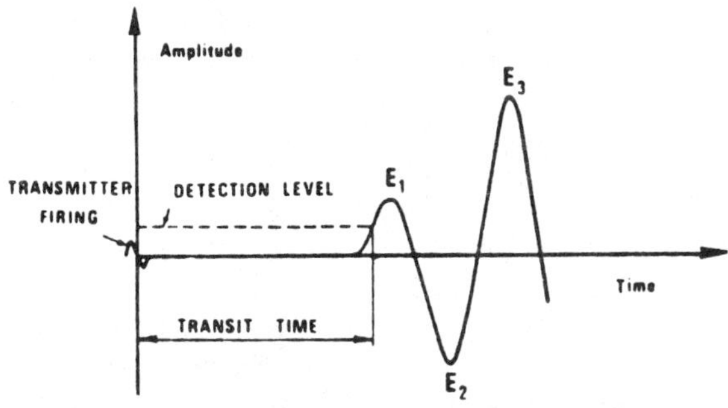

FIGURE 3.22 *Detection of First Arrival. Courtesy Schlumberger Well Services.*

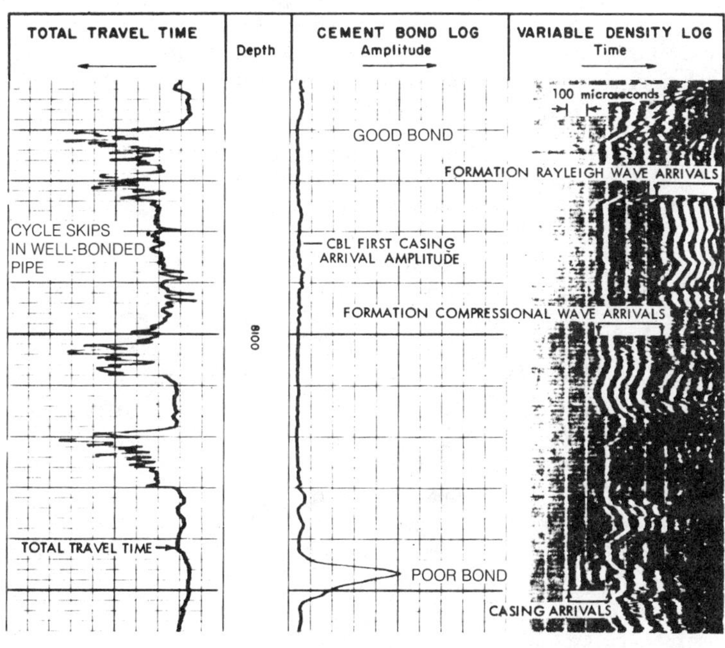

FIGURE 3.23 *Cycle Skipping. Courtesy Schlumberger Well Services.*

In normal logging mode, the system triggers on the first arriving (E_1) half-cycle, measuring both its single-receiver travel time (time from transmitter to receiver) and its amplitude. But two things can prevent that: weak signals in well-bonded pipe can go below the detection threshold; and in hard-rock country it is possible for formation signals to arrive ahead of casing signals. In the first case, cycle skips appear on the log (fig. 3.23), and the amplitudes recorded in the skip intervals are not interpretable. In the second case, the transit-time curve will depart from the fairly straight-line value of casing transit time, and begin to follow formation variations. The scale is not directly correlatable, since the CBL transit time is a 3-ft single-receiver measurement and is not borehole compensated. Normal casing transit time is three 3 ft × 57 μs/ft, plus the travel time from tool to casing and back again, usually around 250 to 260 μs.

CENTERING. Most CBL tools assume in-phase arrivals through all sides of the casing, meaning that the tool must be centered. The degree of centering can be judged from the transit-time curve. A poorly centered tool will produce shorter transit times. The amplitude signal is apt to be weak, causing excessive cycle skipping in normal logging mode, or too-low readings in T_x mode. Centering may be virtually impossible in deviated holes or large casings.

CHECK LIST.* Specific log quality checks for the CBL are as follows (see the complete CBL survey in fig. 3.24):

1. Recording speed: 3000 ft/hr maximum. 1800 ft/hr maximum if gamma ray is recorded as a Perforating Depth Control Log.
2. Correctness of readings:
   a. In unbonded pipe, single-receiver Δt records the correct value considering tool diameter and casing ID. This value does not change with depth except at casing collars.
   b. Amplitude curve and single-receiver Δt-curve recording mode is noted on the heading (either T_0 Delay or T_x mode).
   c. Cycle skipping of single receiver Δt is evident only in zones of low CBL amplitude and/or fast formation signals.
   d. Transmitter-to-receiver distance is noted on the heading.

*The calibration information that follows has been provided, with permission, by Schlumberger Well Services.

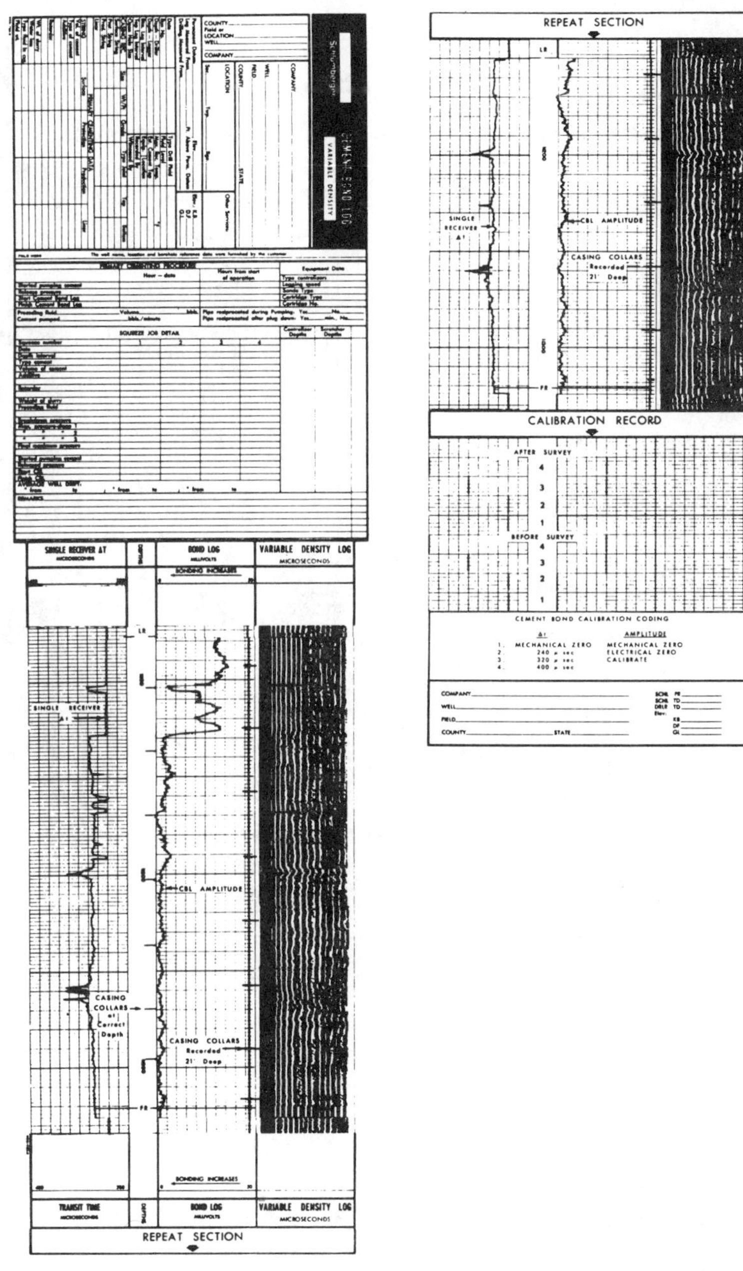

FIGURE 3.24 *Complete CBL Survey. Courtesy Schlumberger Well Services.*

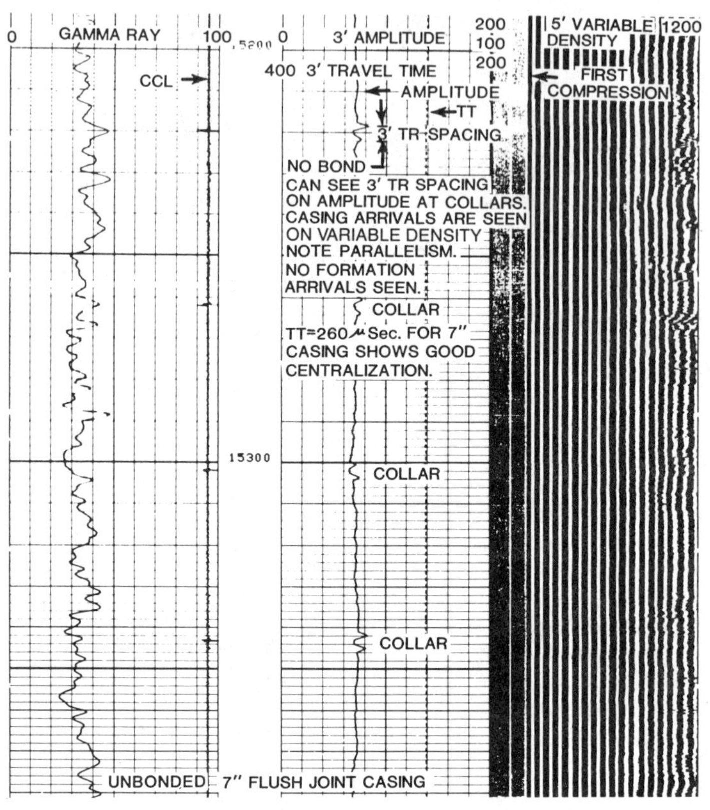

FIGURE 3.25 *CBL Amplitude and VDL in Unbonded Pipe.*

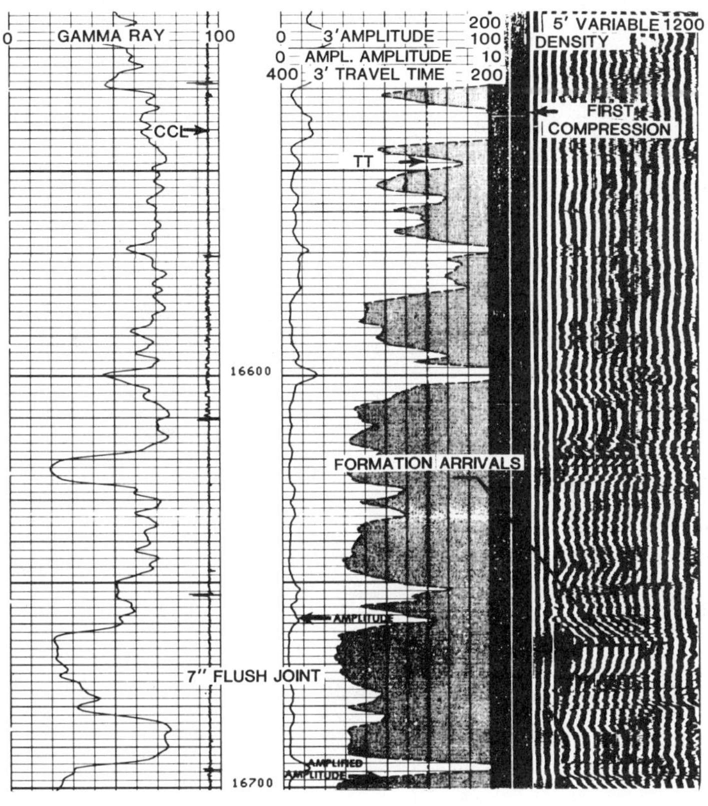

FIGURE 3.26 *Fair-to-Good Bond Shown by Amplitude and VDL Display.*

   e. Casing collars at the correct depths are marked in the depth track. The distance from the collar locator to tool zero is indicated on the heading and the log.
   f. All information concerning primary cementing data and procedures and squeeze-job details is indicated on the heading.
   g. Complete information about centralizers and/or standoffs is provided on the heading.
3. Test films indicate no change between before and after survey calibrations.
4. Test films to be attached:
   a. 200 ft of repeat section (If a gamma ray is recorded, the first 100 ft are unmemorized and the next 100 are memorized.)
   b. downhole calibrations before and after survey.
5. Variable-density log (VDL) quality checks:
   a. Transmitter-to-receiver distance is noted on heading.
   b. White grid lines are clearly visible in well-boned sections.

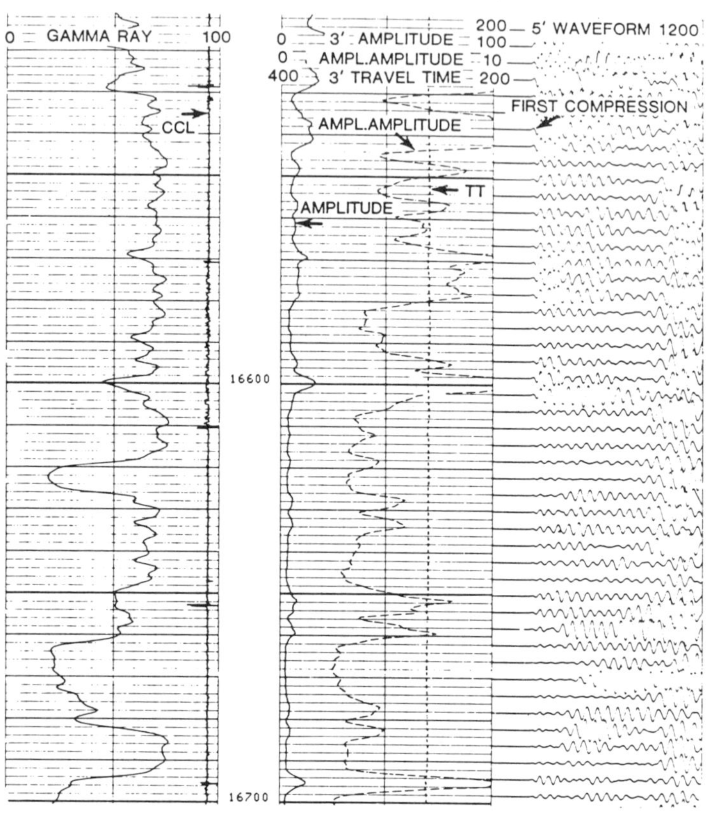

FIGURE 3.27 *Same Section as Shown in Figure 3.26 But with Signature Display.*

c. The VDL trace is centered in Track 3 from 0.5 to 9.5 divisions.
d. The intensity of the trace is such that a gray area exists between the beginning of the trace (0.5 division Track 3) and the first arrival. In well-bonded sections, the first arrival evident on the VDL may occur near the middle or end of the trace.
e. In unbonded sections the various arrivals are sharp and clear. There are no variations in travel time of the first arrival with depth.

CBL Examples

Amplitude and variable-density displays of a CBL for a section of unbonded 7-in. casing are shown in figure 3.25. Fair-to-good bonding over a productive zone are shown by the CBL amplitude and VDL of figure 3.26. The same section is shown in figure 3.27 with a signature log (waveform signature) in Track 3.

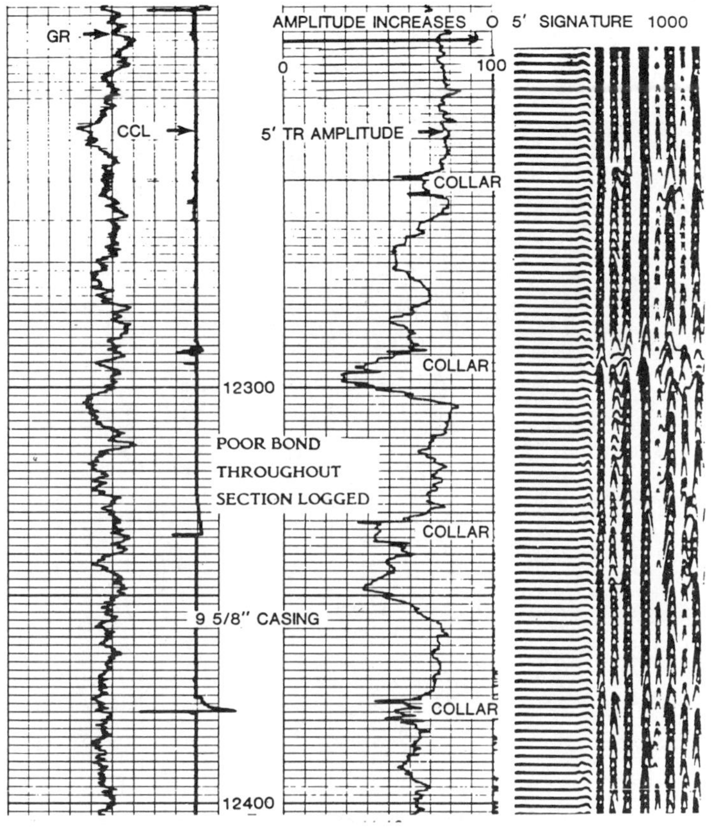

FIGURE 3.28 *Poor Bonding in 9⅝-in. Casing Shown by CBL Amplitude.*

Figure 3.28 shows a CBL for a section of 9⅝-in. casing with poor bonding. Note that the gamma ray log is substandard and no transit-time curve was recorded. A squeeze was performed and 3 months later the well was relogged as shown in figure 3.29. In the last figure, the casing collar log (CCL) is still substandard, and the transit-time curve is showing cycle skips in those places where bonding is good. The cycle skips are normal, but since collars can still be distinguished on the amplitude curve, bonding is still not perfect.

Figure 3.30 shows a case where squeezes were indicated to isolate productive intervals A and B. Note the free-pipe signature shown at the top of the figure.

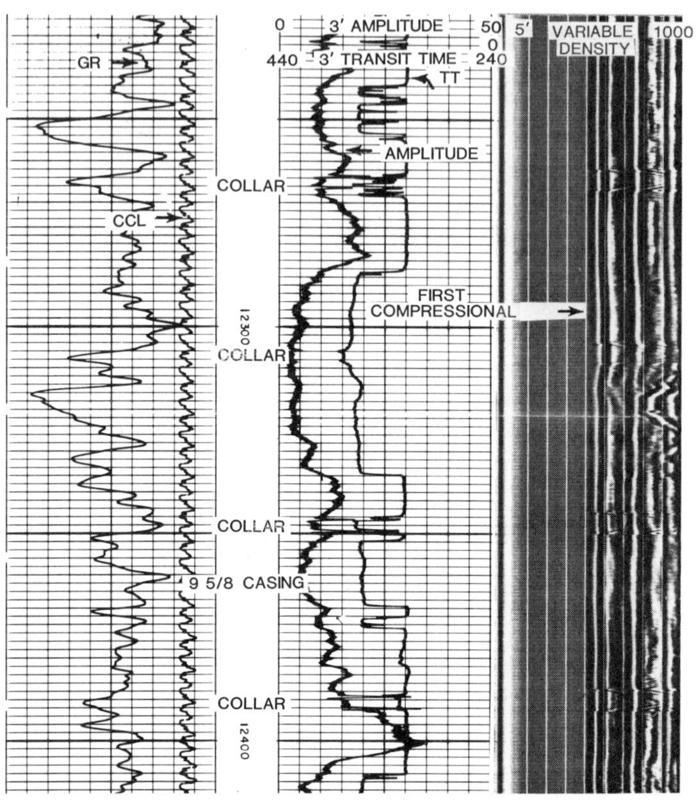

FIGURE 3.29 ***Postsqueeze CBL/VDL for Section Shown in Figure 3.28.***

THE BOREHOLE TELEVIEWER (BHTV)

The borehole televiewer (BHTV) is an acoustic device that scans the surface of a wellbore or casing by rotating an acoustic source (transducer) in the horizontal plane while the tool is moved vertically along the wellbore axis (figure 3.31). The amplitude of the acoustic signal reflected from the borehole or casing wall is displayed as a photograph of the section logged. With the help of a flux-gate compass, an oriented acoustic picture of the inside of the wellbore is provided as if it were split vertically along the north axis and laid flat. The acoustic picture appears in shades of gray and is a record of the amount of acoustic energy that is reflected from the borehole wall. A smooth surface reflects better than a rough one, a hard surface better than a soft, and a normal surface produces larger reflections than an oblique or slanted surface.

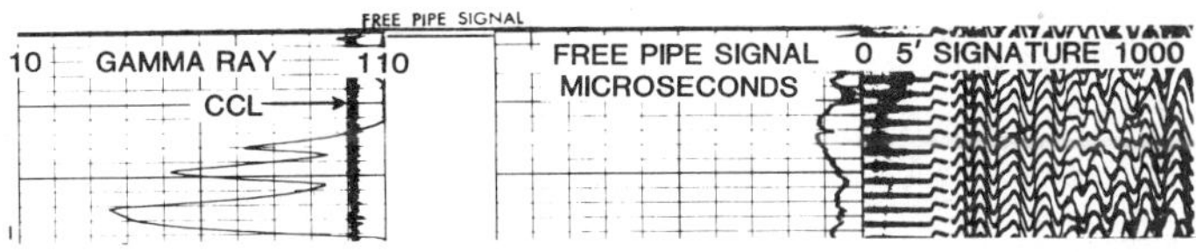

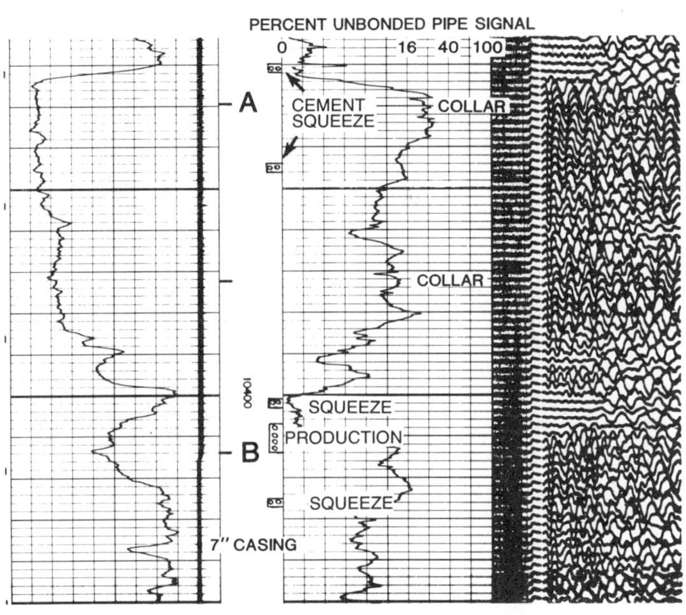

FIGURE 3.30 *Choice of Squeeze Depths from Signature Log.*

When a smooth, normal borehole wall is scanned, maximum energy is reflected and the resulting image will be a series of bright lines. However, when a feature such as a fracture with its attendant discontinuities is surveyed, a minimum amount of energy is reflected and that feature will appear as a dark line (dark represents reduced reflected energy). In addition to fractures, features such as vugs, bedding planes, changes in lithology and perforations, ruptures, or pits in casing can be seen on the televiewer log.

Openhole Applications

The borehole televiewer (BHTV) is used to detect and measure the dip of fractures and bedding planes. An isometric sketch of a wellbore intersected by a nonvertical fracture or bedding plane and a cor-

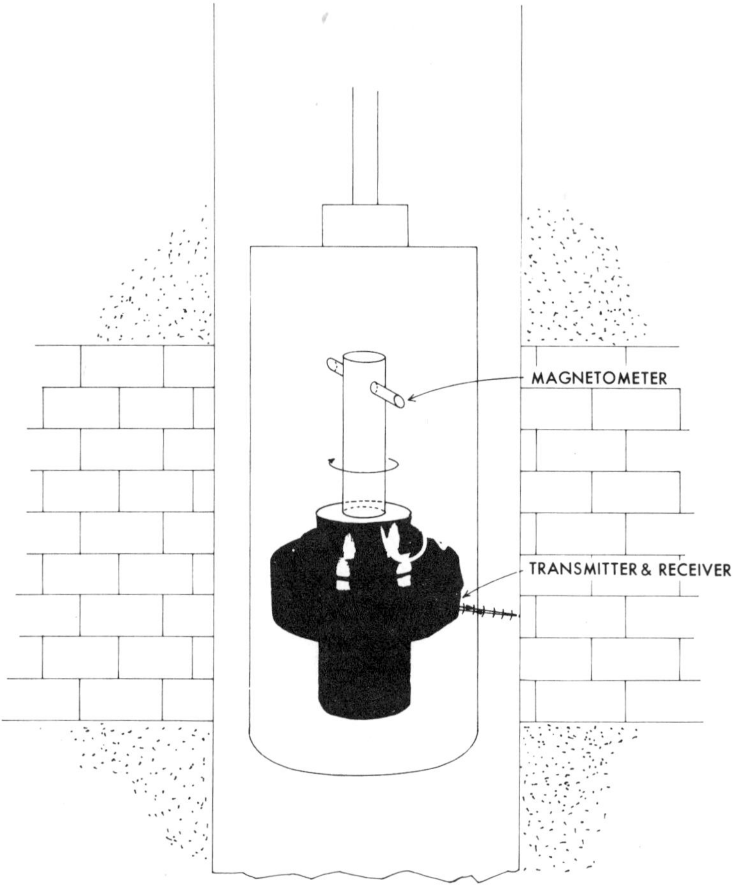

FIGURE 3.31 *Schematic of the BHTV Tool.*

responding BHTV log are shown in figure 3.32. To determine dip, one merely finds the minimum of the sinusoid (indicated by the arrow) and reads the direction from the azimuth scale at the bottom of the log. Dip angle is determined by measuring the peak-to-peak amplitude, h, of the sinusoid and combining it with the diameter, d, of the wellbore:

$$\text{dip angle} = \tan^{-1}\frac{h}{d}$$

Figure 3.33 is a view of a high-angle fracture or bedding place intersecting the wellbore with north dip. If a high-angle fracture intersects

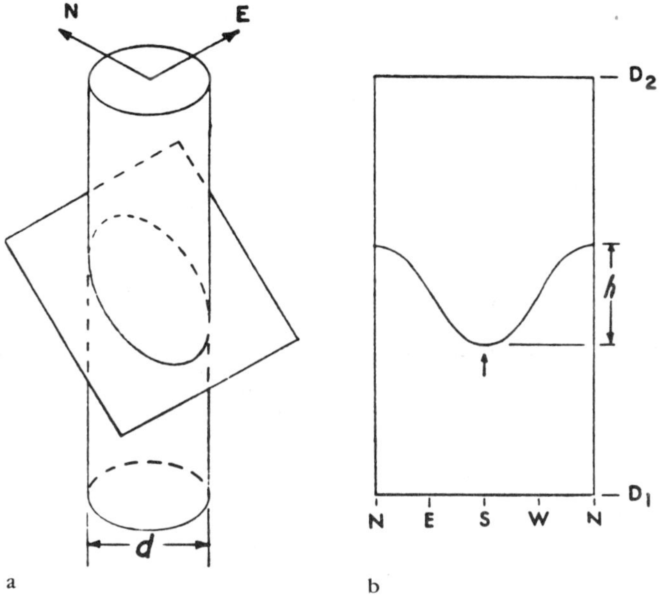

FIGURE 3.32 *(a) Isometric Drawing of Nonvertical Fracture of Bedding Plane Cutting Borehole, and (b) Televiewer Log.*

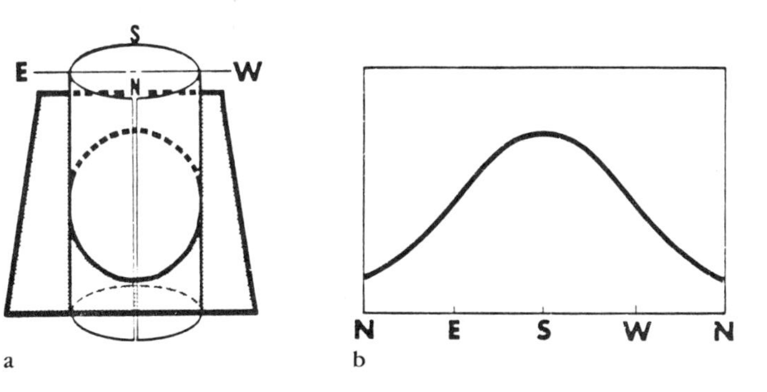

FIGURE 3.33 *(a) North-Dipping Plane Cutting Wellbore and (b) BHTV Log.*

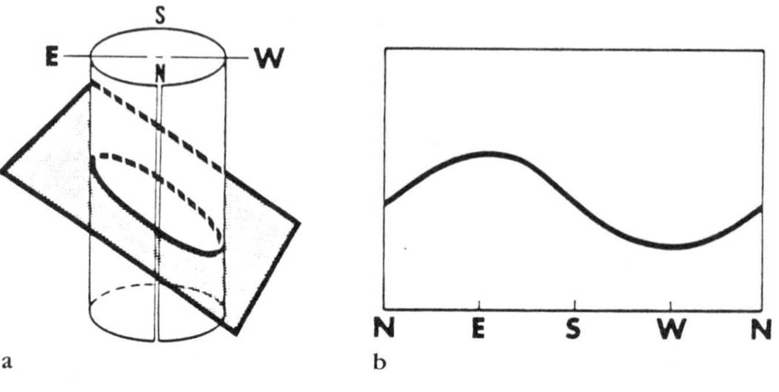

FIGURE 3.34 *(a) West-Dipping Plane Cutting Wellbore and (b) BHTV Log.*

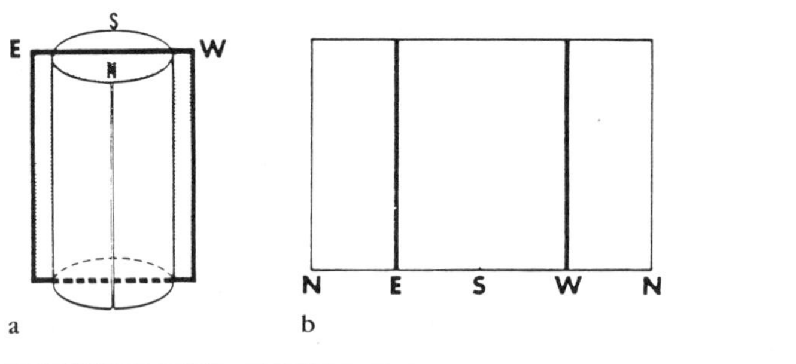

FIGURE 3.35 *(a) Vertical Fracture (East-West) and (b) BHTV Log.*

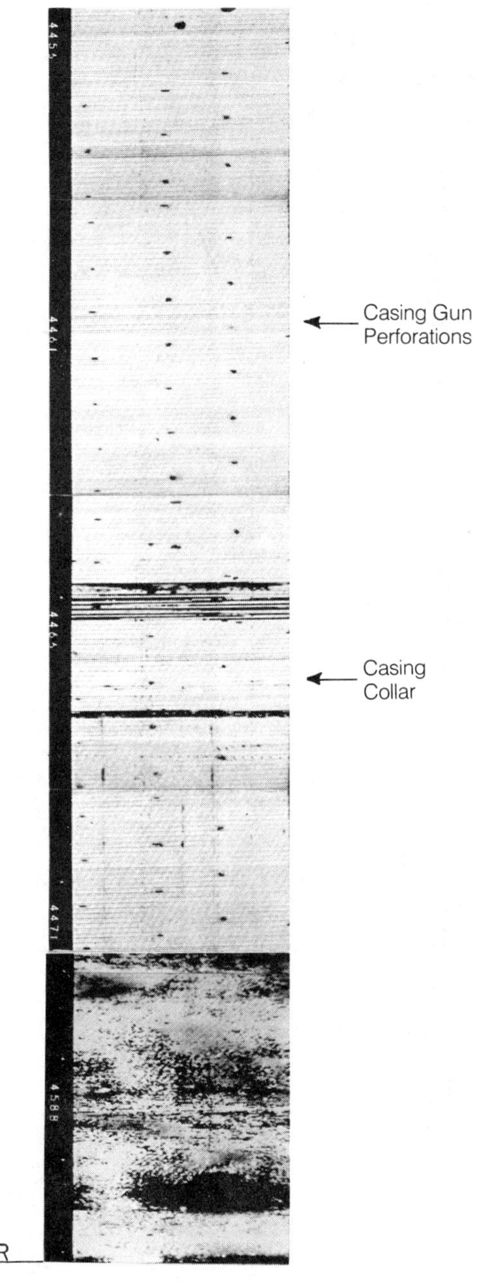

FIGURE 3.36 *BHTV Log Run in Cased Hole over a Section Shot with a Casing Gun.*

FIGURE 3.37 *Collapsed Pipe.*

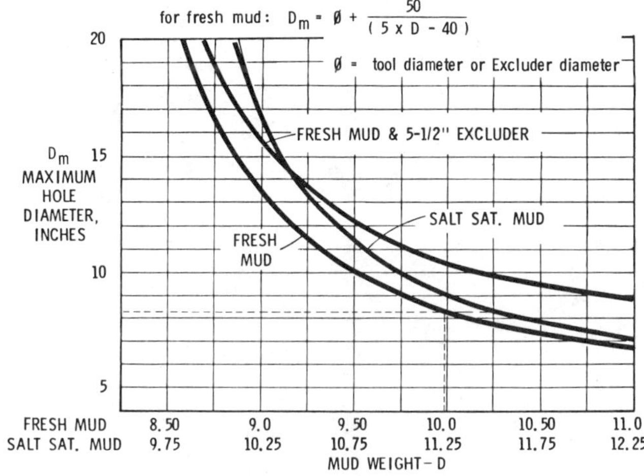

FIGURE 3.38 *Maximum Hole Diameter as a Function of Mud Weight.*

the wellbore with west dip, the BHTV log is a full sine wave with a minimum to the west and a maximum peak occurring to the east as shown in figure 3.34. For the case of a fracture or bedding plane dipping to the east, the minimum of the sine wave would be to the east and the maximum to the west. For a vertical fracture intersecting the wellbore in an east-west direction, the corresponding BHTV log appears as two vertical dark lines 180° apart (see fig. 3.35).

Cased-Hole Applications

The BHTV has two cased-hole applications: (1) to detect perforations (fig. 3.36), and (2) to evaluate damaged casing (fig. 3.37).

General Guidelines

The BHTV can be run in any gas-free liquid such as fresh water, saturated brine, crude oil, or drilling mud. Operating limits for various mud weights and hole sizes may be determined from figure 3.38. For example, the operating limit for an 8¼-in. hole is 10 lb/gal fresh mud and 10.25 lb/gal salt mud. Mud excluders are available where hole size and mud weights exceed normal operating limits.

Prerequisites for a top-quality log are (1) a centered tool in (2) a round hole. In a correctly adjusted scale, all shades of gray from dark to bright are visible. The log shown in figure 3.39 meets these requirements most of the time. There is a dark area on the left side of the log due to the tool being slightly off center. Otherwise, the symmetrical intensity from left to right indicates a centered tool in a round hole.

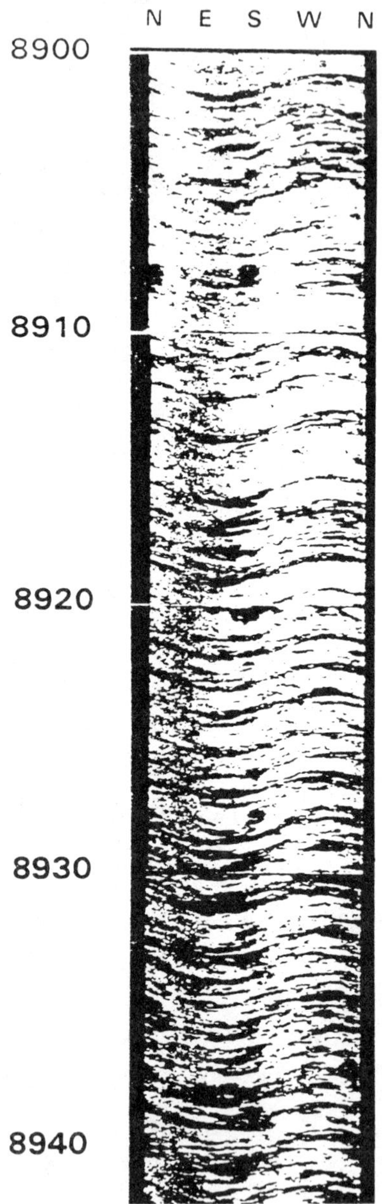

FIGURE 3.39 *A Good BHTV Log.*

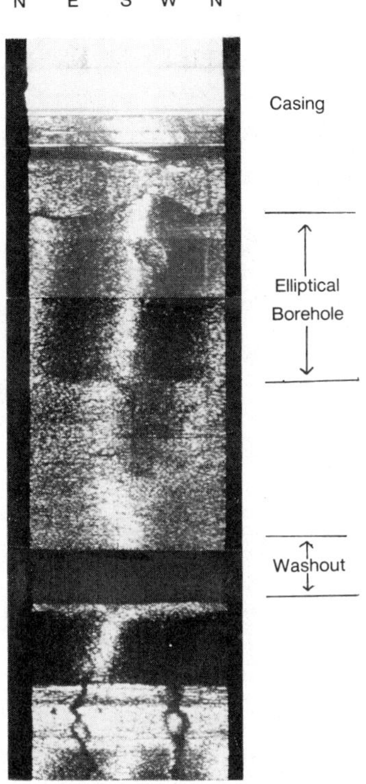

FIGURE 3.40 ***BHTV Log in Elliptical Hole.***

A BHTV presentation of an elliptical borehole is shown in figure 3.40. The wide dark bands marked "elliptical hole" are due to a weak reflected signal from the long axis of the hole. (The long axis is oriented east-west.) The dark interval labeled "washout" is an indication of either a very weak or no reflected signal. This interval should correlate with a caliper log.

CHAPTER FOUR

RADIATION LOGGING

GENERAL RADIOACTIVITY

Logging methods based on radioactive processes began as a way to get correlatable logs in cased holes. The early tools were unable to produce quantitative data in a practical sense. But since the early 1960s, continuing progress in logging technology has brought forth a family of radiation-measuring tools that produce analytical data of high accuracy. Such tools have brought about a major revolution in the science of openhole log evaluation.

In chronological order, and roughly of increasing sophistication, the important radioactive logging tools are:

natural gamma ray (correlation)
single-detector neutron (correlation, relative porosity)
compensated sidewall density (absolute porosity)
sidewall epithermal neutron (hydrogen-filled porosity)
pulsed neutron (lifetime, thermal decay time) (water-filled porosity)
dual-spacing compensated neutron (hydrogen-filled porosity)
compensated epithermal neutron (hydrogen-filled porosity)
natural gamma ray spectroscopy (shale analysis)
neutron activation logging (hydrocarbon detection)

A summary of these devices and measurements is given in table 4.1.

Gamma rays and neutrons are the two types of radioactive energy with enough penetrating power to be significant in well logging. Though they differ from each other in some ways, and are used in widely varied applications, they have several characteristics in common.

Penetration

The most obvious characteristic shared by gamma rays and neutrons is their ability to penetrate cemented casing. This has always been a valuable and almost unique property of the radioactive tools. Some tools are designed mainly for cased holes, some for openholes, and others for either type. A few of these tools are so specialized for one environment that they cannot be run usefully in the other; the sidewall epithermal neutron tool, for example, will not produce a usable log inside casing.

Statistical Fluctuations

All radioactive processes involve random, unpredictable events in time. Measurements of these events are useful, therefore, only in the

TABLE 4.1 *Common Types of Radiation Tools*

Type of Device	Radiation Source	Unit Measured	Uses
Gamma ray device	Gamma rays emitted by the natural decay of radioactive elements in the formation	Gamma rays per unit time (counts/s) expressed in terms of API units	Lithologic determination Correlation Bed boundaries
Neutron (neutron-gamma) device	Neutron sources (radium beryllium, and plutonium beryllium mixtures) or accelerators	Gamma rays created by the neutron capture expressed in API units, counts/s, or porosity units	Porosity determination Lithologic determination Fluid type determination Correlation
Neutron (neutron-neutron) device (includes epithermal neutron)	Same sources as above	Thermal (slow) neutrons and epithermal (fast) neutrons in counts/s or porosity units	Same as Neutron Gamma
Gamma gamma (density) device	High-energy gamma rays from cesium source	High-energy gamma rays not absorbed by the formation presented as bulk density in g/cc	Density determination Porosity determination Lithologic determination
Pulsed neutron	Pulsed neutron generator	High-energy gamma rays	Fluid type and saturation
Inelastic neutron scattering and capture gamma ray spectroscopy	Pulsed neutron generator	Counts/s in selected portions of gamma spectrum	Fluid type and saturation, presence and relative amount of specific elements (i.e., silicon, oxygen, and etc.)
Natural gamma ray spectroscopy	Gamma rays emitted by the natural decay of radioactive elements in the formation	Thorium and uranium in ppm and potassium in %	Lithologic determination Correlation Fracture identification Possible clay-type determination

statistical sense; if there are enough random events to count, meaningful and useful average values can be established. A prime objective of radiation logging is to provide a practical count rate. To this end, and in general, the controllable factors are:

source strength (gamma rays/neutrons per second)
source-detector spacing, coupled with needed shielding
detector efficiency
measuring time

The more known about these factors, the ways in which they interact, and the tradeoffs that become available, the better the evaluation of the quality of radioactive logs. The discussions here are intended to serve as a starting point, and will introduce some of the jargon.

Source Strength

Since gamma rays and neutrons cause biological damage in proportion to their intensity, there is no limit imposed by safety on the size of sources. The sources are shielded when not in use, but they have to be taken out of the shields when logs are to be run. At this writing, the largest source in general use is 16 curies of radioactive material. (A curie is the amount of any material that will produce radiation equivalent to that of a gram of pure radium.)

Pulsed neutron sources have been in use for a couple of decades. They are free of the environmental hazard presented by chemical sources, since they emit radiation only upon command. Their "size," in terms of neutrons emitted per second, is limited not by safety but by technological constraints. Their expense and fragility pretty well confine them to applications where chemical sources cannot be used: "lifetime" or "decay time" logs, primarily.

The gamma ray sources used in density tools are directional; the radiation intensity in the focused direction is much greater than in other directions. This permits compromises in shielding and handling techniques; but these compromises are acceptable only if the special handling requirements are followed rigidly. One company requires, for example, that the source may be transferred from shield to tool or vice versa *only* by the logging engineer (party chief).

Source–Detector Spacing

The distance between source and detector is a design problem, constrained by source strength, logging speed, and required tool response. At this writing, it can be safely generalized that all radioactive-type tools are using shorter source–detector spacings than the theoretical ideal, in order to achieve realistic logging speeds.

Detector Efficiency

Much progress has attended logging-tool detectors since the early proportional counters. First and simplest were gamma ray detectors: Geiger-Mueller tubes of successively better resolution; scintillation counters, which had to be insulated against the heat of boreholes; and now the 3He counters, which are the field standard. The latter can be used to detect thermal neutrons. They are efficient enough that detector sizes can be kept small; they are rugged and thermally stable; and they are not excessively expensive. From the log analyst's point of view, these detectors allow excellent logs: good thin-bed resolution, reasonable logging speeds, and relatively low statistical fluctuations.

Measuring Time

In radiation logging, as with any statistical counting process, time is the factor that heals all ills, in theory anyhow. No matter how low the count rate, if measured long enough whatever accuracy wanted can be achieved. This theory gets compromised when a tool is in motion. An averaging time interval—a time constant—must be chosen by the logging engineer; this time constant is usually one or two seconds with modern equipment. There is a time lapse, or lag, between the observed events and the averaged measurement; i.e., during the time taken to accumulate, and average counts from a given depth interval, the tool will have moved to a shallower interval in the well. This distance (the lag) is thus a function of both the logging speed and the time constant. In order to record the logging measurement at the correct depth on the log, it is necessary to know the lag distance and correct for it. The amount of this lag is determined by the combination of time constant and logging speed; a tool moving at 1800 ft/hr, with a time constant of 2, will move one foot in two seconds. This lag is not seen, because it is taken into account by the engineer in his depth-control technique; but it needs to be understood. Most logging companies make a general practice of holding the lag at one foot for detail logs and possibly two feet for correlation logs.

Failure Modes

Perhaps more than other tools, radiation logs are subject to insidious failures that may be hard to detect. All types of detectors are capable of failures by degree, in which a more or less plausible-looking curve appears. Such logs can range from slightly erroneous to meaningless. No matter how knowledgeable you may be, there is still the possibility that a defective log will slip by. Hence, it is definitely desirable practice to develop a trusting relationship with the logging engineers.

Calibration records and well-chosen repeat sections are very important in validation radiation logs. Every trick should be ready to hand: comparison with offset wells, checking against known minerals, integrating the data with other logs in quantitative log analysis.

Geiger-Mueller counters normally operate on a voltage plateau, where variations in operating voltage produce negligible variation in count rate. But if the G-M tube or the power supply drift off that plateau, voltage changes may appear as log anomalies, sometimes masking real formation signals.

Scintillation counters are more resistant to voltage drifts, but are notoriously temperature sensitive. Normal borehole temperatures are too high for them; they tend to start counting too high as the detector heats up, finally going into a continuous counting mode that has no connection with variations in the formation. This is called *corona*, and the only cure for it is to turn the tool off and cool it down.

Scintillation counters are routinely housed in Dewar-flask insulation, and should perform well up to their rated temperature. Remember, however, that insulation only delays the heating process. Tool temperature ratings do not mean that a tool can operate indefinitely at the rated temperature, but that the tool can descend into the rated temperature, run a log, and come back out without failure. Any extended delay at bottomhole, or any time spent with the tool above its rated temperature, causes the failure probability to rise sharply.

3He counters operate with helium gas under pressure, and of course the gas can leak out. When the pressure drops, the counter efficiency drops disproportionately.

All these detector problems can usually be spotted by reference to repeat logs and wellsite calibration records.

Calibration

Another area of potential problems is the calibration technique itself. Radiation tools are characteristically hard to calibrate, for several reasons:

Statistical variations make the measurement a constantly changing value.
Deep penetration means massive primary-calibration standards.
External radiation affects some tools.

To begin with, the units of measurement of some tools are not standard engineering units, but arbitrary *API gamma ray units* and *API neutron units,* which are defined by a physical installation at the University of Houston. The logging companies take their tools to Houston and establish the primary calibration of a given type of tool by adjusting its response to a standard set of artificial formations to agree with a series of API values. Then a portable secondary standard is calibrated and subsequently used to calibrate field tools. In many cases, even this secondary standard is not handy enough for field use, necessitating a tertiary standard—and inevitably each step brings a little slop into the system.

Newer tools are getting away from API units for quantitative logs: neutron tools are calibrated in *porosity units*; density tools in grams per cc; and pulsed neutron tools in *capture units* or *sigma units* (different names for the same thing). All the computer units have virtually eliminated slop in the calibrations by means of digital recording techniques.

Primary calibration standards now in use vary according to tool types. They are described in the individual tool sections.

Hazards

Biological damage results from excessive exposure to high-energy radiation. The amount of damage depends on source strength, intervening material (shielding), time of exposure, and distance from source. The effects, moreover, are cumulative over a long period of time. It is, therefore, no simple thing to establish safeguards that will protect personnel under every set of conditions. The service companies, under the surveillance of various regulatory agencies, have established normal working practices that insure personnel safety if followed. These practices are based on radiation dosages below those that produce the first observable evidence of damage, and incorporate a safety factor of about four.

The responsibility of the operator's representative may be divided into two parts: safety under normal conditions, and safety in emergencies.

Normal Conditions

When things are going according to plan, the representative should be satisfied that the service-company engineer knows and follows the safety procedures laid down by his company. No compromises should be tolerated. If the service company gives lip service to rules or procedures that are considered impractical or unnecessary, the question should be raised before the competent authority, and the rules or procedures changed where justified.

Ordinarily, the more dangerous sources are those used in neutron and density logging tools. They are kept in radiation-absorbing shields except when in use; the shields provide safe operating conditions even for people who work with the sources routinely. The greatest inherent risk of radiation overexposure comes during transfer of the source from shield to tool and the reverse. Some service companies require the logging engineer to do this task himself; in any case, there should be consultation about the safety procedures recognized by each service company.

The case of stuck source-carrying tools is definitely beyond the scope of this document. First rules to be followed are: do not authorize pulling loose from the tool; and call for experienced supervisory assistance from the well operator and from the service company.

There are many smaller sources that are used for calibration, etc. They are normally safe to handle except in cases of fire, damage to the source housing, etc.

Emergency Conditions

Radioactive sources that are normally benign can become very dangerous if exposed to sustained fire, or otherwise damaged in a way that could release the active material. The principal risk is ingestion of radioactive particles, smoke, or liquid into the body. In event of fire or other accident involving radioactive sources, every precaution must be taken to protect the rig crew and other personnel.

Normally the logging engineer and his crew will be schooled in emergency procedures. Their advice is probably the best guidance until a qualified expert is summoned. For further guidance, refer to table 4.2. If in doubt observe the twin rules of time and distance: stay as far away as possible from radioactive sources; but if you must be close, spend as little time as possible in the vicinity.

GAMMA RAY LOGS

Gamma ray logs are used for three main purposes: (1) correlation, (2) evaluation of the shale content of a formation, and (3) mineral analy-

TABLE 4.2 *Radiation Health and Safety*

A. PERMISSIBLE OCCUPATIONAL[a] DOSES (IN REMS) IN SIGNIFICANT VOLUMES OF CRITICAL ORGANS UNDER VARIOUS CONDITIONS OF EXPOSURE (Rules and Regulations, Division of Radiation Protection Standards, U.S. Atomic Energy Commission, August 1, 1969.)

Part of Body Irradiated	Maximum Quarterly Dose
Whole body, gonads, blood forming organs, and trunk (Radiation of sufficient penetrating power to affect critical organ)	1.25 rems[b]
Skin of whole body (Radiation with a half value layer less than one millimeter in soft tissue)	7.5 rems
Hands, feet, forearms, and ankles (Radiation of any energy)	18.75 rems

[a]Permissible dose for non-occupational personnel and all persons under 18 years of age shall be no more than 1⁄10th of that permitted for radiation workers.
[b]One rem of neutron radiation is generally considered equivalent to 14 million neutrons per cm^2 incident upon the body.

B. PERMISSIBLE NEUTRON FLUX DENSITIES BASED ON A 40-HOUR/WEEK EXPOSURE (Consistent with the recommendations of the National Committee on Radiation Protection)

Neutron Energy	Neutron Flux ($n/cm^2/s$)
0.025 eV	2,000
10 eV	2,000
10 keV	1,000
0.1 meV	200
0.5 meV	80
1.0 meV	60
2.0 meV	40
3–10 meV	30

sis. Other, less-frequent, uses include tracer surveys, monitoring water-flooded intervals, and radioactive sand studies, etc.

The gamma ray log measures the natural gamma ray emissions from subsurface formations. In other types of radioactive logging, induced gamma rays are measured (e.g., in pulsed neutron logging), but that will not be discussed in this section.

Gamma ray tools consist of a gamma ray detector and the associated electronics for passing the gamma ray count rate to the surface. A typical gamma ray log (fig. 4.1) is normally presented in Track 1 on a linear grid and is scaled in API units, which will be defined later. Gamma ray activity increases from left to right. Modern gamma ray tools are in the form of double-ended subs that can be sandwiched

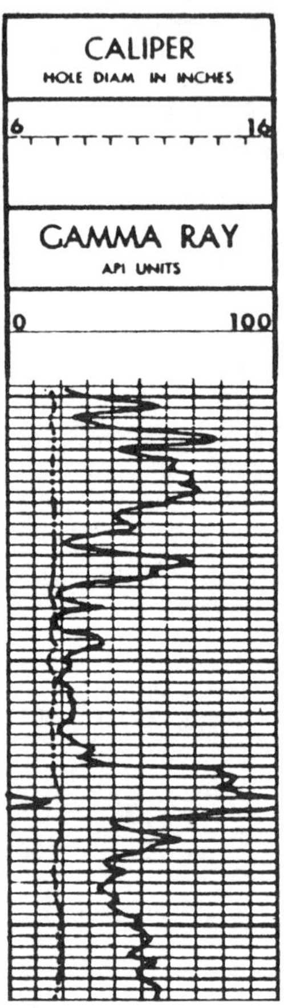

FIGURE 4.1 *Typical Gamma Ray Log. Courtesy Schlumberger Well Services.*

into practically any logging tool string; thus, the gamma ray log can be run with practically any tool available.

Origin of Natural Gamma Rays

Gamma rays originate from three sources in nature: the radioactive elements in the uranium and thorium groups and potassium. Uranium 235, uranium 238, and thorium 232 all decay, via a long chain of daughter products, to stable lead isotopes. An isotope of potassium, K40, which also gives off a gamma ray, decays to argon. It should be noted that each type of decay is characterized by a gamma ray of a specific energy (wave length) and that the frequency of occurrence for each specific energy is different. Figure 4.2 shows this relationship between gamma ray energy and frequency of occurrence. This is an important concept since it is used as the basis for measurement in the natural gamma ray spectroscopy tools.

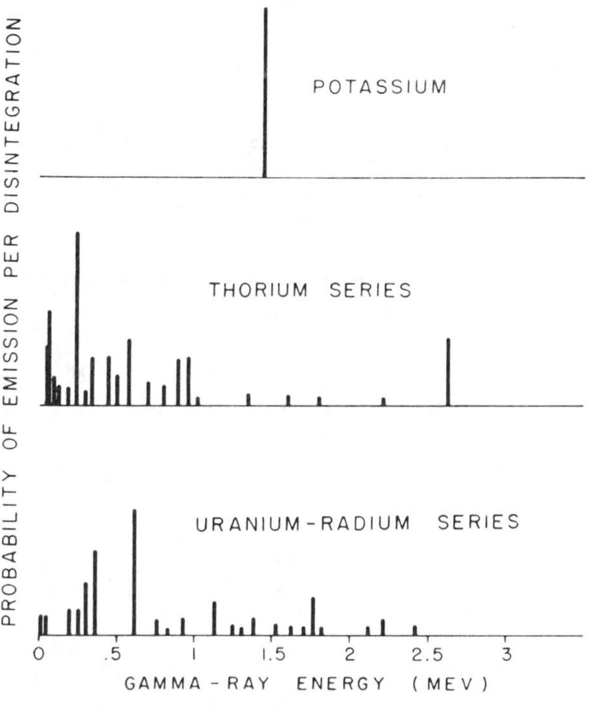

FIGURE 4.2 ***Gamma Ray Emission Spectra of Radioactive Minerals. Courtesy Schlumberger Well Services.***

Abundance of Naturally Occurring Radioactive Minerals

An average shale contains 6 ppm uranium, 12 ppm thorium, and 2% potassium. Since these various gamma ray sources are not equally effective, it is more informative to consider this mix of radioactive materials on a common basis; for example, by reference to potassium equivalents (i.e., the amount of potassium that would produce the same number of gamma rays per unit of time). Reduced to a common denominator, the average shale contains uranium equivalent to 4.3% potassium, thorium equivalent to 3.5% potassium, and 2% potassium. An "average" shale is hard to find. Since a shale is a mixture of clay minerals, sand, silts, and other extraneous materials, there can be no standard gamma ray activity for shale. Indeed, the main clay minerals vary enormously in their natural radioactivity. Kaolinite has no potassium, whereas illite contains between 4 and 8% potassium. Montmorillonite contains less than 1% potassium. Occasionally, natural radioactivity may be due to the precipitation of dissolved potassium or other salts in the water contained in the pore space. Shales with high organic content also tend to have high uranium content.

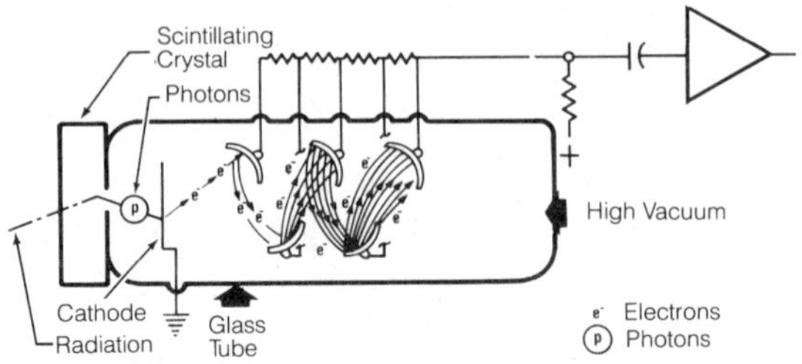

FIGURE 4.3 *Scintillation Counter. Courtesy Gearhart Industries, Inc.*

Operating Principle of Gamma Ray Tools

Traditionally, two types of gamma ray detectors have been used in the logging industry: Geiger-Mueller and scintillation detectors. Today, practically all gamma ray tools use scintillation detectors (those interested in the G-M detectors should refer to the literature). The scintillation counter (fig. 4.3) contains a sodium iodide crystal; and when a gamma ray strikes the crystal, a single photon of light is emitted. This tiny flash of light then strikes a photocathode made from cesium antimony or silver magnesium. Each photon, when hitting the photocathode, releases a bundle of electrons. These in turn are accelerated in an electric field to strike another electrode producing an even bigger shower of electrons. This process is repeated through a number of stages until a final electrode conducts a small current through a measure resistor to give a voltage pulse that signals that a gamma ray struck the sodium iodide crystal. The system has a very short "dead time" and can register many counts per second without becoming swamped by signals.

Calibration of Gamma Ray Detectors and Logs

One of the problems of gamma ray logging is the choice of a standard calibration system since all logging companies use counters of differ-

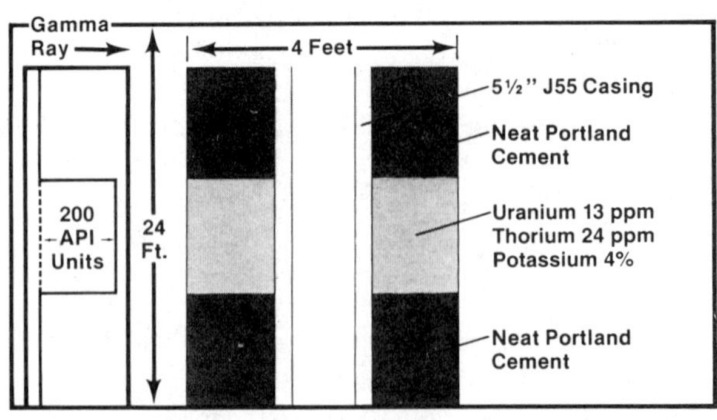

FIGURE 4.4 *API Gamma Ray Standard.*

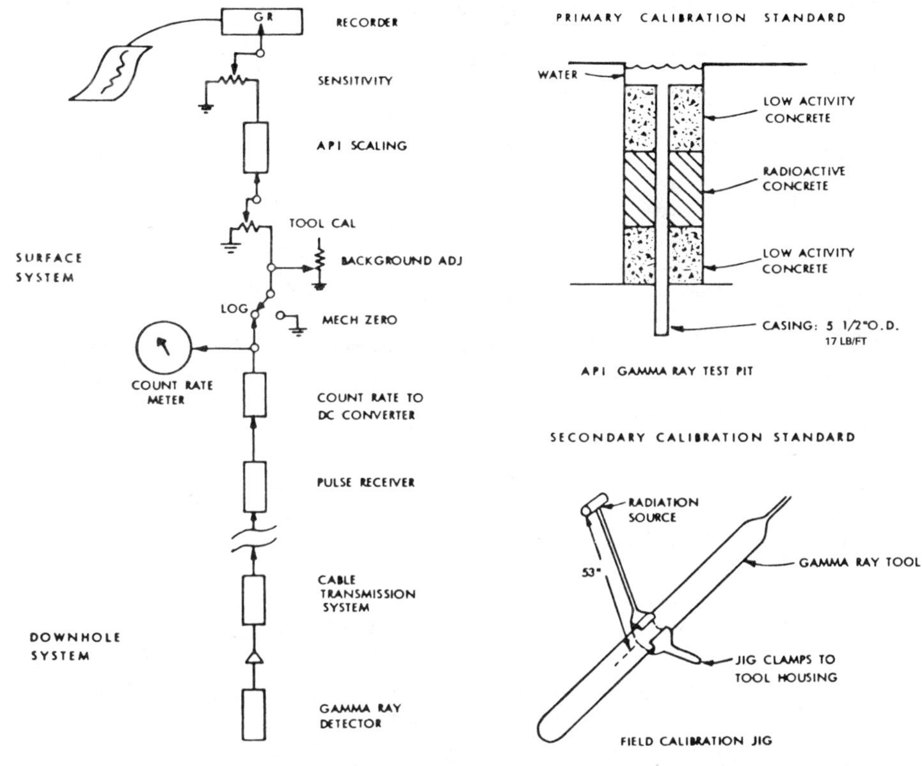

FIGURE 4.5 *Gamma Ray Calibration System. Courtesy Schlumberger Well Services.*

ent sizes and shapes encased in steel housings of varying characteristics. On very old logs, the scale might be calibrated in micrograms of radium per ton of formation. For many reasons this was found to be an unsatisfactory method of calibration for gamma ray logs. So an API standard was devised. A test pit (installed at the University of Houston) contains an "artificial shale," as illustrated in figure 4.4. A cylinder 4 ft in diameter and 24 ft long contains a central 8-ft section consisting of cement mixed with 13 ppm uranium, 24 ppm thorium, and 4% potassium, sandwiched by 8-ft sections of neat portland cement on either side. This 24-ft sandwich is cased with a 5½-in. J55 casing. The API standard defines the difference between tool readings in the neat cement and the radioactive cement mixture as 200 API gamma ray units. Any logging-service company may place its tool in this pit to make a calibration.

Field calibration is performed using a portable jig that contains a radioactive *pill*. The pill is a small radioactive source. When placed a fixed distance from the center of the gamma ray detector it produces a known increase over the background count rate. This increase is equivalent to a known number of API units, depending on the tool type and size and the counter it encloses.

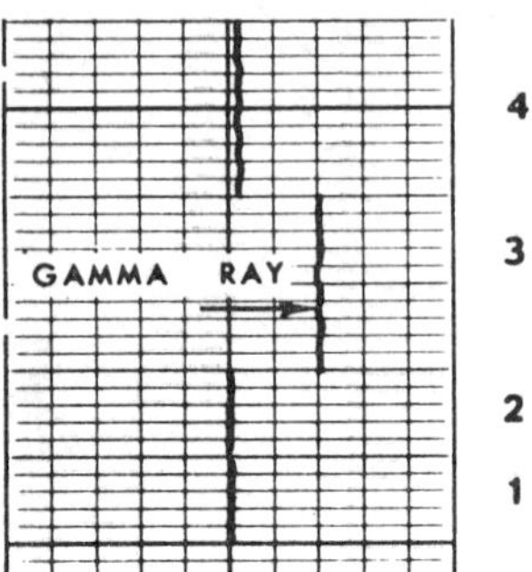

FIGURE 4.6 *Gamma Ray Calibration Record. Courtesy Schlumberger Well Services.*

Calibration of gamma ray logging tools at the API test facility in Houston really amounts to calibration of the secondary field standard (test-pill gamma ray source placed at a fixed distance from the tool). The specific tool under test is used to log the test pit, then the tool is removed and the test-pill source is positioned at a fixed distance from the tool. By comparing the response of the tool to the gamma ray source with its response in the API pit, the strength of the source in API units is determined. This procedure must be repeated for each tool size, each source spacing, and each detector type (Geiger-Mueller, scintillation, etc.).

A graphic representation of the gamma ray system and its calibration standards is given in figure 4.5. For example the Schlumberger secondary calibration for gamma ray tools consists of a 0.1-millicurie radium 226 source positioned 53 in. from the detector. Since errors introduced by incorrect source positioning can be substantial, a mechanical fixture is used which maintains the desired radial spacing. This system produces an increase in radiation field strength of 165 API units for standard-size tools and 135 API units for $1\frac{11}{16}$-in. tools.

Typically, a gamma ray calibration record (fig. 4.6) includes the following steps (from Schlumberger):

1. Mechanical zero: Gamma ray galvanometer at 5 divisions, Track 1 (shifted for calibration; normal zero point for logging is at 10 divisions Track 1, or off-scale left, as indicated by log scale on heading).
2. Electrical zero: Gamma ray galvanometer is functionally on mechanical zero, Track 2.
3. Background: Gamma ray galvanometer reads API GR-unit value of background (2 divisions deflection in example).
4. Gamma ray calibration: Gamma ray calibration jig is in place; galvanometer reads 8.25 divisions higher than in Step 3. The primary galvanometer is off-scale right; the backup spot reads 4.75 divisions of Track 1 (8.25 divisions deflection above background).

Time Constants

All radioactive processes are subject to statistical variation. For example, if a source of gamma rays emits an average of 100 gamma rays each second over a period of hours, the source will emit 360,000 gamma rays per hour (100 × 60 seconds × 60 minutes)/hour. However, if the count is measured for 1 second, the actual count might be less than 100 or more than 100. Thus, a choice must be made. Gamma rays can be counted for a very short interval of time (logging fast), resulting in a poor estimate of the real count rate, or the gamma rays can be counted for a long time resulting in a more accurate estimate of the count rate, but at the expense of an inordinately long time period (logging slowly). In order to average out the statistical variations, various time constants may be selected according to the level of radioactivity to be measured. The lower the count rate, the longer the time constant required for adequate averaging of the variations.

In the logging environment, gamma rays can be counted for a short period of time (e.g., 1 second) with the recognition that during that time period, the detector will have moved past the formation whose activity is being measured. Thus, the logging speed and the time interval used to average count rates are interrelated. The following rules of thumb are generally recognized:

Logging Speed (in feet/hour)	*Time Constant (in seconds)*
3600	1
1800	2
1200	3
900	4

At very slow logging speeds (900 ft/hr = 1.5 ft/s) and long time constants, a more accurate measurement of absolute activity is obtained at the expense of good bed resolution. At high logging speeds and short time constants, somewhat better bed resolution is obtained at the expense of absolute accuracy. At some future time, when the efficiency of gamma ray detectors and their associated electronics improve by one or two orders of magnitude, the use of a time constant will be obsolete except in the cases of very, very inactive formations with intrinsically low gamma ray count rates.

Perturbing Effects on Gamma Ray Logs

Gamma ray logs are subject to a number of perturbing effects:

sonde position in the hole (centering/eccentering)
hole size
mud weight
cement and casing thickness

Since there are innumerable combinations of hole size, mud weight, and tool position, an arbitrary set of standard conditions is defined as

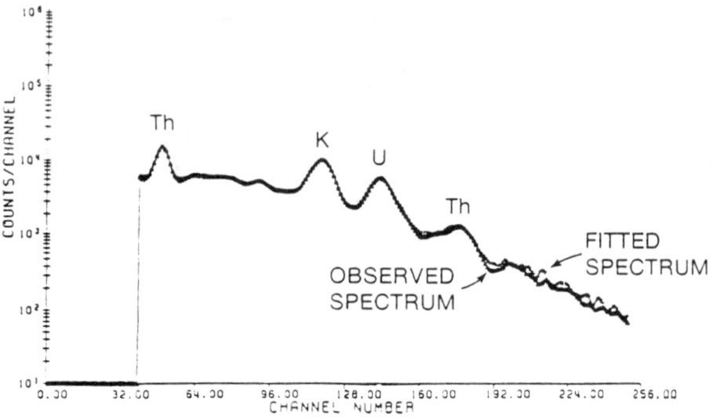

FIGURE 4.7 *GR Spectroscopy Principle.*

a 3⅝-in.-OD tool eccentered in an 8-in. hole filled with 10-lb/gal mud. A series of charts for making the appropriate corrections are available in service-company chart books.

Log Quality Control

Gamma ray logs should be checked for several easily detectable errors. The check list should include the following:

a. Insure that the GR curve is on depth with the other curves; if it is not, there may be some memorization error.
b. Make sure the time constant used was appropriate for the logging speed and requirements of the log (i.e., detail or correlation).
c. See that the GR curve is free from spikes, dead spots, or sections off-scale without an appropriate backup trace.
d. Be particularly watchful for excessive count rates in deep, hot holes where counters may exceed their temperature ratings.

GAMMA RAY SPECTRAL LOGGING

Measurement Principle

Naturally occurring gamma radiation in earth formations consists of individual gamma rays of many different energies (wave lengths). These rays originate from the natural radioactive decay of elements in the uranium and thorium groups and potassium. Uranium and thorium decay, via a long chain of daughter products, to stable lead isotopes. Potassium decays into argon. Each link in the decay chains of these original isotopes gives rise to a gamma ray of distinct energy. By analyzing the distribution of the observed gamma ray energy spectrum, the relative concentration of U, Th, and K may be inferred. These concentrations may be interpreted to indicate properties of the formation that cannot easily be measured in any other way.

The instrument used is a scintillation spectrometer that effectively sorts the incident gamma rays by energy and then uses a computer to

TABLE 4.3 *Applications for Natural Gamma Ray Spectral Logging*

Lithology	Applications
Argillaceous sediments (shales, clays)	Lithology; clay typing; recognition of clay diagenisis; cation exchange capacity (CEC); organic carbon content estimates
Fractured shale reservoirs	Lithology variations (shale, chert, silt, carbonate)
Sandstone reservoirs	Shaly versus clean reservoirs; recognition of radioactive minerals; clays versus feldspar, mica and accessory radioactive minerals; cation exchange capacity (CEC) and Q_v estimates for Waxman-Smits equation. Clay typing, clay mineral distribution, formation damage estimates. Natural open and sealed fracture systems. Apparent correlation to pyrite and other sulfides, presence of sour gas. Flags depleted reservoirs in stratified strata
Carbonate reservoirs (limestone, chalk, dolomite)	Shaly versus clean reservoirs. Natural open (limestone, chalk, dolomite) and sealed fracture systems, presence of stylolites; permeable radioactive dolomites; flags depleted reservoirs in stratified intervals
Igneous and metamorphic rocks	Lithology; hydrothermally altered rocks; open natural fracture systems
Evaporites (salts)	Lithology; ore grade evaluation; differentiation between evaporites and shales
Coals	Lithology; ash content estimates
Uranium	Lithology; ore grade assessment
Miscellaneous	Recognition of radioactive sealing on casing and around perforations; recompletion guide in partially depleted stratified reservoirs; tracing channeling behind pipe, well stimulation attempts, cementing (e.g., tagging squeeze operations), etc.

Source: Dresser Atlas.

fit a variable mixture of U, Th, and K to the observed energy spectrum. Figure 4.7 illustrates the process schematically.

Applications

The main applications of natural gamma ray spectroscopy are (1) clay typing, (2) ore grading, and (3) fracture finding. Other applications are given in table 4.3.

Tools Available

Three tools are currently available for gamma ray spectral logging: the Spectralog (rated at 400°F and 20,000 psi), the Natural Gamma Ray Tool (rated at 350°F and 20,000 psi), and the Compensated Spectral Natural Gamma Ray Tool. Both tools are 3⅝-in. OD and can be run in either an open or a cased hole. Logging speed is slow. Plan on 900 ft/hr or less and a 4-second time constant.

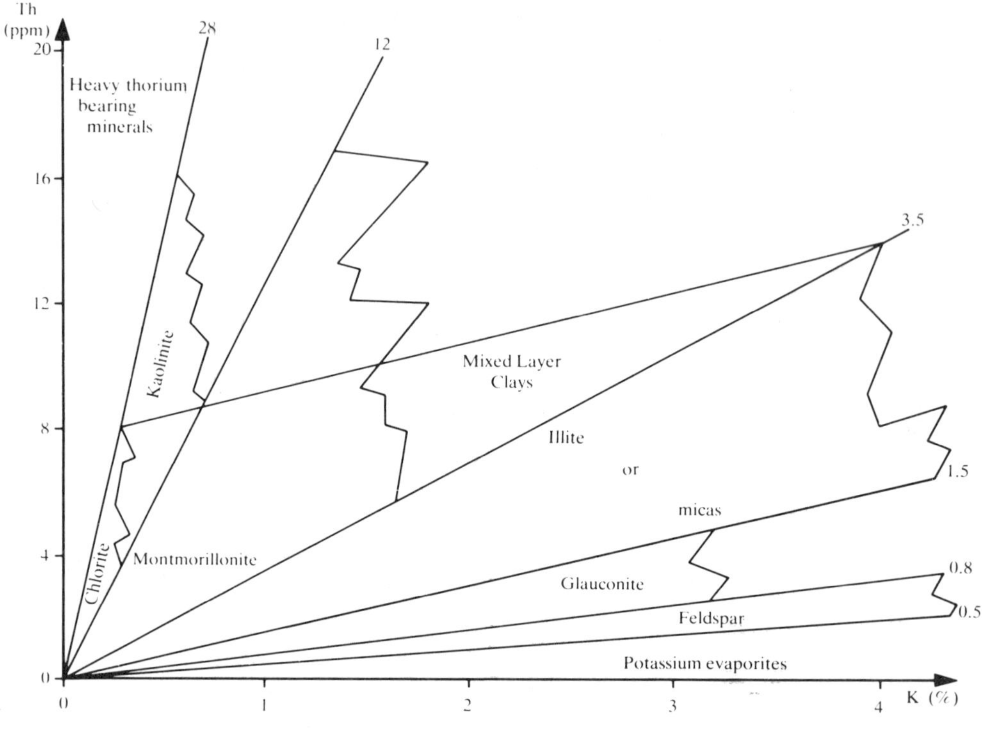

FIGURE 4.8 *Clay Typing from Th and K Content. Courtesy Schlumberger Well Services.*

Log Quality Control

Natural gamma ray spectra logs should be carefully checked:

a. Be sure the logging speed is indicated on the log. Look for the 1-minute timing marks. At 900 ft/hr, the timing marks should appear every 15 ft.
b. Check that no curve reads less than zero. If the "fitting" technique used to compute the concentrations of U, Th, and K fails to find the correct mix, an apparently negative amount may appear on one of the curves.
c. Curves should repeat fairly well in an openhole and less well in a cased-hole log (where count rates are lower).
d. Be aware of the calibration system used by the service company and keep up to date with reports on calibration problems.

Field Examples

For clay typing, figure 4.8 is particularly useful. Fracture finding is illustrated in figure 4.9.

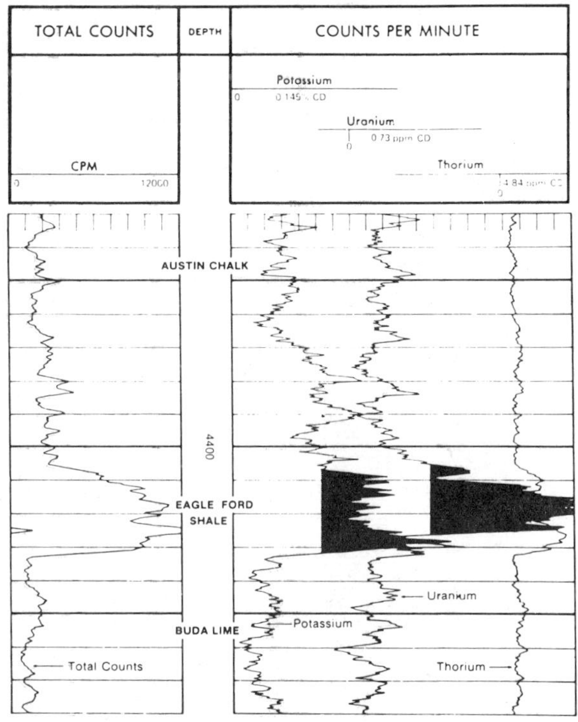

FIGURE 4.9 *Fracture Finding from High U Content. Courtesy Dresser Atlas.*

DENSITY LOGGING

The density of a formation is one of the most important pieces of data in formation evaluation. It is used in the majority of the wells drilled as the primary indicator of porosity. In combination with other measurements (e.g., sonic and neutron), it may also be used to indicate lithology and formation-fluid type.

A conventional compensated formation density log (FDC) is shown in figure 4.10, with the value of formation bulk density (ρ_b) in Tracks 2 and 3. The most frequently used scales are either 2.0 to 3.0 g/cc or 1.95 to 2.95 g/cc across two tracks. A correction curve, $\Delta\rho$, is sometimes displayed in Track 3 and less frequently in Track 2. The gamma ray and caliper curves usually appear in Track 1. An alternative presentation shows density porosity computed from ρ_b.

Figure 4.11 illustrates the articulated device (skid) that carries a gamma ray source and two detectors, which are referred to as the short-spacing and long-spacing detectors. This tool is a pad-type tool; that is, the skid is forced against the side of the borehole.

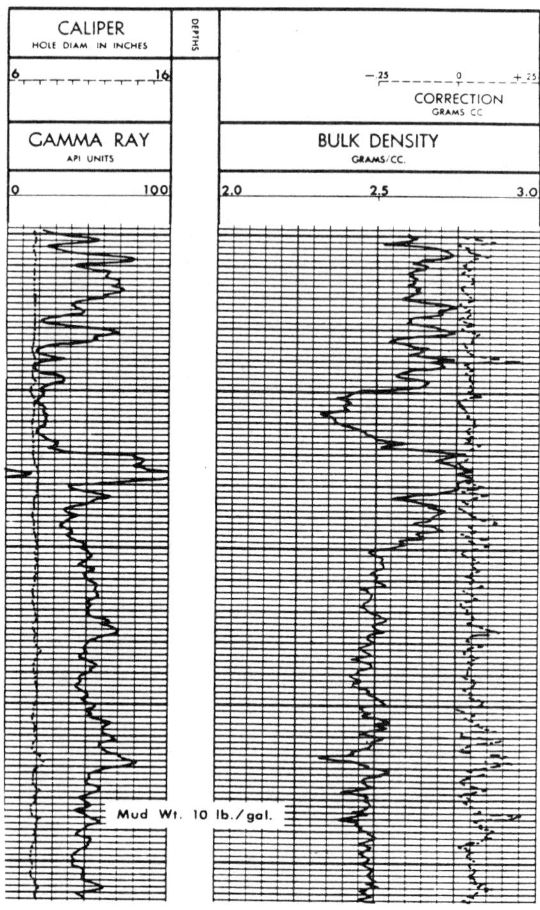

FIGURE 4.10 *Compensated Formation-Density Log. Courtesy Schlumberger Well Services.*

Operating Principle

Gamma rays are continuously emitted from the source. They pass through the mudcake (if present) and enter the formation, where they progressively lose energy until they either are completely absorbed by the formation or return to one or the other of the two gamma ray detectors in the tool. Dense formations absorb many gamma rays, while low-density formations absorb fewer gamma rays. Thus, high count rates at the detectors indicate low-density formations and low count rates at the detectors indicate high-density formations.

Gamma rays can interact with matter in three distinct ways:

1. *Compton scattering*: A gamma ray collides with an electron orbiting some nucleus. In this case, the electron is ejected from its orbit and the incident gamma ray loses energy.

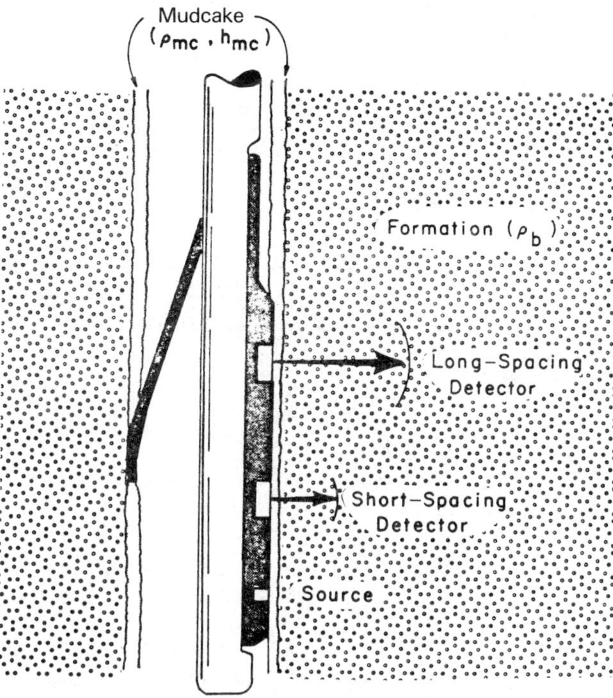

FIGURE 4.11 *Schematic of a Density Tool. Courtesy Schlumberger Well Services.*

2. *Photoelectric effect*: A gamma ray collides with an electron and loses all its energy. In this case, the electron is ejected from the atom.
3. *Pair production*: A gamma ray interacts with an atom to produce an electron and a positron; these will later recombine to form another gamma ray.

The conventional gamma ray source used in logging tools is made either of cesium or cobalt and the emitted gamma rays have an energy of 0.63 MeV. At this level, it is impossible to expect any form of pair production, since this type of interaction only occurs at energies higher than 2 or 3 MeV.

The detectors used in conventional density tools have a practical lower limit to the gamma ray energy level they can detect. This lower limit is about 0.2 MeV. Compton scattering, therefore, becomes the only possible form of interaction that conventional density tools can monitor.

The net effect of gamma ray Compton scattering and absorption is that the count rate seen at the detector is logarithmically proportional to the formation density (fig. 4.12). Both the near and the far spacing detectors behave in this way, so a plot of far count rates against near count rates will also produce a straight line, as shown in figure 4.13. Note that formation density increases as count rates decrease.

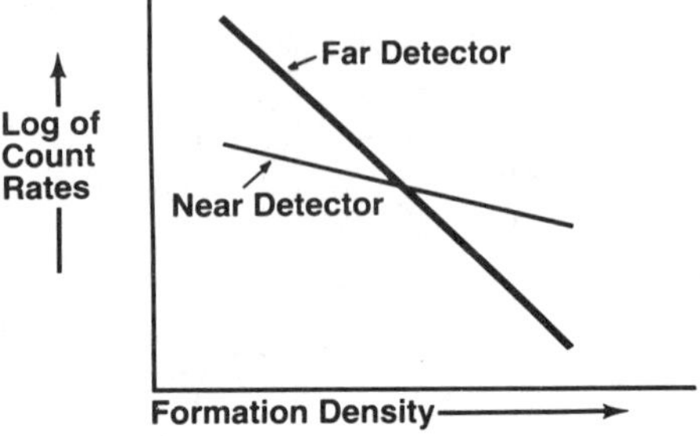

FIGURE 4.12 *FDC Count Rates vs. Formation Density.*

Compensation of the FDC

If the density of the mudcake is different from that of the formation, then both the near and far count rates will change. Figure 4.14 shows where a plotted point would fall if a formation with a bulk density of 2.7 g/cc were to have an ever-increasing amount of mudcake of density 1.5 g/cc placed between it and the density tool. In the extreme case of "infinite" mudcake thickness, both detectors will see only mudcake and read a value of 1.5 g/cc.

The arc describing the locus of the points is referred to as a *rib*. The zero mudcake line is referred to as a *spine*. A complete set of spine and ribs can be drawn for various thickness and densities of mudcakes, as shown in figure 4.15. Note that ribs also extend to the left of the spine for mudcakes having a density greater than the formation density (e.g., in barite muds).

The surface equipment associated with the formation density tool computes the position of the point on the spine-and-ribs chart and then moves the point down the rib to intersect the spine. At this point,

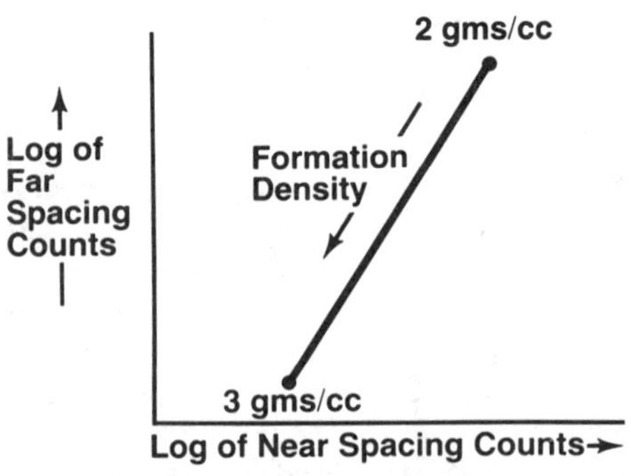

FIGURE 4.13 *FDC Near and Far Count Rates.*

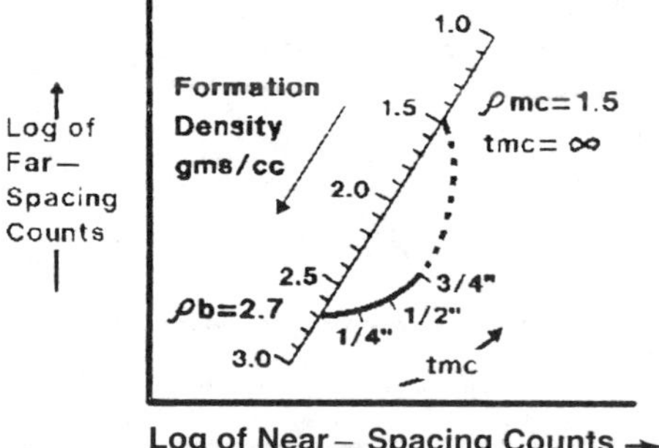

FIGURE 4.14 *FDC Mudcake Compensation.*

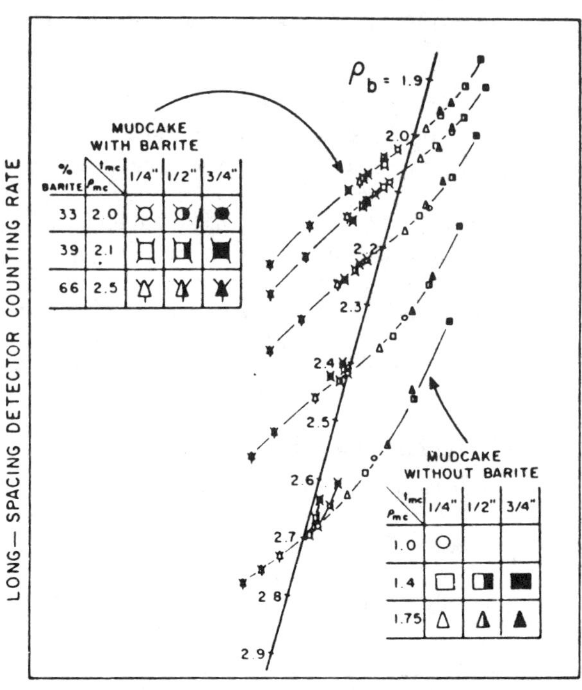

FIGURE 4.15 *"Spine-and-Ribs" Plot, Showing Response of FDC Counting Rates to Mudcake. Courtesy Schlumberger Well Services.*

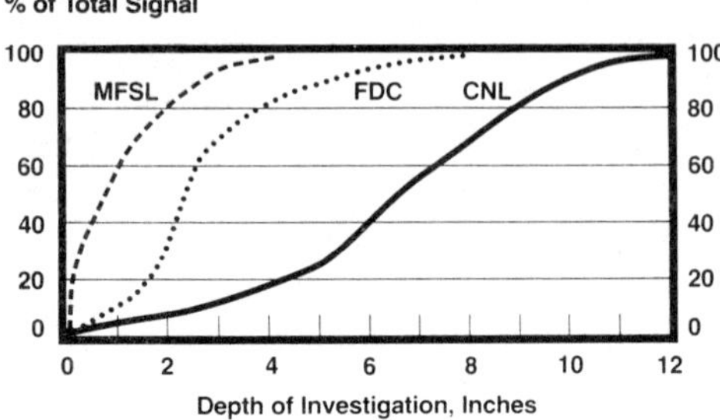

FIGURE 4.16 *Comparison of Depth of Investigation of Density and Neutron Tools.*

a corrected value of ρ is recorded for the log. The value of Δρ is calculated as the difference between ρ from the long spacing and corrected ρ. Thus, Δρ is positive in light muds and negative in heavy muds.

Depth of Investigation

The density tool has quite a shallow depth of investigation. Figure 4.16 illustrates the depth of investigation of the density tool as compared with the MFSL and CNL (compensated neutron log; see next section, Neutron Logging) tools. Most of the density tool signal comes from a region less than 8 in. from the borehole wall. The CNL tool, by contrast, gathers most of its signal from the region within 12 in. of the borehole wall. Thus, the density tool is less affected by light hydrocarbons than is the CNL tool. In situations where deep invasion has occurred, there may be very little hydrocarbon effect on the density tool.

The Lithologic Density Tool (LDT)

The lithologic density tool (LDT) uses the photoelectric effect as a basis for its measurement. This tool must use special shielding for special detectors since the energy level of the gamma rays (resulting from the photoelectric-effect collisions) is much lower than can normally be detected by the conventional FDC tool. In practice, special plastic windows are used. This tool makes a direct measurement of Z, the average atomic number of atoms in the formation. The log itself records a function of Z called P_e, the photoelectric factor, and P_e is defined as $(Z/10)^{3.6}$. Figure 4.17 shows a typical log presentation.

The value of knowing P_e is that it is a direct indication of the lithology and is almost independent of the porosity. For example, the P_e values for

sand = 1.81,
limestone = 5.05,
dolomite = 3.14.

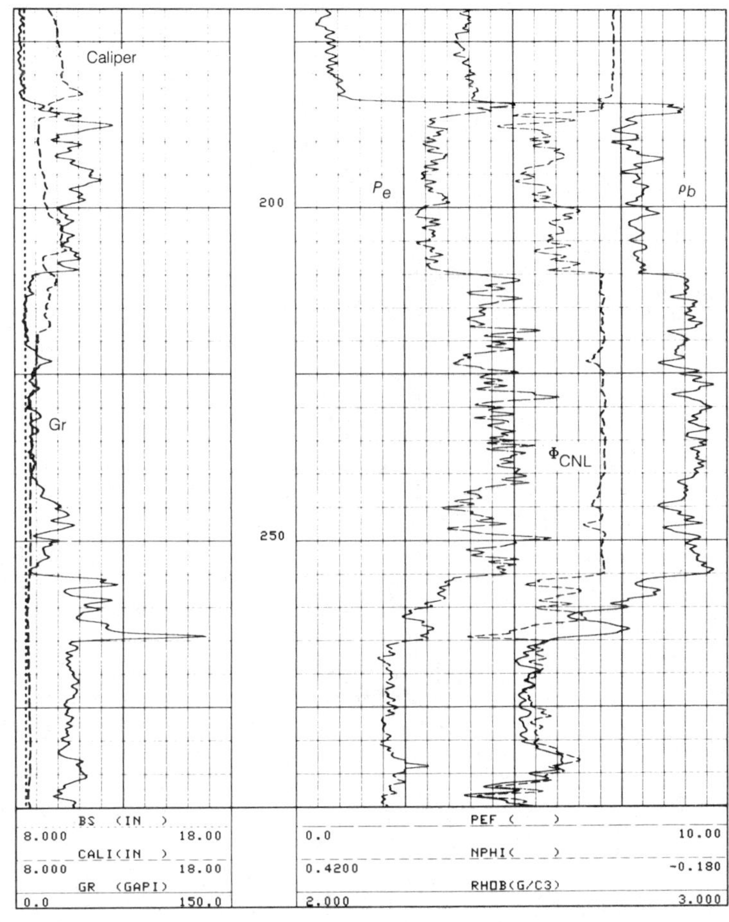

FIGURE 4.17 ***Typical Presentation of a Lithodensity (LTD) Log. Courtesy Schlumberger Well Services.***

Calibration

Practical calibration of the density tool is accomplished by a series of standards. The primary standard (fig. 4.18) is made by using laboratory formations. Since these cannot be transported, a set of secondary standards are available at logging-service company bases in the form of aluminum, magnesium, and/or sulphur blocks of accurately known density and geometry. These blocks, weighing about 400 lb, are not easily transportable either, so a field calibrator containing two small gamma ray sources is used to reproduce the same count rates as those found in the secondary standard block. Checks for the mudcake compensation can be made by insertion of special spacers in these blocks.

Wellsite calibration should be performed before and after each log

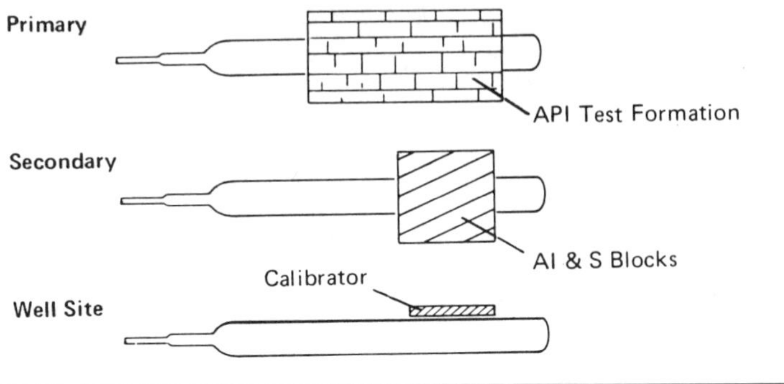

FIGURE 4.18 ***Density Tool Calibration.***

is run; the shop calibration should be run at least every 60 days and a copy of it attached to the main log. It is important to note that the field calibrator, the skid with the detectors, and the source form a matched set. If any item of the three does not match the serial numbers on the master calibration, then the log should be rejected.

As a result of the calibration procedures used for density tools, and the physics of Compton scattering, log densities may be slightly different from the bulk densities (see table 4.4).

Natural benchmarks for checking the validity of a density log are salt, which has a ρ_a of 2.032 g/cc, and anhydrite, which has a ρ_a of 2.977 g/cc. These minerals may not appear in the wellbore being logged, and even if they do, they may not be 100% pure. So even these benchmarks should be used with caution. In general, density logs are either well calibrated (and therefore correct) or very noticeably bad.

Apart from the natural benchmarks, the next-best quality check is a review of the $\Delta\rho$ curve. If the short-spacing detector fails, then the whole compensation mechanism is thrown out of kilter. So if $\Delta\rho$ is generally within the limits of ± 0.05 g/cc, the log may be assumed to be correct. However, if in light muds the $\Delta\rho$ is negative, something is wrong. Likewise, positive values for $\Delta\rho$ in heavy (barite) muds is a danger signal.

BEFORE SURVEY CALIBRATION—FDC.* A complete FDC calibration record is shown in figure 4.19.

1. Mechanical zero: Galvanometer circuits are open.
   a. ρ galvanometer is at 10 divisions Track 2.
   b. $\Delta\rho$ galvanometer is at 5 divisions Track 3.
2. Recorder sensitivity: Panel signals produce one-track deflection.
   a. ρ galvanometer is at zero Track 2.
   b. $\Delta\rho$ galvanometer is at 5 divisions Track 2.

*The calibration information that follows has been provided, with permission, by Schlumberger Well Services.

TABLE 4.4 *Actual and Apparent Density Values for Various Minerals and Fluids*

Compound	Formula	Actual Density ρ_b	$\frac{2\Sigma Z}{\text{Mol. Wt.}}$	ρ_e	ρ_a (as seen by tool)
Quartz	SiO_2	2.654	0.9985	2.650	2.648
Calcite	$CaCO_3$	2.710	0.9991	2.708	2.710
Dolomite	$CaCO_3MgCO_3$	2.870	0.9977	2.863	2.876
Anhydrite	$CaSO_4$	2.960	0.9990	2.957	2.977
Sylvite	KCI	1.984	0.9657	1.916	1.863
Halite	NaCl	2.165	0.9581	2.074	2.032
Gypsum	$CaSO_42H_2O$	2.320	1.0222	2.372	2.351
Anthracite Coal		1.400 1.800	1.030	1.442 1.852	1.355 1.796
Bituminous Coal		1.200 1.500	1.060	1.272 1.590	1.173 1.514
Fresh Water	H_2O	1.000	1.1101	1.110	1.00
Salt Water	200,000 ppm	1.146	1.0797	1.237	1.135
"Oil"	$n(CH_2)$	0.850	1.1407	0.970	0.850
Methane	CH_4	ρ_{meth}	1.247	1.247 ρ_{meth}	1.335 ρ_{meth}—0.188
"Gas"	$C_{1,1}H_{4,2}$	ρ_g	1.238	1.238 ρ_g	1.325 ρ_g—0.188

Source: Schlumberger Well Services.

3–7. Panel test positions: With panel function-former switch set on FDC-Liquid, galvanometer reads as shown by the tabulation in figure 4.19.

8. Mechanical zero-caliper: Caliper galvanometer reads 8 divisions Track 1 (8-in. diameter on 6- to 16-in. scale).
9. 8-in. ring: Tool in measure condition. Caliper galvanometer reads 8 divisions Track 1.
10. 12-in. ring: Caliper galvanometer reads 4 divisions Track 1.
11. Tool calibration No. 1: Set ρ = 2.50 g/cc. Count rate of "far" detector recorded in calibration data on heading. ρ galvanometer reads 10 divisions Track 2.
12. Tool calibration No. 2: Set $\Delta\rho$ = zero. Count rate of "near" detector recorded in calibration data on heading. $\Delta\rho$ galvanometer reads 5 divisions Track 3.
13. Jig response in log position:
    a. ρ galvanometer reads 2.59 g/cc (1.8 divisions Track 3).
    b. $\Delta\rho$ galvanometer reads 0.015 g/cc correction (5.3 divisions Track 3).

    Note: Steps 11, 12, and 13 are affected by statistical fluctuations and are recorded for one minute or more to permit valid averaging.

AFTER SURVEY CALIBRATION—FDC. Steps 1, 2, 11, 12, and 13 are recorded as in the before survey calibration, without changing the panel settings used for logging.

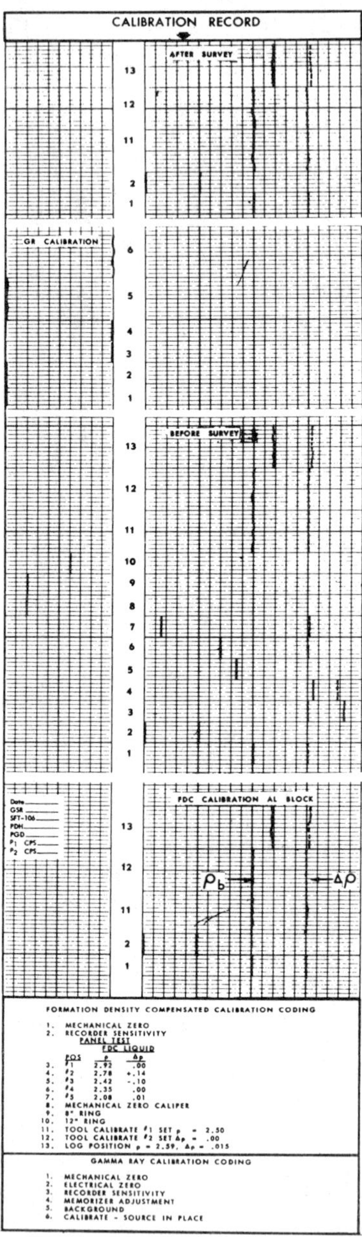

FIGURE 4.19 *Density Calibration Record. Courtesy Schlumberger Well Services.*

FDC CALIBRATION IN ALUMINUM BLOCK. Steps 1, 2, 11, 12, and 13 are recorded as in the before survey calibration. Galvanometer deflections in Steps 11, 12, and 13 are produced by tool response in the aluminum calibrator with logging source, rather than by the field calibration jig.

Log Quality Control

A complete density log is shown in figure 4.20. Specific log quality checks should include the following:

1. Correct recording speed and time constant (TC):

   1800 ft/hr $TC = 2$
   1200 ft/hr $TC = 3$
   900 ft/hr $TC = 4$

2. Correctness of readings:
   a. Repeat section should be identical to log except for statistical variation and variation due to changes in skid orientation between runs, as in out-of-round holes.
   b. Porosity values should be the same as for other porosity tools in shale-free zones.
3. Test films indicate no change during survey.
4. Test films to be attached:
   200 ft of repeat,
   calibration before survey,
   calibration after survey, and
   shop calibration in aluminum block.

Field Examples

Figure 4.21 shows a combination CNL/FDC log on which the density measurement progressively failed as the survey progressed. One or other of the counters undoubtedly failed. Below 7150 ft, reasonable values of ρ_b (and hence apparent density porosity, ϕD) are obtained. Above 7150 ft, the apparent density porosity reads too high.

Figure 4.22 shows a density log that exhibits very little character. Deflections run from 2.35 g/cc to 2.5 g/cc and $\Delta\rho$ values are excessively high. This log can be rejected on sight and should be rerun with a different set of equipment.

Figure 4.23 illustrates a poor comparison between a main density log run and the repeat section. An additional run with backup equipment is indicated in a case like this to determine the correct formation density.

NEUTRON LOGGING

Neutron porosity devices respond to the hydrogen content of the formation. In clean reservoirs, hydrogen is present in liquid-filled

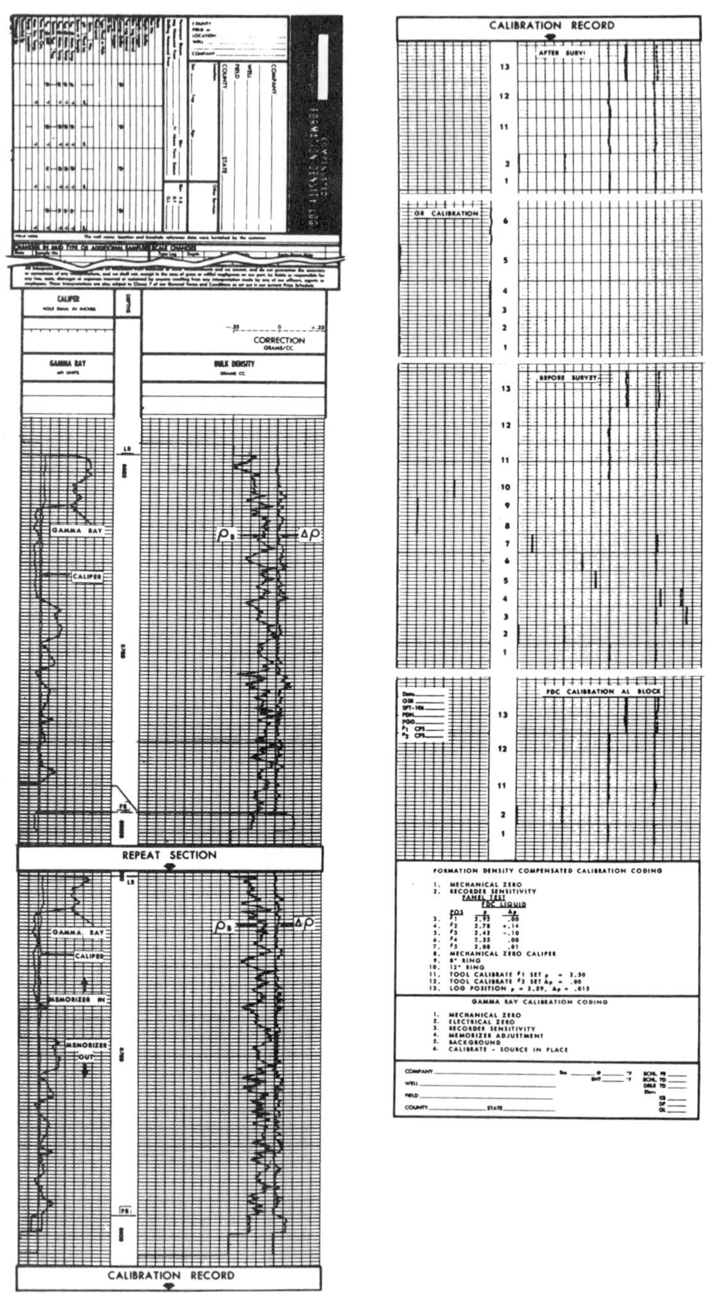

FIGURE 4.20 *Complete Density Log Presentation. Courtesy Schlumberger Well Services.*

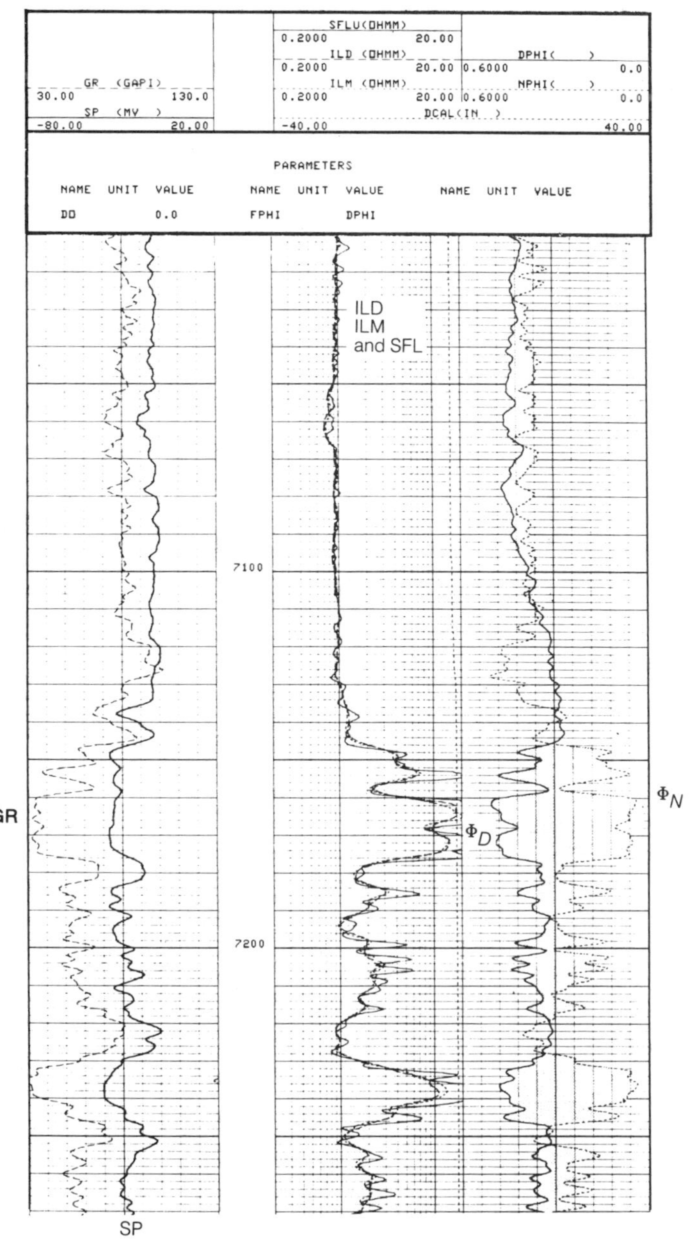

FIGURE 4.21 *Progressive Failure of Density Tool.*

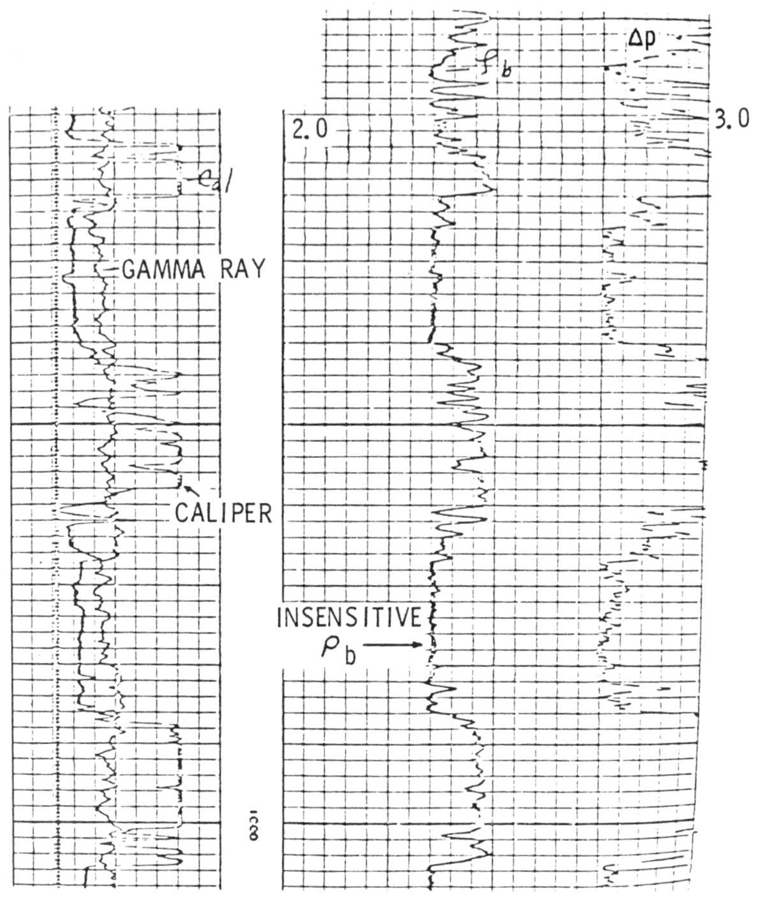

FIGURE 4.22 *Bad Density Log.*

pore space. Neutron logs thus measure the amount of liquid-filled porosity. Neutron logging devices contain a neutron source that continuously emits energetic (fast) neutrons, and one or more detectors. Neutrons collide with formation nuclei causing them to lose energy. After a sufficient number of collisions, the neutrons will reach a lower energy state referred to as epithermal, and continue to lose energy until they reach a still lower energy state (thermal), whereupon they are captured by formation nuclei. When a nucleus captures a thermal neutron, a gamma ray is emitted to dissipate the excess energy. Porosity (or hydrogen index) can be determined by measuring either the flux of capture gamma rays, or by measuring the epithermal or thermal neutron populations, or by any combination thereof. Neutron logs based on the detection of epithermal neutrons are referred to as SN (Sidewall Neutron) Logs. The Compensated Neutron Log (CNL), in

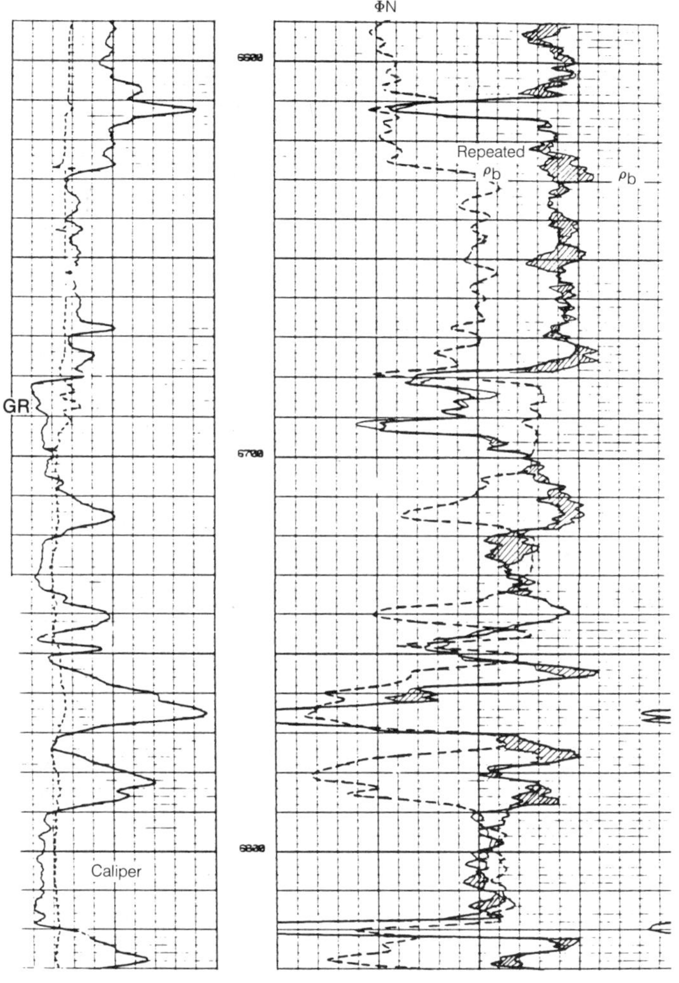

FIGURE 4.23 *Bad Repeat on Density Log.*

widespread use today, uses two thermal neutron detectors to reduce borehole effects. Single thermal-neutron-detector tools, although poorer in quality, are also available in many areas of the world. Capture gamma rays are also used to determine porosity, and logs of this type are referred to as neutron–gamma tools. The responses of these devices are dependent upon porosity, lithology, hole size, hole rugosity, fluid type, temperature, etc. CNL and SNP (sidewall neutron porosity) logs have corrections included in the electronic panels to account for some of these variables, while neutron–gamma logs require corrections from departure curves.

The main applications for neutron tools are, therefore, determination

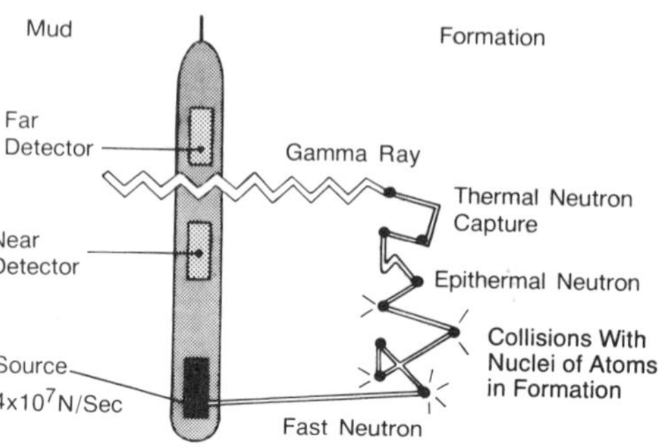

FIGURE 4.24 *Generalized Neutron Tool.*

of porosity; determination of formation-fluid type; and determination, in combination with other devices, of lithology. Depending on the device, measurements with these tools may be made in either open or cased holes.

Measurement Principle

A generalized neutron logging tool is illustrated in figure 4.24. Historically, a number of devices have been used that rely on one portion or another of a neutron's progress from a chemical source to eventual capture by formation nuclei (see table 4.5).

All neutron tools respond to porosity; but they are also influenced by other parameters and certain environmental factors. Neutron devices are affected by the lithology and the formation-fluid type (e.g., gas). Environmental factors that can affect neutron logs include borehole-fluid type, density, and salinity; borehole size; mudcake;

TABLE 4.5 *Neutron Logging Tool Principles*

Nuclear Process	Fast Neutron	Epithermal Neutron	Thermal Neutron	Capture Gamma Ray
Encountered	Near source	After collisions	Just before capture	At capture
Population dominated by	—	Hydrogen	Hydrogen	Chlorine
Energy	Several MeV	A few eV	0.025 eV	—
Tool type	—	SNP, CNT-G	CNT	GNT
Generic name	—	Sidewall or epithermal	Compensated neutron log	Neutron-gamma
Detection mode	—	Single or dual epithermal neutron detectors	Dual thermal neutron detectors	Single-capture gamma detectors
Calibrated in	—	Porosity units	Porosity units	Count/s (API units)

TABLE 4.6 *Neutron Tools by Service Company*

	Gearhart	Dresser	Schlumberger	Welex
Neutron Gamma	GR/N/CCL	GR-N Log	GNT	
Sidewall	SNL	SWN	SNP	SWN
Compensated	CNS	CN Log	CNL	DSN

standoff; temperature; and pressure. To some extent modern neutron tools incorporate corrections for many of these environmental factors, but older (neutron-gamma type) tools do not. Service-company correction charts should be consulted for the particular tool used to assess the magnitude of these effects.

Tools Available

The commercial designations for currently available neutron logging tools are given in table 4.6.

Neutron Gamma tools may infrequently be found in use for cased-hole correlation logging and perforation-depth control.

Sidewall Neutron tools are still available from some service companies but have mostly been replaced with compensated neutron tools.

Compensated Neutron tools are widely in use and are frequently run in combination with compensated density tools.

Dual Epithermal Neutron tools are in the prototype stage and may become more widely available in the future.

Choice of a neutron logging tool should be dictated by well conditions and the tool's ability to combine with other required services. Figure 4.25 shows a schematic of a CNL tool eccentralized in a borehole together with two typical tool-string arrangements, one (combined with density and gamma ray) for openhole and one (combined with a collar locator) for cased-hole logging.

The presentation of neutron logs can vary considerably from tool to tool and service company to service company. Figure 4.26 illustrates some typical presentations.

Calibration

The standard for all neutron logging tools is the *API Neutron Calibration Pit* maintained at the University of Houston. Figure 4.27 illustrates this pit, which contains three limestone formations of different (1.9%, 19%, and 26%) porosity. Any service company may test its tool in this pit and thus relate detector count rates to porosity.

Secondary shop calibrations are carried out once every 60 days in a water-filled neutron calibration tank, which appears to the tool as a formation of known porosity. Wellsite calibration is effected by use of a calibration jig containing small radioactive sources. When the jig is

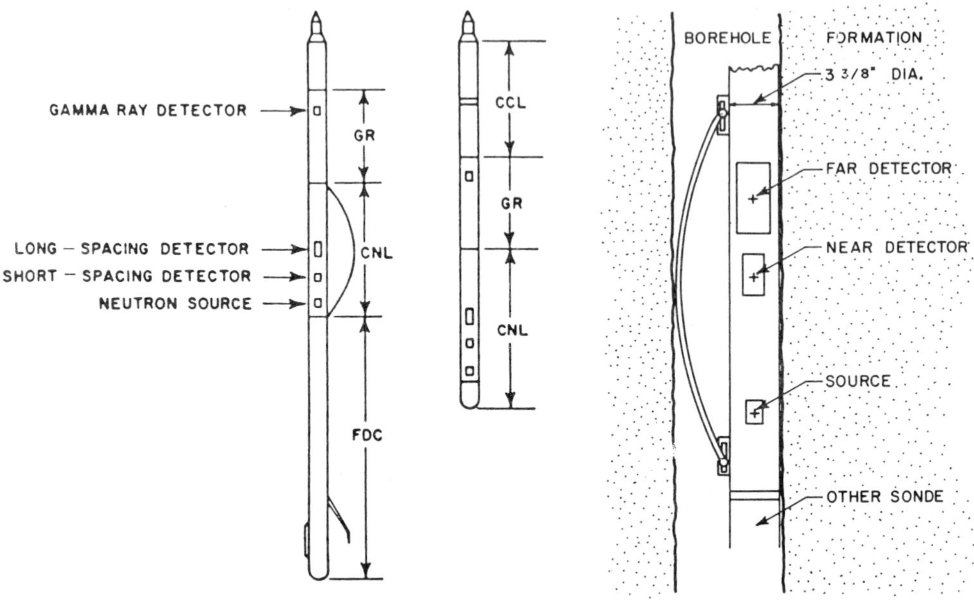

FIGURE 4.25 *Compensated Neutron Tool Strings. Courtesy Schlumberger Well Services.*

clamped over the detector(s) it reproduces the count rate(s) observed in the calibration tank.

The complete calibration chain is shown in figure 4.28. The calibration tail for the CNL is reproduced in figure 4.29; and a complete CNL survey is shown in figure 4.30.

BEFORE SURVEY CALIBRATION.* The following shop calibrations are performed monthly, or any time the ratios observed during wellsite calibrations do not agree with those recorded at the previous shop calibration. The test films are dated, and the serial numbers of the downhole cartridge and the field calibrator are also included.

1. Mechanical zero:
   a. Porosity galvanometer (ϕ) is set at zero porosity, 5 divisions of Track 3.
   b. Ratio galvanometer (dashed) is set at ratio of zero, 10 divisions of Track 3.
2. Recorder sensitivity:
   a. An internal signal produces one-track deflection of the ϕ gal-

*The calibration information that follows has been provided, with permission, by Schlumberger Well Services.

vanometer and a ratio of 2.30. ϕ galvanometer reads 5 divisions of Track 2.

b. Ratio galvanometer reads 0.8 divisions of Track 3.

3.–6. Panel test: Artificial ratio values of 1, 2, 3, and 4 are fed into the porosity computer. The porosity reading depends upon which function former is being used; the test film must include this information. Ratio and ϕ values are given in figure 4.29. (*Note*: ϕ values over 30 will be off-scale if the usual -10%, $+10\%$, $+30\%$ scale is used. The backup galvanometer must then be read—it will be scaled 30%, 50%, 70% right to left.)

7. Field calibrator installed: Porosity response normalized to give one-track deflection.
   a. ϕ reads 5 divisions Track 2.
   b. Ratio galvanometer reads near 2.17 (1.15 divisions Track 3). The exact figure varies with different tools and is not critical; the variation is taken into account by the calibration.

7A. Tool in calibration tank: Readings are the same as Step 7.

8. Log position—field calibrator installed: ϕ galvanometer reads the same as Step 7. Both galvanometers are now affected by statistical fluctuations. (See figure 4.29 for ϕ values.)

8A. Log position—tool in calibration tank: Readings are the same as Step 8.

AFTER SURVEY CALIBRATION. Steps 1, 2, 7, and 8 of the shop calibration are repeated without changing the sensitivity controls. This verifies that the system has remained in calibration. Steps 3, 4, 5, and 6 are made immediately following the completion of the log to verify porosity scale and matrix settings.

Log Quality Control

Specific log quality checks include:

1. Correct recording speed and time constant:

   1800 ft/hr $TC = 2$
   1200 ft/hr $TC = 3$
   900 ft/hr $TC = 4$

2. Correctness of readings:
   a. Repeat section should be identical to log except for statistical variation and variations due to changes in skid orientation between runs, as in out-of-round holes.
   b. Porosity values should be the same as for other porosity tools in shale-free water zones.
3. Test films should indicate no change during survey.
4. Test films to be attached:
   a. 200 ft of repeat,
   b. calibration before survey,
   c. calibration after survey, and
   d. shop calibration.

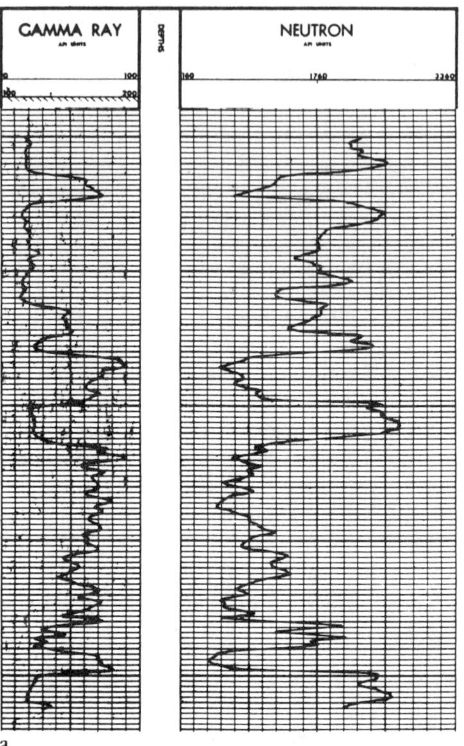

FIGURE 4.26 *Neutron Log Presentation: (a) Gamma Ray/Neutron, (b) Sidewall Neutron, and (c) Compensated Neutron. Courtesy Schlumberger Well Services.*

5. Natural benchmarks for compensated neutron tools are evaporites such as salt and anhydrite, which should indicate close to 0% limestone porosity.
6. Carefully inspect all neutron logs for scaling information. Most compensated neutron logs may be run on a variety of matrix settings (e.g., limestone, sandstone, etc.). Depending on the matrix setting and the actual lithology the log may or may not reflect the true formation porosity.

Field Examples

Figure 4.31 is a schematic of a combination neutron/density/GR log run in complex lithology. Note the relative positions of the curves in different rock types. Figure 4.32 shows an unsatisfactory repeat section on a neutron/density log. Logs with poor repeatability should be rejected.

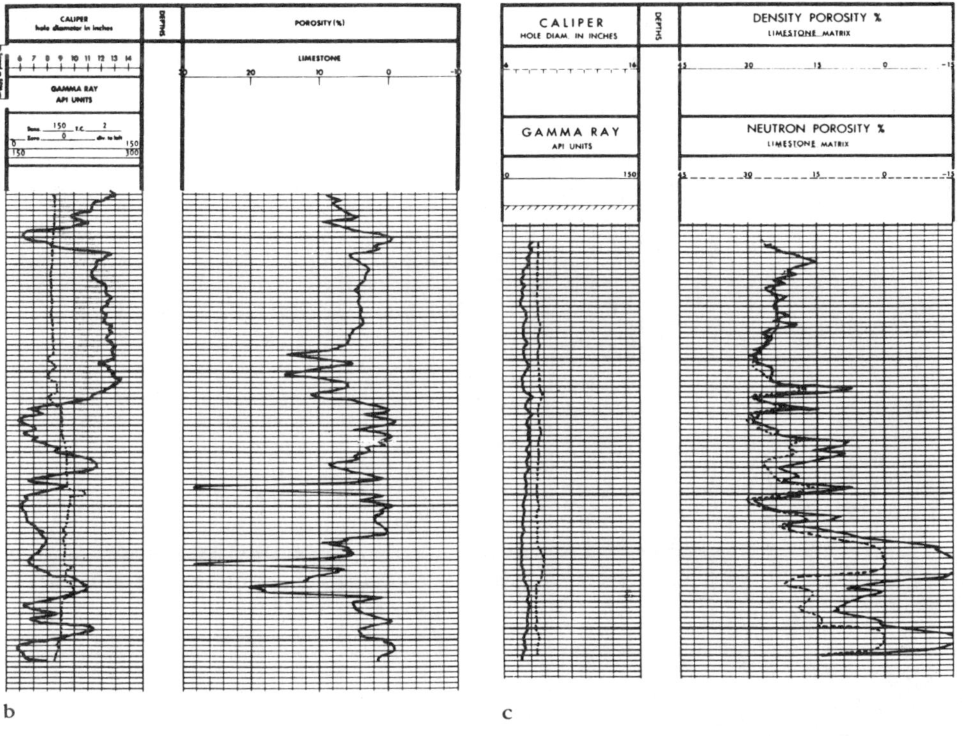

ACTIVATION LOGGING AND THE C/O LOG

Theory of Measurement

Neutrons can interact with matter in two distinct ways to create gamma rays: by *inelastic scattering* with nuclei at high energies (>5 MeV) and from *capture* or *absorption* by nuclei at low energies (≤0.025 eV). The gamma rays produced from each of these reactions have unique energies that depend on the type of nucleus with which the neutron reacts. By measuring the number and energy of gamma rays produced by neutron bombardment, the elemental composition of the formation can be inferred. Thus, the process of what we have called *activation logging* consists of directing high-energy neutrons (14 MeV) produced by a pulsed neutron source into a formation and sampling the energy spectrum of gamma rays produced by the neutron bombardment at various times during and after the neutron burst.

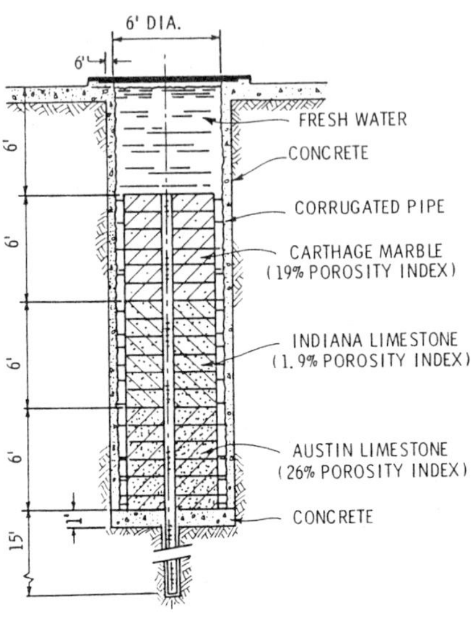

FIGURE 4.27 ***API Neutron Calibration Pit. After Belknap et al. 1959.***

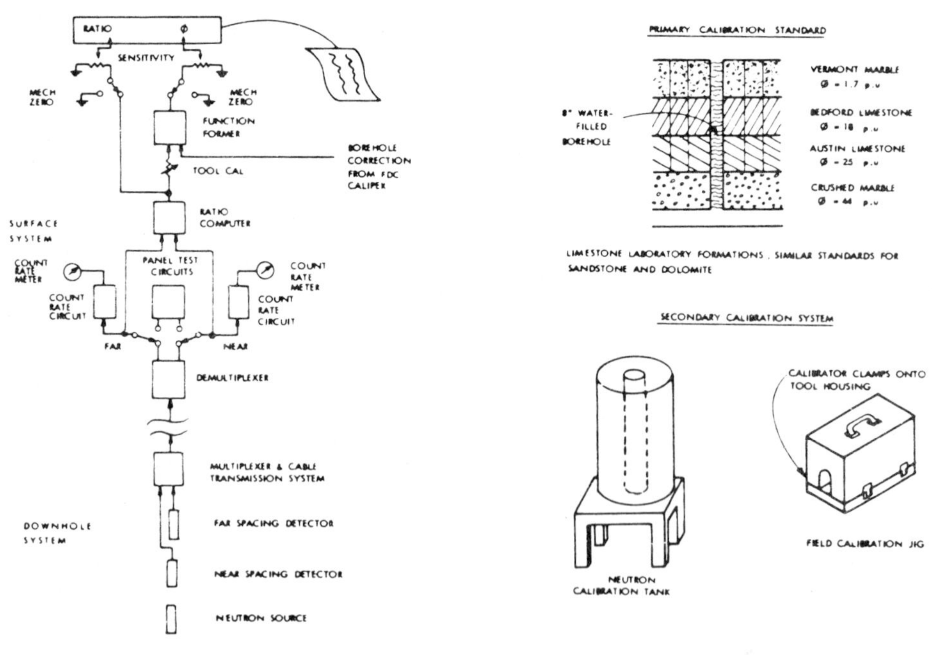

FIGURE 4.28 ***Compensated Neutron Calibration. Courtesy Schlumberger Well Services.***

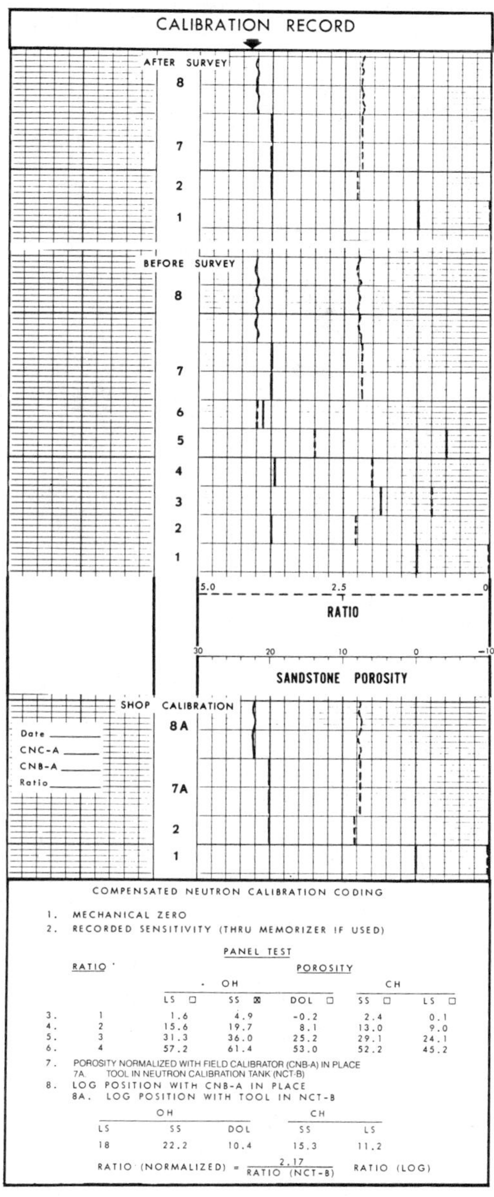

FIGURE 4.29 *Compensated Neutron Calibration Tail. Courtesy Schlumberger Well Services.*

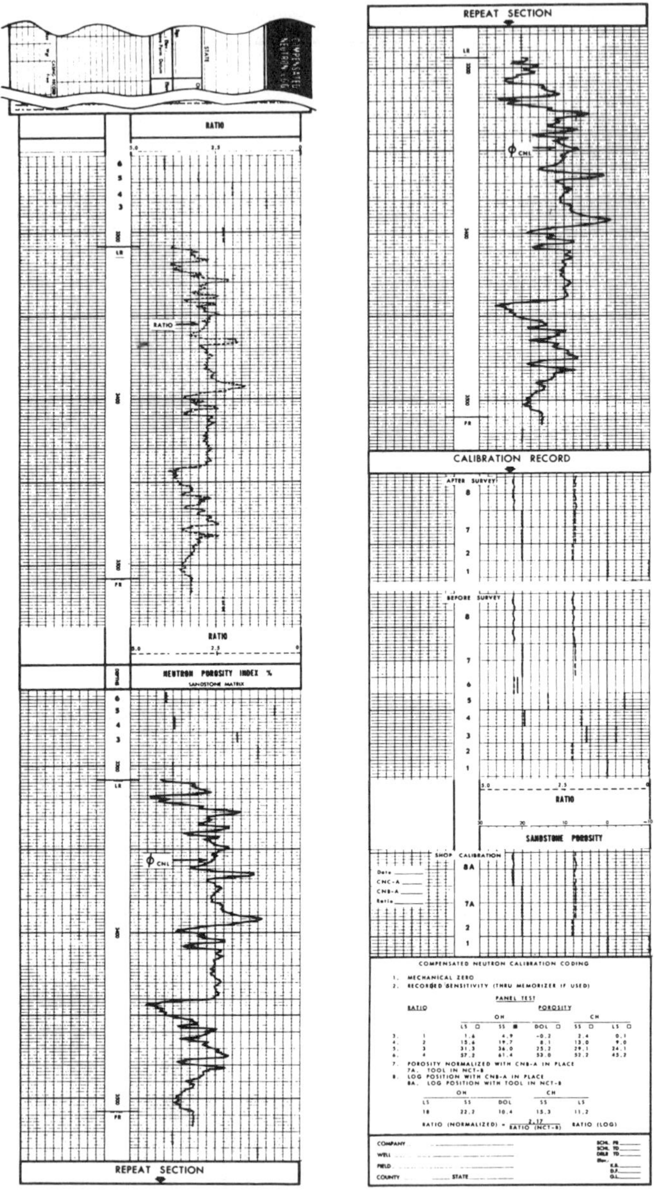

FIGURE 4.30 *Complete Compensated Neutron Log. Courtesy Schlumberger Well Services.*

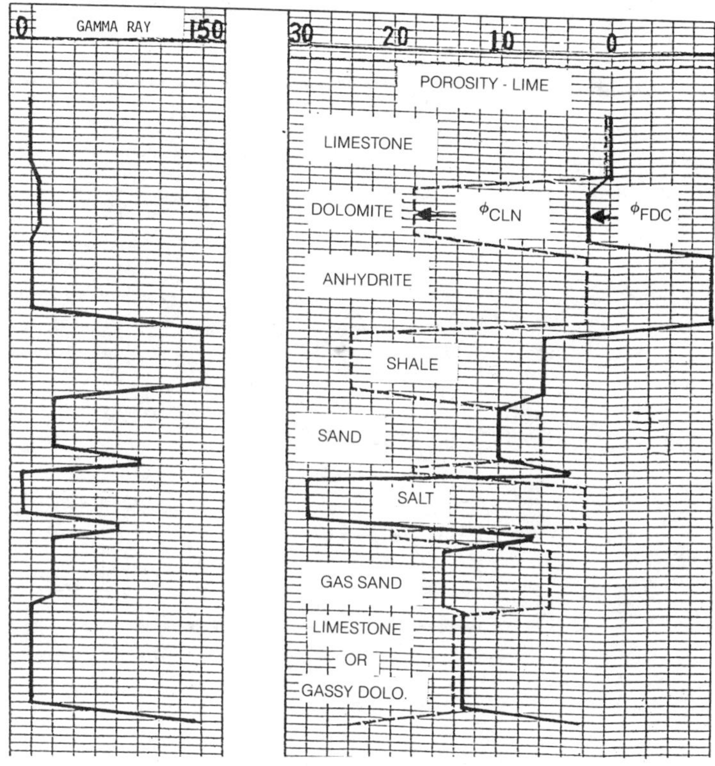

FIGURE 4.31 *Combination Neutron/Density Responses to Variations in Lithology. Courtesy Schlumberger Well Services.*

Applications

Logging tools based on neutron activation provide a measure of oil saturation, C/O ratio; lithology, Si/(Ca + Si) ratio; porosity, H/(Ca + Si) ratio; shale, Fe/(Ca + Si) ratio; and salinity, Cl/H ratio in open or cased holes. The method is used to determine the presence of hydrocarbons behind casing regardless of formation-water salinity.

At present, reliable measurements can be made only with optimum borehole and formation conditions and even then, the measurement should be used qualitatively. The major interpretative uncertainty stems from the inability of the measurement to distinguish between carbon associated with carbonates (e.g., limestone, $CaCO_3$) and carbon associated with hydrocarbons.

Ratios of elemental yields (C/O, Si/Ca, Cl/H, etc.) are normally presented. Given a constant porosity and lithology, an increase in the carbon/oxygen ratio indicates an increase in oil saturation. It should be noted that, by taking elemental ratios, any variations in neutron output from the source are normalized.

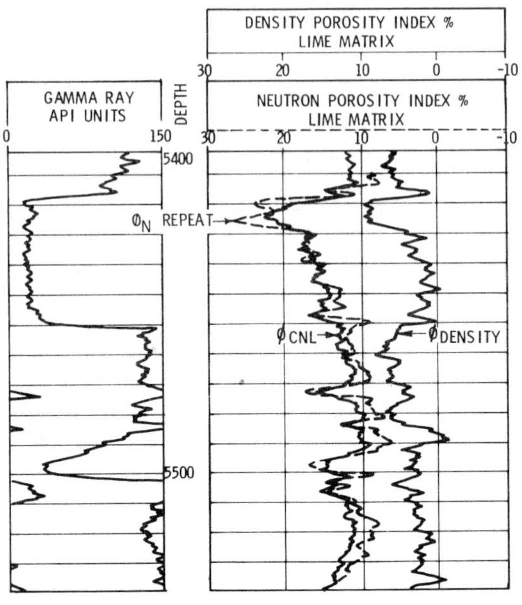

FIGURE 4.32 *Poor Repeat on a Neutron/Density Log.*

Tools Available

The Dresser Atlas Carbon/Oxygen Log measures the number of gamma rays in two energy "windows" centered around the expected carbon and oxygen inelastic-scattering energies during the burst and around the silicon and calcium thermal-capture energies after the burst. The Schlumberger Induced Gamma Ray Log (IGT) employs a *spectral-fitting analysis* to determine the relative concentrations of carbon, oxygen, calcium, silicon, and several other elements. This spectral-fitting analysis uses three gates. The burst gate, the background gate, and the capture gate. The burst gate is at the source, the background gate cuts down on borehole interference, and the capture gate gives capture readings. The burst gate minus the background gate gives the inelastic spectrum, and the capture gate gives the capture spectrum of the IGT Log as shown in figure 4.33.

Log Quality Control

Insufficient field experience has been gained to effectively prescribe a quality control procedure for the C/O log. Data should still be used qualitatively. Disagreement between fluid content of rock suggested by the log as compared to test results cannot necessarily be attributed to log quality, but could possibly be related to interpretation technique. The following considerations are relevant to log quality control:

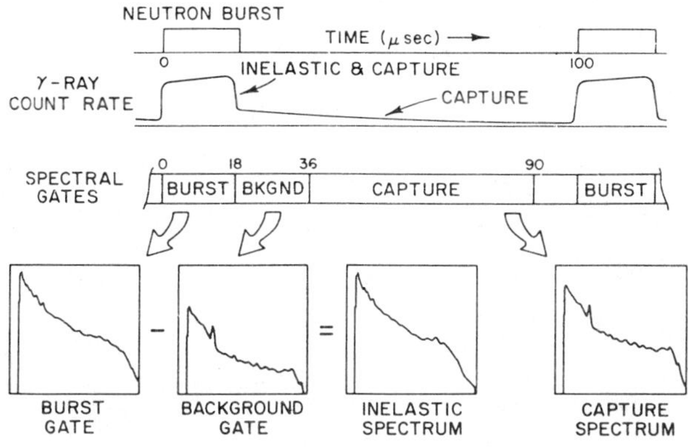

FIGURE 4.33 *IGT Measurement Principle. Courtesy Schlumberger Well Services.*

1. The log is generally run in cased holes where conditions are not favorable for TDTs (thermal decay-time logs) or NLLs (neutron lifetime logs) due to low formation salinities.
2. Optimum formation conditions are high porosity (>20%), low water salinity (<50,000 ppm NaCl), and consistent or known lithology. The log can be useful where salinities are unknown or variable.
3. Depth of investigation is very shallow for measurements on the inelastic-scattering spectrum. This limits the tool's openhole use and forces consideration of the effects from the casing annulus.
4. Optimum borehole conditions are a small-diameter hole and constant fluid composition in the casing. If an oil-water contact or varying salinities are expected in the casing, a fluid displacer should be considered.
5. At present, the statistical uncertainty in analyzing the spectrum is the tool's limiting feature. Advances in detector design and spectrum analysis are expected to solve these problems.
6. Current tools are station reading devices (up to 8 minutes) so that the statistics are minimized. Dresser Atlas has a continuous logging tool that can be run at 2 ft/min, but station readings should also be made over the zones of interest.

Field Examples

Figure 4.34 shows a continuous carbon/oxygen log; the curves presented include:

Track 1	monitor
	silicon correlation
Tracks 2 and 3	silicon/calcium ratio (capture spectrum)
	carbon/oxygen ratio
	calcium/silicon ratio (inelastic spectrum)

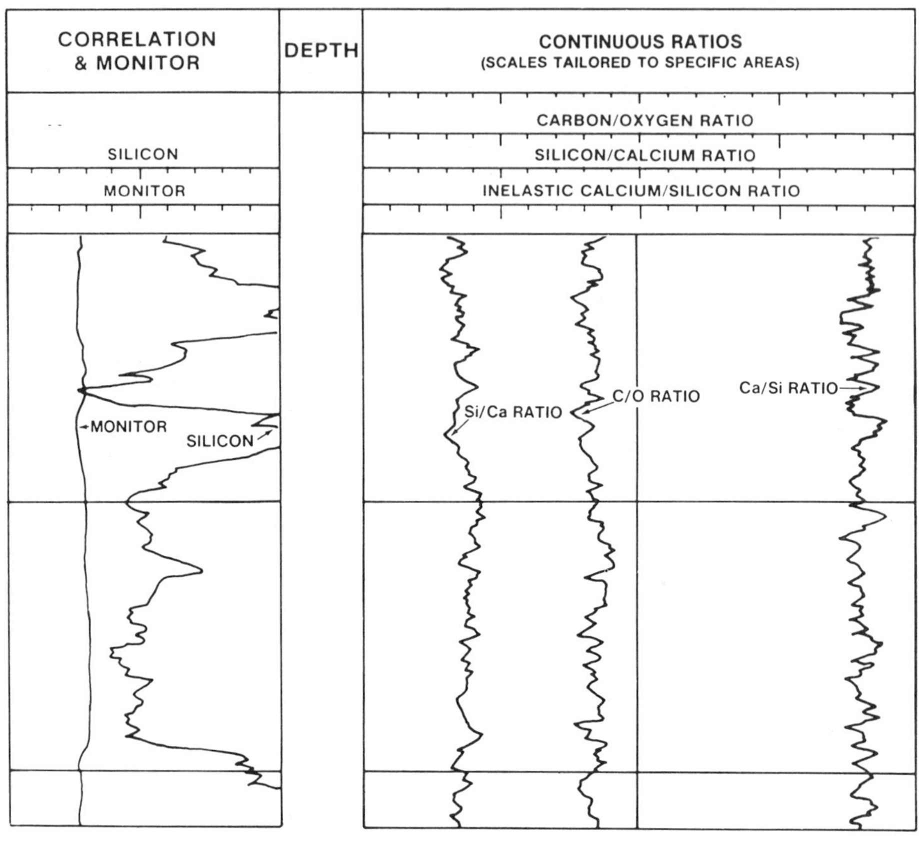

FIGURE 4.34 *Carbon/Oxygen Log. Courtesy Dresser Atlas.*

Figure 4.35 shows a Schlumberger IGT Log, which records the following data:

Track 1	iron-indicating ratio	Fe/(Si + Ca)
	gamma ray	
Tracks 2 and 3	net count rate	
	lithology indicator	Si/(Si + Ca)
	porosity indicator	H/(Si + Ca)
	salinity indicator	Cl/H
	capture cross section (Σ)	

THE PULSED NEUTRON LOG

Measurement Principle

The pulsed neutron log provides a measurement that can be used to determine the formation-water saturation in an open hole or through

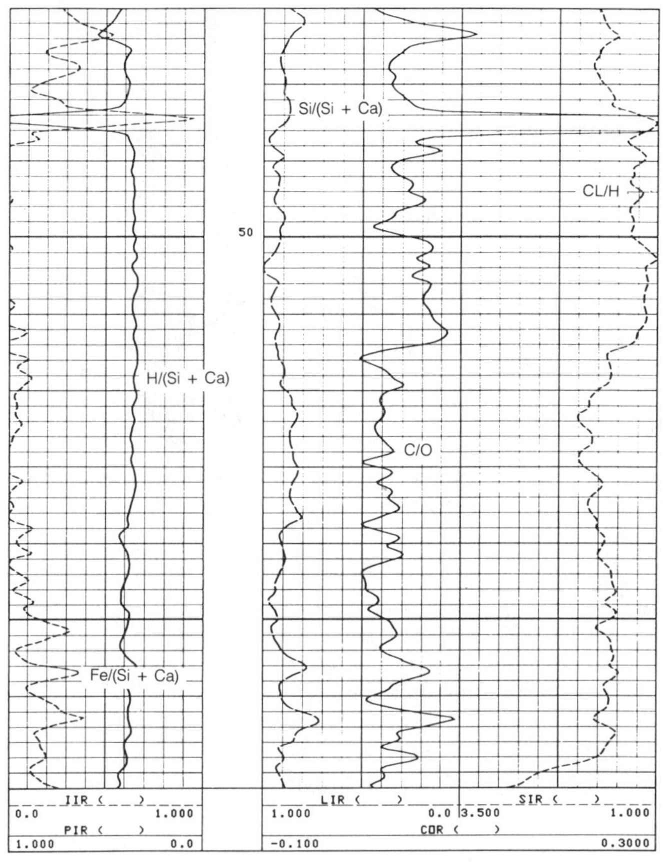

FIGURE 4.35 ***Schlumberger IGT Log. Courtesy Schlumberger Well Services.***

casing. This is accomplished by relating water saturation to the thermal-neutron-capture cross section of the formation (Σ_{log} capture units), as derived from the pulsed neutron log measurements. The thermal-neutron-capture cross section is a basic physical property of the formation and is sensitive to the amount and type of fluid present in the formation.

The system used to provide these measurements includes a pulsed neutron source emitting high-energy (14 MeV) neutrons, and two gamma ray detectors (scintillation counters). Fast neutrons from any one pulse collide with the nuclei of the atoms in the formation and slow down to thermal energies, at which point they are captured and a gamma ray of capture is released. The rate at which neutrons are captured is monitored by observing the gamma ray count rate at the counter(s). Figure 4.36 is a schematic of a generalized pulsed neutron logging tool.

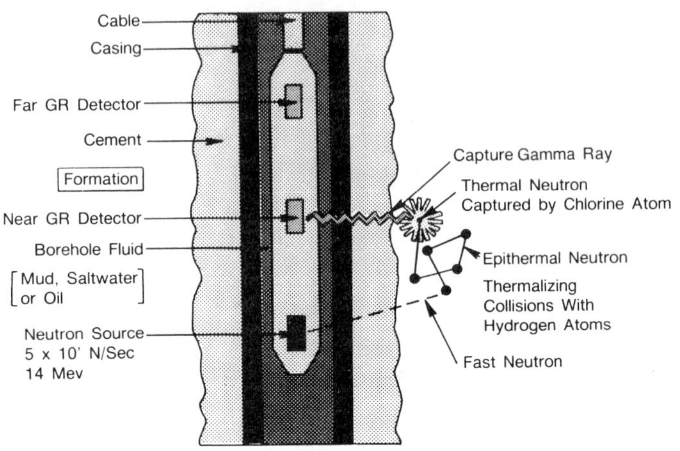

FIGURE 4.36 *Generalized Pulsed Neutron Tool.*

Pulses of high-energy (14 MeV) neutrons are obtained by bombarding a target with high-energy electrons in a neutron generator. The NLL pulse rate is 1000 cycles/s, for a duration of 30 μs. For the TDT-K, the pulse rate and duration is variable, being controlled by the decay time of the formation (τ_{log} microseconds, where $\Sigma = 4550/\tau$). Pulse rates can vary from 200 to 1000 cycles/s for a duration of from 100 to 500 μs.

Neutrons are slowed to a thermal state and captured by elements in the formation, primarily by chlorine in the formation fluids. Capture gamma rays are liberated when thermal neutrons are captured by nuclei of the formation. A scintillation counter measures the emitted gamma radiation during two different time intervals after the neutron emissions. With the NLL device, these time intervals are from 400 to 600 and 700 to 900 microseconds after the neutron burst. With the TDT-K device, the measure intervals are controlled by the decay time (τ) of the formation. The pulse–measure cycle has a duration of 10τ. The neutron generator provides a burst of neutrons for one τ. The first detection gate is positioned in time from 2τ to 3τ and the second detection gate from 3τ to 5τ. The third gate, which measures background, extends from 6τ to 9τ. These background counts are subtracted from the counts measured during Gates 1 and 2 before determining sigma or tau. Background counts are not measured by the NLL. Since only gamma rays with energies above 2.2 MeV are counted with this tool, background effects are reportedly eliminated.

The number of capture gamma rays liberated is proportional to the number of thermal neutrons present. The neutron decay or capture rate is related to the thermal-neutron-capture cross section of the formation using counts received at the near detector.

Figures 4.37 and 4.38 illustrate the pulsed neutron logging technique.

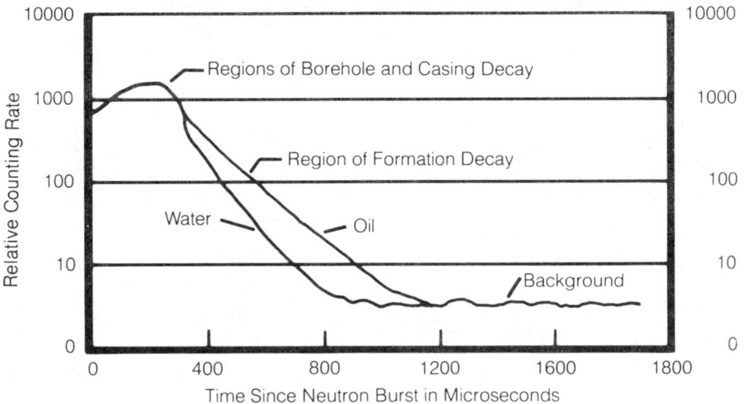

FIGURE 4.37 ***Pulsed Neutron Count Rates vs. Time for Oil and Water-Bearing Formations: Relative Counting Rate.***

Water saturation of the formation is computed from the response of the pulsed neutron log as follows:

$$\Sigma_{\log} = (1 - \phi)(\Sigma_m) + (\phi)(\Sigma_{hc})(1 - S_w) + (\phi)(\Sigma_w)(S_w).$$

Solving for S_w:

$$S_w = \frac{\Sigma_{\log} - (1-\phi)(\Sigma_m) - (\phi)(\Sigma_{hc})}{(\phi)(\Sigma_w - \Sigma_{hc})},$$

where: S_w = formation water saturation, fraction of pore space.
$\Sigma_{\log}$ = neutron-capture cross section of formation (capture units), obtained from pulsed neutron log.

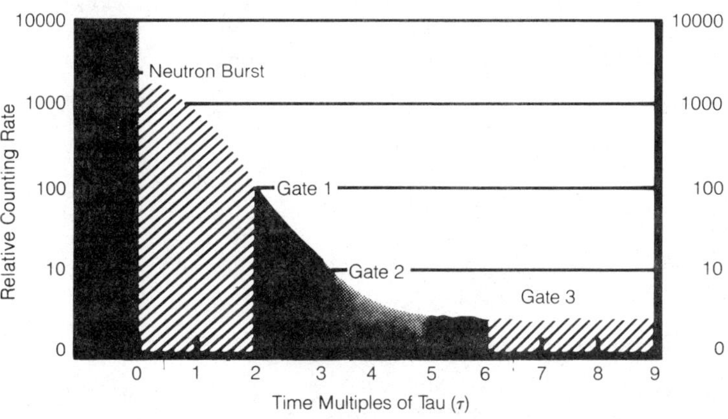

FIGURE 4.38 ***Pulsed Neutron Gating System: Relative Counting Rate.***

Σ_m = neutron-capture cross section of rock matrix (capture units), assumed from known lithology or computed in zones of known water saturation.

Σ_w = neutron-capture cross section of formation water (capture units), derived from knowledge of water salinity, or more accurately determined using water analysis.

ϕ = porosity, fraction of bulk volume. Estimated from the ratio curve (ratio of count rates measured by near and far detectors) as presented on pulsed neutron display or derived from openhole logs or cores.

Σ_{hc} = neutron-capture cross section of hydrocarbons in the pore space (capture units)—a variable, but usually assumed to be approximately 22 for oil and 4 for gas. A cross section for gas can be reasonably estimated using published charts, provided bottomhole pressure, gas composition, and temperature are known. Published charts are also available for estimating the capture cross section of oils when the gravity and gas/oil ratio is known.

Application

The pulsed neutron log is used principally to provide a measure of the formation-water saturation through casing. It can be run openhole or through-tubing. If core porosity data or openhole porosity logs are not available, porosity can be estimated from the ratio curve presented with the pulsed neutron display. This makes the tool valuable in evaluating formations for recompletion after the initial production interval has been depleted; for logging wells that have inadequate openhole logs; for detecting the movement in water-oil and gas-oil contacts resulting from production; and for detecting intervals of water breakthrough in production wells of a waterflood project. It is also used to distinguish salt-filled porosity from gas-productive intervals and to evaluate intervals in high-angle wells when conventional openhole tools encounter problems reaching logging depth. These tools are also run through stuck drillpipe to evaluate the formations penetrated.

Tools Available

Four tools are generally available for pulsed neutron logging: (1) the Dual Detector Neutron Lifetime Log (DNLL), which is available in either a 3⅝-in. OD or a 1$^{11}/_{16}$-in. OD version; (2) the Dual Spacing Thermal Decay Time (TDT) log (mark of Schlumberger), which is available in a 1$^{11}/_{16}$-in. OD version; (3) an updated version of the TDT known as the TDT-M; and (4) the Thermal Multigate Decay tool (TMD) (mark of Welex).

Limitations

Since the depth of investigation of the pulsed neutron device is shallow (generally <12 in.), the device should be run in a cased well after

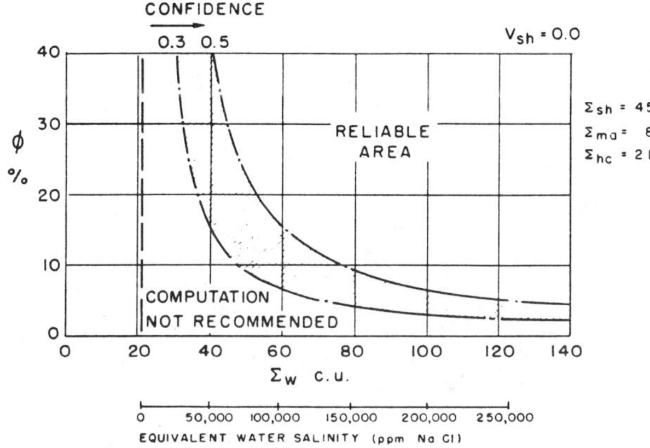

FIGURE 4.39 ***Minimum Formation-Water Salinity Requirements. Courtesy Schlumberger Well Services.***

invasion effects associated with drilling, completion, and cementing have dissipated. The time required for those fluids to dissipate is variable and is controlled primarily by the vertical permeability of the interval (can be as long as several months). If the well is completed with an open hole it is usually necessary to produce the well to establish representative saturations in the vicinity of the well prior to logging.

Since the chlorine content of the formation water is mainly responsible for the difference in log response between oil and water productive formations, there are minimum values of salinity and porosity for reliable pulsed neutron log evaluation. Figure 4.39 shows the minimum formation-water salinity (in ppm NaCl) required to obtain a reliable interpretation as a function of formation porosity. The average capture cross sections of various formation fluids are shown in table 4.7.

Quality Control

Quality control for pulsed neutron logs requires attention to the following details:

1. Pulsed neutron logs from offset wells or wells in the same field should be used to aid in the evaluation of the log.
2. If the formation has been invaded by fresh water or oil, the computed water saturations are probably too low.
3. If the formation has been invaded by salt water, the computed water saturations could possibly be too high.
4. The response of pulsed neutron logs is influenced by neutron diffusion and borehole effects. Correction charts to correct the sigma or tau values of the TDT-K log for these effects are available in the literature. It is *necessary* that these corrections be made to the sigma

TABLE 4.7 *Formation Fluid Neutron-Capture Cross Sections (in capture units)*

Fluid	Capture Cross Section
Fresh water	22
Oil (dead)	22
Condensate	14 to 18
Oil (varying composition and GOR)	16 to 22
Gas (depends on PVT data)	0 to 16
Water (100,000 ppm NaCl)	58
Water[a] (200,000 ppm NaCl)	100

[a]Small amounts of high neutron absorbers such as boron or lithium can significantly increase the capture cross section of the formation water.

and tau values before they are used to determine water saturations, porosities, water salinities, etc. The NLL reportedly does not require corrections. Schlumberger's TDT-M tool also does not require correction.

5. Porosity must be known accurately. Porosities can be obtained from core analysis or derived from openhole logs or cased-hole neutron logs, or computed from the ratio curve. Core analyses from other wells in the field in which similar logs were run can be used to calibrate these logs.
6. The gas saturation must be known or measured. It may be estimated from production data and/or history, or computed from neutron logs.
7. The capture cross section of the formation water can be computed from the response of the pulsed neutron logs in zones of known 100% water saturation, or it can be computed from the mineral analysis of a water sample (charts are also available).
8. The neutron-capture cross section of the rock matrix must be known or assumed from known lithology. This is particularly troublesome in shaly sands where it can vary from 10 to 18. The capture cross section of the rock matrix can also be computed from the response of the pulsed neutron log if there are zones of known 100% water saturation and if the lithology is similar to the intervals of interest. The most common matrix ranges for the major lithologies are as follows (*note*: Small amounts of high neutron absorbers such as boron, salt, or pyrite can significantly increase matrix cross-values):

   Sandstone 8–13 capture units
   Limestone 8–10 capture units
   Dolomite 8–12 capture units

9. Scaling of calibration should allow full presentation of calibration.
10. Accuracy of tau-sigma conversion should be checked:

$$\Sigma_{\text{capture units}} = \frac{4550}{\tau_{\text{seconds}}}$$

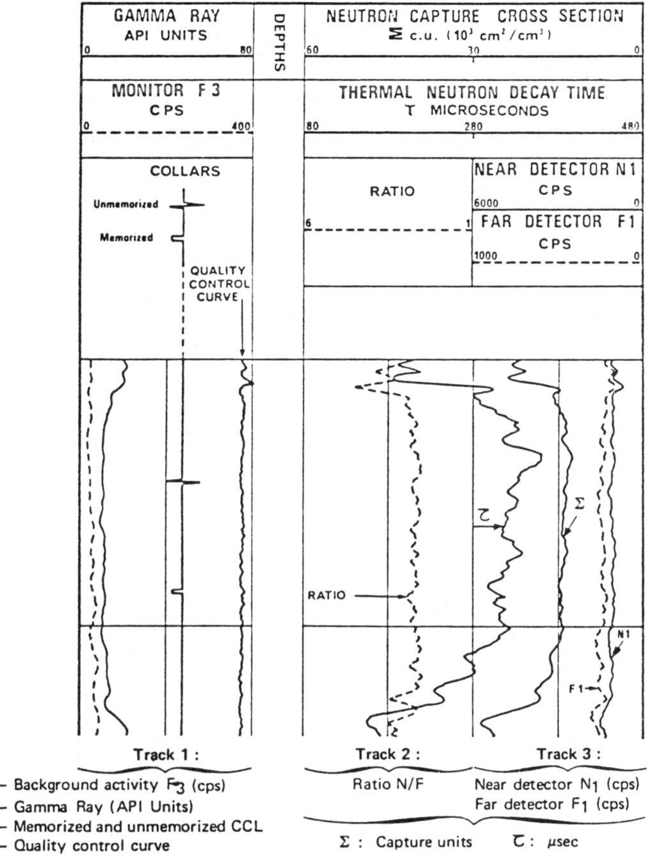

FIGURE 4.40 *TDT-K Presentation. Courtesy Schlumberger Well Services.*

11. Check F_1 and N_1 count-rate in intervals known to have 100% water saturation. If curves do not overlay in these intervals then separation between curves ($F_1 > N_1$) will not be a valid indicator of gas.
12. When quality-curve values deviate significantly from 1.0 the ratio value should not be used.
13. Several passes should be made and the logs overlayed to determine variation in log response. If the variation in sigma is as much as one capture unit the log should be repeated. It may be necessary to make several logging runs through the intervals of interest to obtain a reliable average log.
14. Check accuracy of sigma derived from NLL by:

$$\Sigma = \frac{10{,}500}{\Delta T} \log N_1/N_2$$

Note: On NLL logs the value of ΔT used is reported on the log heading and is normally 300 μs.

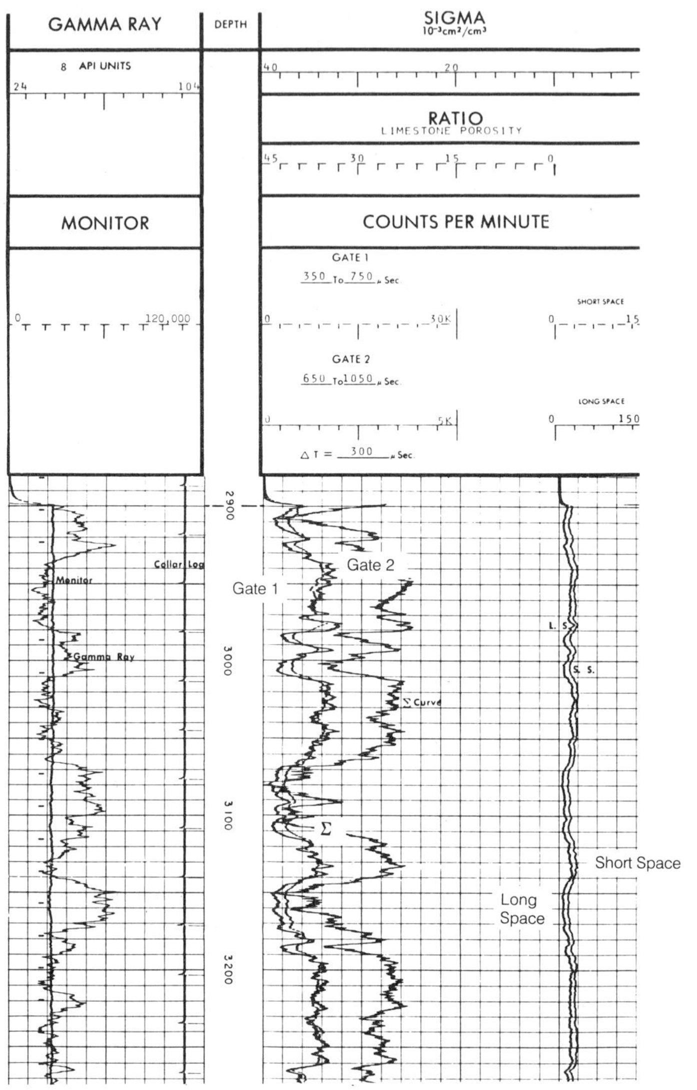

FIGURE 4.41 *NLL Presentation. Courtesy Dresser Atlas.*

Field Examples

THERMAL DECAY TIME LOG—TDT-K. The following curves should be displayed on TDT-K logs (see the example in fig. 4.40):

tau (τ): thermal-decay time of the formation (μs)
sigma (Σ): capture cross section of the formation (capture units)
ratio: obtained from the net count rates of the two detectors, related to porosity

F_3: count rate from Gate 3 of the far detector (counts per second), background
F_1: count rate from Gate 1 of the near detector (counts per second)
N_1: count rate from Gate 1 of the near detector (counts per second)
quality: when quality values deviate significantly from 1.0, counting-rate losses, bed-boundary effects, or incorrect tau determination are indicated

NEUTRON LIFETIME LOG—NLL. The following curves should be displayed on neutron lifetime logs (see fig. 4.41):

Gate 1: count rate from Gate 1 of the near detector (counts per minute), N_1
Gate 2: count rate from Gate 2 of the near detector (counts per minute), N_2
short space: count rate from Gate 1 of the near detector (counts per minute), N_1
long space: count rate from Gate 1 of the far detector (counts per minute), F_1
ratio: obtained from the net count rates of the two detectors
sigma (Σ): capture cross section of the formation (capture units)
monitor: (Source output should be above 18,000 counts per minute for reliable sigma measurements.)

CHAPTER FIVE

DIPMETER APPLICATIONS AND DIRECTIONALS

DIPMETERS

Dipmeters are in a class by themselves. The technique, the purpose, the interpretation, and the quality control technique of dipmeters are entirely different from that of other logging tools.

The purpose of the dipmeter is to measure the dips of formations. To do this, the tool must simultaneously and continuously do two separate jobs: first, it must orient itself in space; and second, it must react to formation bedding planes.

All present dipmeter tools go about this job in the same way. An inclinometer section supplies continuous measurements of deviation, both the amount and the direction, and of the orientation of the tool's electrode array, either with respect to the borehole direction or with magnetic north (a few specialized tools for far-north operations use gyroscopic orientation, nominally with true north). At the same time, an electrode array is maintained in contact with the borehole wall by pressured linkages. The electrodes respond to resistivity variations, while the expanding linkages activate a caliper recording.

The dipmeter pads, normally numbering four, are identical, and so mounted as to remain in a plane normal to the tool axis. When an anomaly is detected by at least three pads, the resistivity deflections plus the caliper reading identify three points in what is assumed to be a plane—the plane of deposition of the formation. This identification is then referred to vertical and true north, giving the true dip of the formation.

Recording correct data with a dipmeter tool is a straightforward, more or less mechanical process—though the tools used to do it are some of the most sophisticated in the industry. Interpretation of the data draws heavily on computer technology.

Tools Available

All major service companies offer four-arm dipmeter tools. Figure 5.1 shows a typical dipmeter tool's mechanical section with the four pads, the electrodes, and the caliper assembly visible.

Most of these tools will operate to 20,000 psi and 350°F in holes with diameter between 6 and 16 in. Different varieties of tools handle low- and high-deviation holes using different methods of measuring deviation and azimuth angles. Figure 5.2 illustrates a monitor log and a computed-answer log.

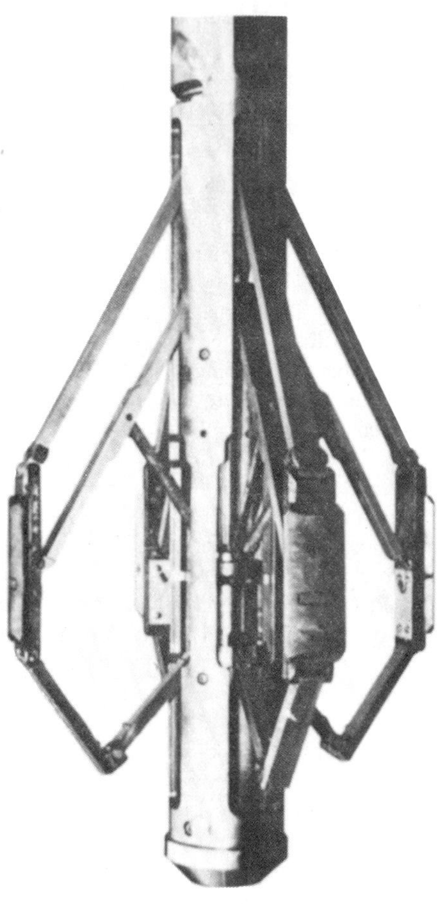

FIGURE 5.1 *Four-Arm Dipmeter Tool. Courtesy Gearhart Industries, Inc.*

Calibration

The dipmeter is essentially a physical tool, and its calibration is physical. The inclinometer section is adjusted to read correctly in a special test jig that rotates the three sensors through their ranges. Special care is given to insure that the deviation sensor registers zero with the tool held vertical. The operation of the inclinometer is checked before and after each log run, first by allowing the tool to hang vertically in the derrick, then by manually rotating the tool through at least one full revolution. The calipers are checked as usual, by calibration jigs of known diameter. Four-arm dipmeters record a separate caliper with

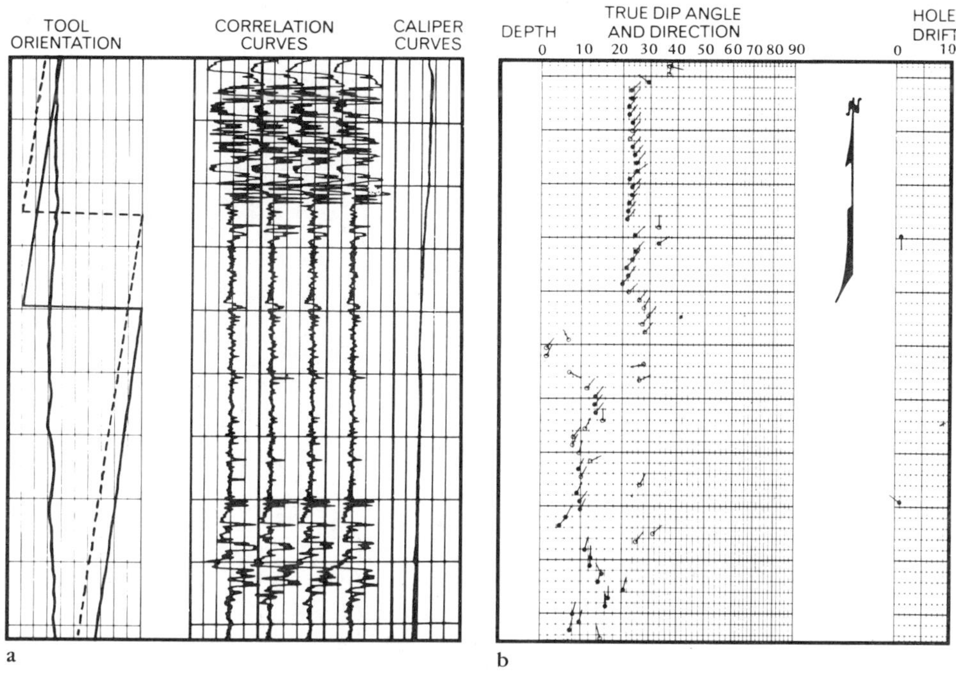

FIGURE 5.2 *Dipmeter Monitor Log and Computed Dipmeter Log: (a) Monitor Log and (b) Computed Dipmeter. Courtesy Gearhart Industries, Inc.*

each opposing pair of pads, which can thus flex independently of each other while remaining in the same plane. The sensitivity of the electrodes is checked by shorting them out in sequence; this also verifies the correct wiring of the electrode array.

Note that these are tool checks, not calibrations. The integrity of the service company is your only routine assurance of proper calibration, though if there is any reason to doubt the veracity of a log, a recalibration of the equipment that was used should be requested.

The following paragraphs consider the general principles of dipmeter calibration and details of the various steps that appear on a typical dipmeter calibration record such as that shown in figure 5.3.

GENERAL PRINCIPLES OF CALIBRATION.* The calibration of any dipmeter tool can be divided into two distinct and unrelated problems: those sensors that define the geometry of the borehole in space, and those that define the intersections of bedding planes with the borehole.

The High-Resolution Dipmeter (HDT) (mark of Schlumberger) differs from contemporary tools in that it was designed primarily to

*The calibration information that follows has been provided, with permission, by Schlumberger Well Services.

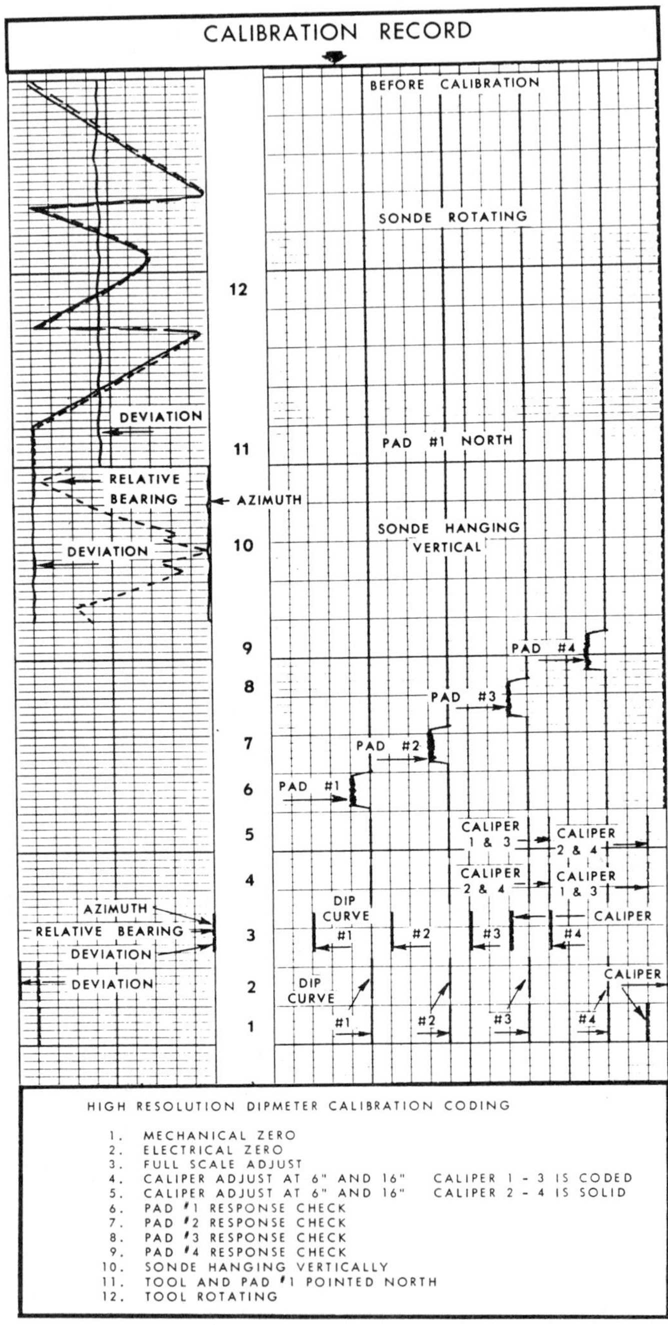

FIGURE 5.3 *Dipmeter Calibration Record. Courtesy Schlumberger Well Services.*

record its data on digital tape, and only secondarily to make an optical recording. This feature enhances the importance of the field calibration film, since there is at present no practical way to validate tape-record calibrations in the field. The optical record consists of analog curves derived from the digital data. Hence, if the analog calibration record is correct, it is reasonable (if not totally infallible) to infer that the digital record is also correct. Current experience with optical playbacks of dipmeter tapes supports this inference.

GEOMETRY CURVES. These include five measurements:

1. azimuth of electrode 1 with respect to magnetic north (when using the high-angle, 72°-inclinometer assembly, a different principle is used. See the discussion of deviation measurements below)
2. relative bearing of electrode 1 with respect to the upper half of the vertical plane passing through the axis of the tool
3. the angle of deviation from vertical of the axis of the tool
4. a caliper measuring hole diameter in the plane of electrodes 1 and 3
5. a caliper in the plane of electrodes 2 and 4

The first two measurements are made by low-torque continuous circular potentiometers. Zero to full-scale movement of the galvanometers is equated to minimum (zero) and maximum voltages on the windings of the pots.

1. The azimuth galvanometer is actuated by a gimbal-mounted bar compass carried in a nonmagnetic pressure housing. It is periodically shop checked against a strong fixed magnet in a special test stand, and checked against a hand compass on each job.

2. The relative-bearing galvanometer is driven by a pendulum pivoting about the longitudinal axis of the tool. Its accuracy is also shop checked in the test stand. On each job, its action is verified by physically leaning the tool in a known direction and rotating it through a turn and a half each way.

3. The deviation measurement is made by a pendulum pivoted transversely within the rotating relative bearing assembly. It uses a circular low-torque pot similar to those above. Two types of assemblies are available, a low-angle (up to 36°), and a high-angle (up to 72°). The high-angle inclinometer uses a different principle than does the standard 36° unit. The compass assembly is suspended in the relative bearing pendulum assembly, permitting direct measurement of hole azimuth. Both are shop checked for accuracy in the jig, and operationally checked during calibration on each job.

4, 5. The two calipers use linear pots within the sonde hydraulic section, driven by the action of the pad linkages. The geometry of this linkage produces a nonlinear response to changing hole diameter which is not resolved by the field recording system. This is no handicap for machine computation, but requires nonlinear scaling of the optical log. Calibration is by means of a two-diameter jig giving 6- and 16-in. readings. Rarely, other calibration sizes will be used.

DIP CURVES. Although the HDT dip sensors respond to resistivity anomalies, their zero and full-scale readings have no quantitative significance. Calibration of the optical curves has three purposes:

1. to insure that the four curves deflect the same distance at the same rate;
2. to insure that the pads respond comparably and without cross talk to a test probe; and
3. to insure that the pads are connected to the correct sensing channels. The fifth dip sensor (the "speed button" on the No. 1 pad) is also checked with the probe, but its response cannot be recorded optically.

CALIBRATION FILM

1. Mechanical zero: galvanometer circuits are open.
   a. Azimuth and relative bearing are zeroed at 9 divisions Track 1, to produce a scale of 40° per division.
   b. Deviation is mechanically zeroed at the same 9 divisions Track 1, permitting scales that are multiples of 9° per 9 divisions.
   c. The caliper galvanometers are mechanically zeroed at 6 in., with the pads open. The caliper galvanometers will stop at approximately 9 divisions Track 3 with newer panels (HDP-D), or at 10 divisions if an HDP-C panel is used.
   d. The dip galvanometers are usually zeroed at 5 divisions Track 2, 9 divisions Track 2, 3 divisions Track 3, and 7 divisions Track 3. These spots are rather arbitrary. In some areas, in order to reduce the interference between no. 4 dip and the calipers, the zeroes used are 4 divisions Track 2, 7 divisions Track 2, 10 divisions Track 2, and 3 divisions Track 3.
2. Electrical zero: All galvanometers except those for deviation and the calipers are connected, but have no signal applied. They read the same as in Step 1.
   a. The deviation galvanometer is electrically offset 1 division (4° with low-angle unit; 8° with high-angle) to the left. The HDT uses digital coding for signal transmission, which cannot be negative. The deviation pendulum is not mechanically damped; its swings are recorded faithfully on tape and averaged in the computer (the optical galvanometer is electrically damped to remove these oscillations). When the tool axis is near vertical, the swings of the pendulum go past zero degrees, so the digital recording must begin at a point below zero degrees; otherwise the negative swings would be clipped, and a serious error introduced. Thus the digital zero level is set at − 4 (or − 8) degrees, and corrected in the computer. To give the optical galvanometer a linear response, it too must be electrically zeroed at − 4 (or − 8) degrees. The zero point on the

optical scale remains at 9 divisions Track 1, but the galvanometer electrical zero is at 10 divisions Track 1.

b. The calipers are now on "measure." The pads are closed; galvanometers will read approximately 10 divisions Track 3.

3. Full-scale adjust:
   a. Azimuth and relative bearing deflect to zero Track 1.
   b. Deviation also deflects to zero Track 1, but much more slowly due to its long damping-time constant.
   c. Calipers deflect to approximately 2 divisions of Track 3.
   d. Dip curves deflect approximately 3 divisions from their respective zero spots. This deflection is 4 divisions for earlier panels HDP-C. Their rates of deflection are indicated here by the amount of overshoot, which may vary considerably in absolute value but must be the same for all four curves.
4. Caliper adjust: Caliper 1 (Pads 1 and 3) at 6 in. (9 divisions Track 3) and Caliper 2 (Pads 2 and 4) at 16 in. (4 divisions Track 3). This sequence may be reversed: Caliper 1 is coded (dashed) while Caliper 2 is solid.
5. Caliper adjust: The reverse of Step 4.
6. Pad No. 1 response check: With the tool operating, a small finite resistance is connected by probe to the no. 1 sensing electrode. The galvanometer deflects to the left. The absolute value of the deflection may vary, but each of the four curves should deflect approximately the same amount when the appropriate electrode is contacted. Check for crosstalk.
7. Pad No. 2 response check: The probe is touched to sensing electrode No. 2. The most important part of Steps 6–9 is to insure that the curves deflect in the same sequence as the electrode contacts are made, i.e., 1, 2, 3, 4.
8. Pad No. 3 response check.
9. Pad No. 4 response check.
10. Zero-deviation check: With tool hanging vertically in the derrick and No. 1 electrode pointing toward rig north (as determined by a hand compass on the rig floor), the three galvanometers in Track 1 read as shown below. The other galvanometers remain as in Step 2.
    a. Deviation galvanometer reads zero (9 divisions Track 1).
    b. Azimuth reads near north (9 divisions or 0 divisions Track 1).
    c. Relative bearing may read anything or wander as tool swings.
11. Azimuth and relative-bearing check:
    a. With No. 1 pad pointing north as above, lean the bottom of the tool in the same direction. Azimuth and relative bearing stay together near north (9 or 0 divisions Track 1).
    b. Deviation reads the angle of the tilt (its scale may be 9°, 18°, 36°, or 72° per 9 divisions; it should be the same as the scale on the log heading).
12. Tool rotating: With the bottom of the tool still inclined toward rig north, the tool is slowly rotated 1½ turns clockwise looking down, then back the other way 1½ turns.
    a. With the 36° inclinometer assembly, azimuth and relative bearing

should track closely, increasing from left to right until they reach zero Track 1, then jumping back and increasing further to about 4.5 divisions Track 1, then reversing.

b. With the high-angle (72°) inclinometer assembly, azimuth should remain constant near the north reading, while the relative-bearing galvanometer behaves as described above.

c. In either case, the deviation galvanometer should read the inclination of the tool, on whatever scale is used for logging. No indication of sticking ("stair stepping") should be evident on any of the three curves.

13. High-angle (72°) azimuth check: With No. 1 pad pointed north, the bottom of the tool is inclined and rotated 360°. The relative-bearing galvanometer should remain constant near the north reading while the azimuth galvanometer should move through all four directions as the tool is rotated through 360°. This step does not apply to the 36° inclinometer assembly. The example log and calibration (fig. 5.3) was made with a dipmeter using the low angle (36° inclinometer), hence this step is not included in the calibration.

Log Quality Control

The primary quality-control tool is the 5- or 6-in. log film. (This is the only film available on location unless a computer logging unit is used for a quick-look interpretation of dips.) This check film should contain before and after tool checks, at least one repeat section, and the main body of the log. Since virtually all dipmeters are taped, and the footage per reel of tape is limited, the main log will often include several segments. Each such segment should overlap at least 200 ft of the previous one. Figure 5.4 shows a complete dipmeter survey example.

Until the advent of computer logging units, it was standard practice to record dipmeters on 60-in. film in addition to the 5- or 6-in. film. This was done first for manual dip computation, later as a backup to be used if the tape record was lost or defective. However, the computer logging units with wellsite dip computations have rendered the 60-in. film obsolete.

With a little experience a lot can be told about correct functioning of the dipmeter tool by watching the galvanometer spots. The dip curves are so compressed vertically on the 5-in. log that you cannot tell much from the film, but the eye can detect similar motions while the log is running.

The log quality control film will detect a dead dip curve, or one with serious noise. The four dipmeter curves should have good activity. *Good activity* is at least ½ in. (2 large divisions) movement on all curves. In addition, this movement should be very similar in character and amplitude. Noise problems are best detected when the resistivity measurements are being recorded with the tool stopped in the hole. The curves should be drawing straight lines. Any curve movement at all should be suspected of being noise. Noise will usually appear on all four resistivity curves as virtually identical in both character and amplitude, with no depth displacement between the anomalies.

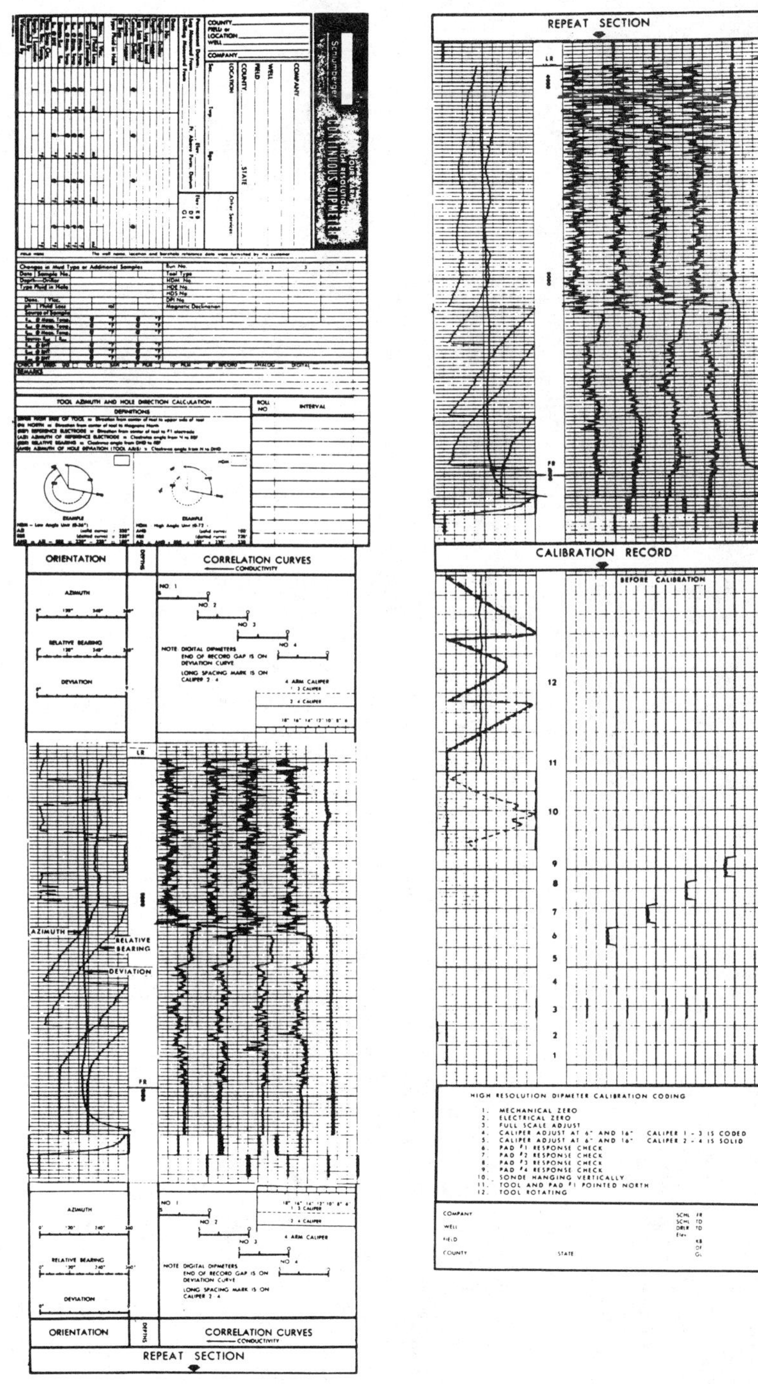

FIGURE 5.4 *A Complete Dipmeter Survey. Courtesy Schlumberger Well Services.*

Although it is true that a four-arm dipmeter can still be computed with the loss of one dip curve, *do not settle for this* except where there is no alternative—such as loss of the hole. Present dip-computation techniques make full use of all four dip curves, and the quality and usefulness of the final product are greatly reduced if one curve is missing.

The check film will usually verify correct operation of the inclinometer and calipers, particularly in straight holes where the tool will tend to rotate counterclockwise as it comes up the hole (see fig. 5.2). The azimuth and relative-bearing curves follow the rotation of the tool, which normally happens much faster than changes in the direction of the borehole. At the same time, the deviation curve holds nearly constant. All the inclinometer and deviation curves should move slowly and smoothly. Any abrupt changes or intervals of no curve movement should be examined closely. Discuss any trouble indicated by the curves with the logging engineer.

Caliper indications from a four-arm dipmeter are also helpful in quality control. They should overlay in round holes, which are usually at or under bit size (the latter where there is mudcake); but often they will show different diameters. It is not unusual that one will show every large washout, sometimes exceeding the limit of expansion. This is a common condition, particularly in deviated holes where the drill string wears itself into the formation. If other caliper indications, and the tool checks, show correct caliper operation, there should be no difficulty accepting the log. Sometimes the dip curves will provide support; one or both of the curves from the pair of pads spanning the large diameter will sometimes be suppressed, due to alteration of the borehole wall and/or mud between the pad and the wall.

Caliper action should be continuous, even over long gauge sections. Be suspicious of log intervals in which a caliper (or any other curve) is completely stationary. If the straight-line caliper is below bit size, be doubly suspicious—the caliper linkage may be jammed, or choked with shale. A routine caliper check recorded both before and after logging in the same casing interval can be of assistance in evaluating the caliper calibration and stability during the log.

Many dipmeter tools are very flexible in operation. They can be opened and closed at will, and the pad pressure can be changed while logging. These options should be tried any time the log suggests caliper linkage problems.

Sometimes the dipmeter tool will want to rotate excessively. Since the effect of tool rotation is one of degree, and the degree of error that can be accepted varies with the intended use of the log, a fixed rule is hard to make. However more than one complete rotation in 60 ft of the hole should be discouraged. One reasonable criterion sets the maximum allowable rotation at ten times the correlation window being used for interpretation. The correlation window (interval) is the length of curve that is correlated with the other curves to determine the displacement of one curve with another, thus setting up the dip calculation at each level. Figure 5.5 shows a correlation interval and also illustrates the terms *step* and *search angle*.

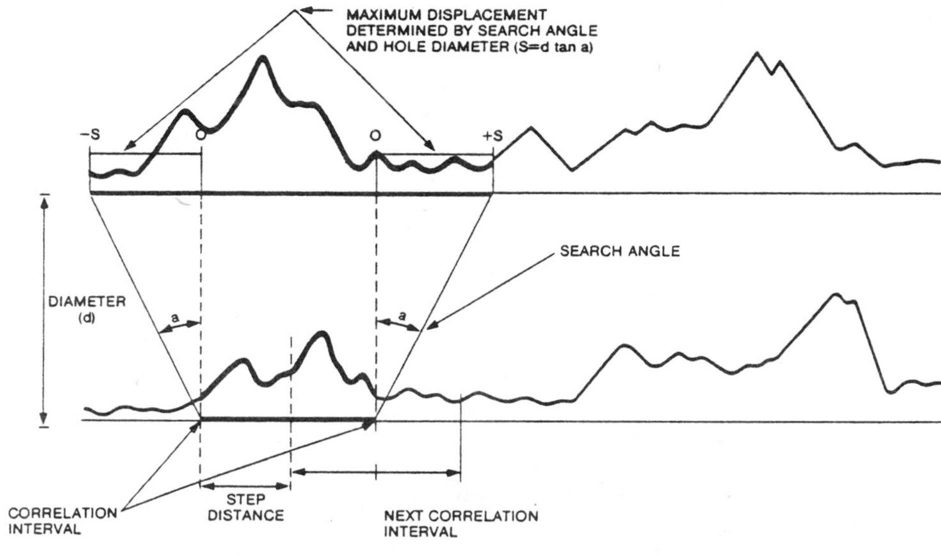

FIGURE 5.5 *Illustration of the Terms* Correlation Interval, Step, *and* Search Angle.

Usually correlation windows of 2 or 4 ft are used: 2 ft for detailed (stratigraphic) work, 4 ft for more general work. Thus, by the "ten times" criterion, a rotation once in 20 or 40 ft of hole is acceptable.

Tool rotation causes error because the orientation potentiometers, being mechanical devices, lag behind changes in tool rotation; and the interpretation software now in use assumes constant orientation for the length of each correlation interval.

Most tool rotation is thought to result from cable torque, which changes drastically when the logging crew starts the tool uphole. But if the tool is free to rotate, the up torque can usually be worked out by several fast pickups off bottom, perhaps for 200 or 300 ft.

If rotation problems can be anticipated, such as in small and/or deviated holes, ask the logging company to run a *swivel adapter head.* This device allows the tool to rotate independently of the cable while maintaining electrical contact. In tough cases, the pad pressure and logging speed may be varied.

Oil-Base Muds

Dipmeters run in oil-base muds present a special problem. Because the oil-base mud will not electrically connect the resistivity electrodes to the formation, it is necessary to use electrodes with a special knife-edge blade. These blades mechanically cut into the formation and make contact with the waters in the formation. This method does not give the quality of data that is obtained with the conventional system. The reason for this is the considerable amount of noise introduced into the resistivity recording, due to the knife blade sliding along the borehole wall. The better the contact made between the knife edge and the formation, the better the quality of the resistivity measurements. Two important things can be done to improve this contact:

1. Make sure that the knife edge blade is sharp and not worn. Demanding new blades is the best way to insure sharp edges. New blades are also less likely to have electrical insulation problems.
2. Have the logging company dress and adjust their dipmeter tool to apply the maximum arm pressure. This will mechanically force the knife-edge electrodes into the formation. This adjustment may be made with different spring mechanisms or by applying greater pad pressure through a hydraulic linkage. Either way this is very important to obtaining good resistivity data.

Because the oil-base-mud dipmeter is a low-usage tool, the service company should be given maximum notice so they can prepare for the job. All special instructions and stipulations should be given at the same time. It should be remembered that the final data on the recorded log can be expected to be only 10 to 20% as good as that from an ordinary dipmeter. It therefore is advisable to consider multiple repeats over critical zones. The data will usually only be valid for general structural use so particular attention should be paid to shale zones.

All resistivity curves should have good character, although their similarity will not extend down to small details. A slow or dead curve usually indicates a faulty knife-edge electrode. The orientation curves are unchanged from an ordinary dipmeter so the same comments apply to both.

The Computed Dipmeter Log

The dipmeter log as recorded is not easily evaluated for quality; it has to be interpreted by a computer program before the measurements can be related to formation dips. An appraisal of the computed log should be included in dipmeter quality control routine. This log may be available on location, with a computer logging unit; but it is normal to wait for days or even weeks for the results, if the processing is done at a central computer office.

Working with the computed log, look for defects at two distinct levels: undetected problems with the recorded data, and problems with the computation. It is beyond the scope of this manual to discuss details of dipmeter computation. Suffice it to say that a computed dipmeter log should model real-life geology. If it seems not to do that, an investigation is in order.

If the problem is in the computation, it can normally be solved by repeating the computation job. Even if the problem is with log measurements, the logging company's computer experts can often solve it by special handling.

Field Examples

Figure 5.6 shows a dipmeter with excessive rotation. Note that the orientation curve in Track 1 shows a complete rotation about every 12 ft. Reliable computation of dips from this log are impossible.

DIRECTIONAL SURVEYS

Directional surveys are used to determine the location of the hole path with respect to the surface location. Some of the uses for this information are:

1. legal proof that the bottomhole location is under the correct surface property
2. to ascertain the bottomhole location on purposely deviated wells
3. to determine the radius of curvature of the hole as it affects the ability to run casing or tools
4. to differentiate between measured depth and true vertical depth when using formation elevations for structural mapping

There are two basic types of directional surveys: continuous surveys and station surveys. The station surveys can be either single-shot or multishot. (Single and multi refer to the number of stations recorded.) Single-shot surveys are normally recorded during the drilling operation and are recorded at given depth or time intervals. Single-shot records are accumulated and used to plot the hole path. Multishot surveys are the result of several shots run at given depth intervals after the hole has been drilled. Continuous surveys are run after a portion of the hole has been drilled and are recorded continuously over the selected interval.

Tools Available

Continuous, multishot, and single-shot instruments are all different; and there is still another classification of instruments to be considered when choosing a survey—the gyroscopic survey tool.

Openhole directional surveys normally use magnetic-compass orientation to fix the hole direction. (This requires input of the deviation between magnetic and true north.) But, since surveys run inside metal casing cannot use the magnetic compass for hole direction, gyroscopes are normally used whenever a survey is needed inside casing. These gyroscopes must be aligned on the surface before proceeding with the survey. They should also have their alignment verified as part of the after-survey checks. A gyroscopic survey tool incorporating an accelerometer is illustrated in figure 5.7.

There are several new systems being made available which are able to sense the Earth's magnetic field through steel pipe. These systems have no moving parts and are expected to replace both gyroscopes and compasses. At present very few surveys use this system.

All directional devices use a pendulum system for determining the angle of hole deviation. All continuous dipmeter surveys measure

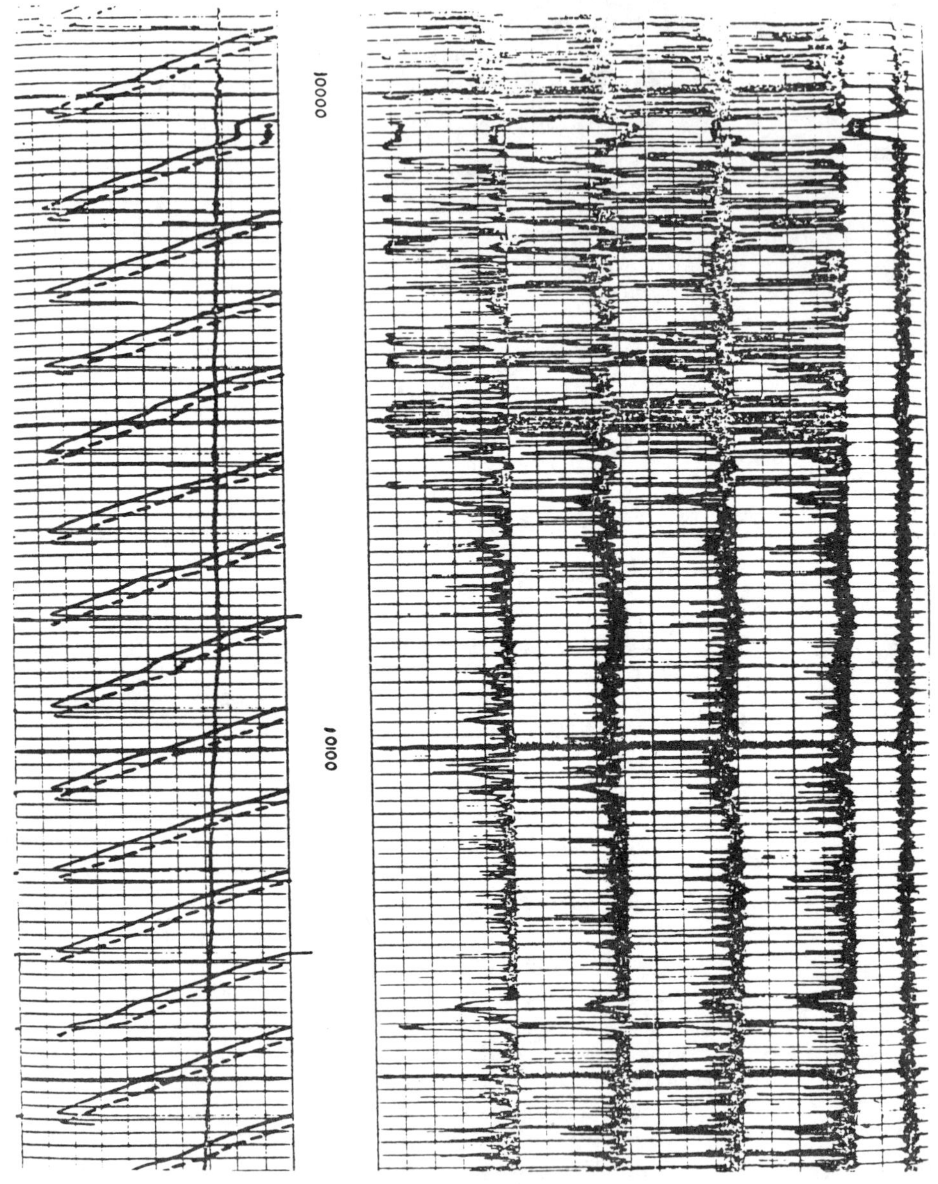

FIGURE 5.6 *Dipmeter with Excessive Tool Rotation.*

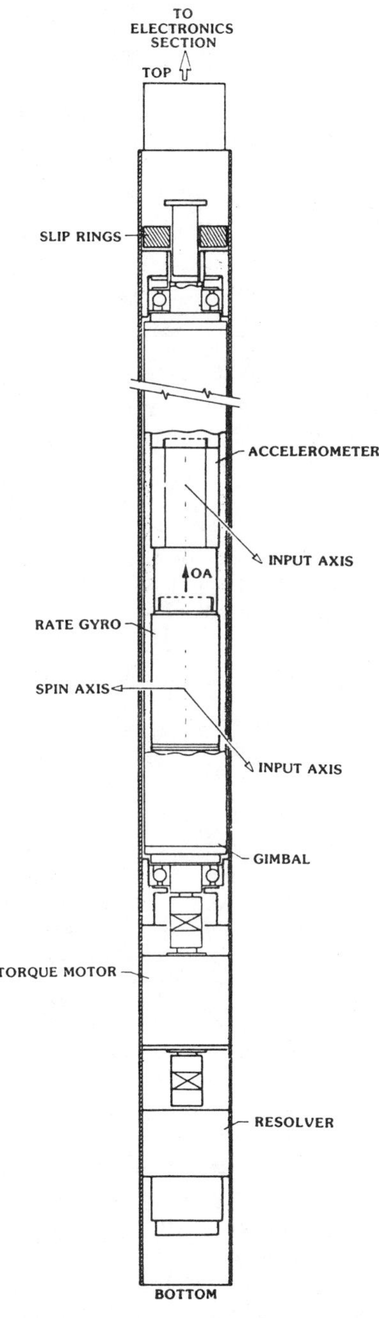

FIGURE 5.7 *Eastman Whipstock Seeker-1. From Scott 1982. ©1982 SPE-AIME.*

data that can yield a directional survey. This directional survey is available either as part of the dipmeter or as a separate survey over portions of the hole where information about formation dips is not desired. This instrument does not at present work in a cased hole, so the survey must be tied in to known coordinates at the bottom of the casing.

Any device using a magnetic compass to fix direction will be affected by metal in or near the borehole. This effect must be considered when a survey is run in an open hole that has been whipstocked past abandoned drillpipe or may be near a cased wellbore.

Legal Requirements

Each state has a separate definition of what constitutes a *legal* directional survey. These definitions may include the following criteria—which must be considered when choosing a service company and a type of survey:

length of downhole sensor
whether or not such assembly is centered
method of calculating station surveys
professional qualifications of person supervising or certifying the results
documentation, presentation, and distribution of results

Computation of Results

Directional surveys are available in any area where directional drilling is done or where dipmeters are available. The field log may be a series of station readings or a continuous curve showing hole direction and deviation. The computed results will include a well plot showing the vertical projection of the wellbore. Additional plots may show wellbore projections on a vertical plane passing through the surface location. A tabulated listing will show the wellbore coordinates and deviation angle.

METHODS OF CALCULATION. There are many methods of calculating directional surveys. Most companies use one of the following five basic methods:

Tangential Method. Directional calculations with the tangential method use the inclination and azimuth angles at the bottom of each course length. This is usually the most common and most inaccurate method. The error introduced increases with the inclination angle and the course length (distance between readings or stations). This method is not recommended.

Balanced Tangential Method. The balanced tangential method uses the inclination and azimuth angles at both the top and bottom of each course length to tangentially balance the two sets of measurements over the course length. This method is more accurate than the tangential method but is still sensitive to the course length.

Angle-Averaging Method. The angle-averaging method uses a simple mathematical average of the inclination and azimuth angles at the top and the bottom of the course length to compute the survey using the tangential method. This is more accurate than the tangential method but still simple enough for hand field calculations. Course length should be kept as short as feasible.

Radius-of-Curvature Method. The inclination and azimuth angles at the top and bottom of the course length can be used to generate a space curve representing the curve path. This space curve passes through the measured angles at the top and the bottom of the course length. This radius-of-curvature method is usually considered the most accurate but is still sensitive to course length.

Mercury Method. Used by the U.S. government at the Mercury Test Site in Nevada, the mercury method is a combination of the tangential and balanced tangential methods. The portion of the course length defined by the length of the surveying instrument is treated by the tangential method. The remainder of the course length is treated by the balanced tangential method.

In General. All these methods are critical to the course length or separation between stations. As the course length increases, their inaccuracies and deviations one from the other increase. As the course length decreases, they all become more accurate. On very short course lengths (10 ft or less) there is very little to choose between the methods. For this reason, directionals computed from continuously measuring devices such as dipmeter tools may be more accurate than station reading devices. Dipmeter devices normally compute every 1 or 2 ft, although data listings may be accumulated and listed only every 50 ft.

PRESENTATIONS AND FIELD EXAMPLES. Directional data is normally presented both in well sketches and tabulated data. Well sketches include (1) a planar view and (2) vertical sections.

The planar view is a vertical projection of the wellbore path on a horizontal plane (fig. 5.8). This projection shows the separation between the wellbore and the surface location. The wellbore path is marked with measured depths.

Vertical sections are of two types. The first is a projection of the wellbore on a vertical plane through the surface location and aligned at various azimuths (fig. 5.9). The second is plot of depth against closure where closure is the horizontal distance of the wellbore from the surface location (fig. 5.10). The tabulated data listing will show the measured depth, vertical depth, hole azimuth, deviation angle, X and Y distances, and closure distance (fig. 5.11).

Log Quality Control

When a directional survey is computed from a dipmeter survey, the same log quality control checks should be made as noted under the dipmeter surveys. Station surveys should be checked for repeatability.

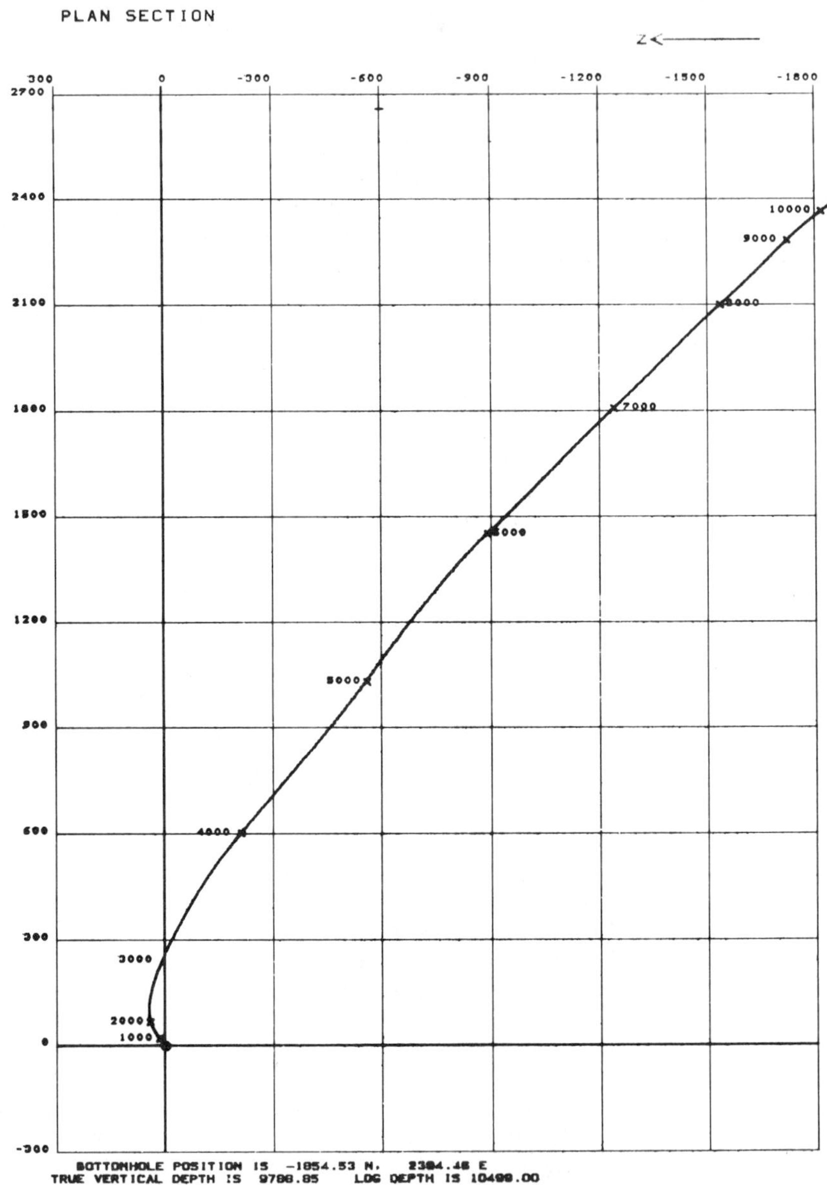

FIGURE 5.8 *Deviation Survey Planar View. Courtesy Vizilog, Inc.*

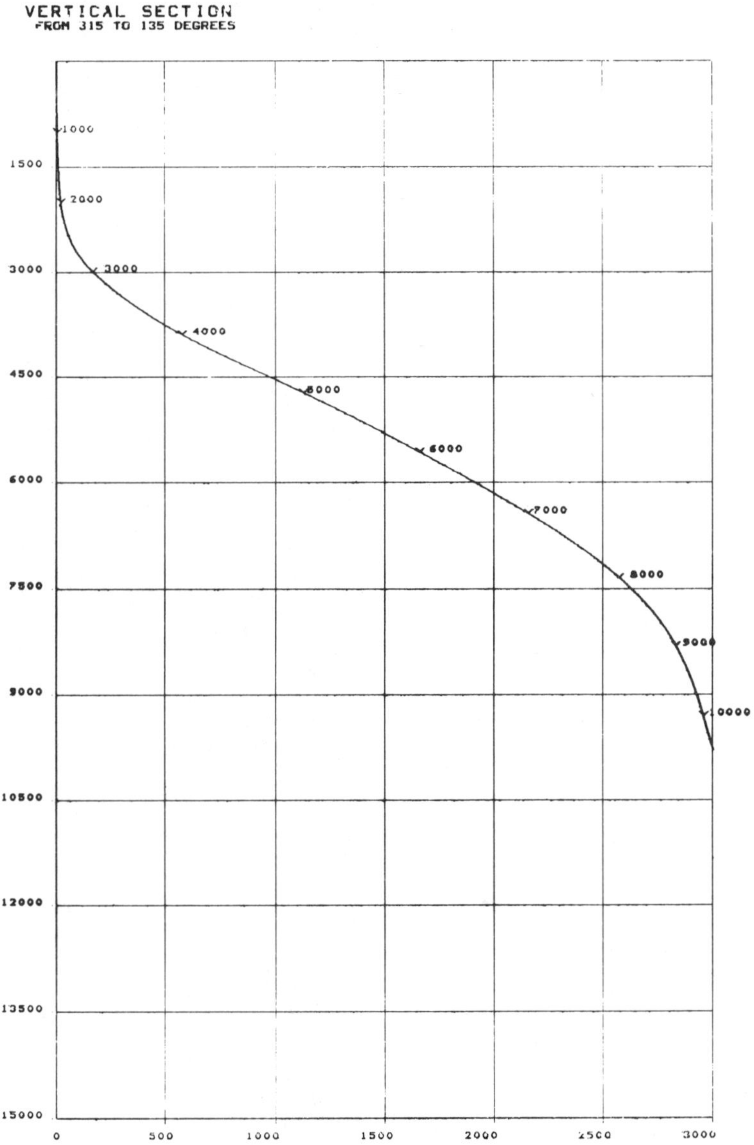

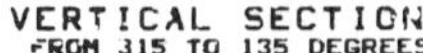

FIGURE 5.9 *Deviation Survey Vertical Section. Courtesy Vizilog, Inc.*

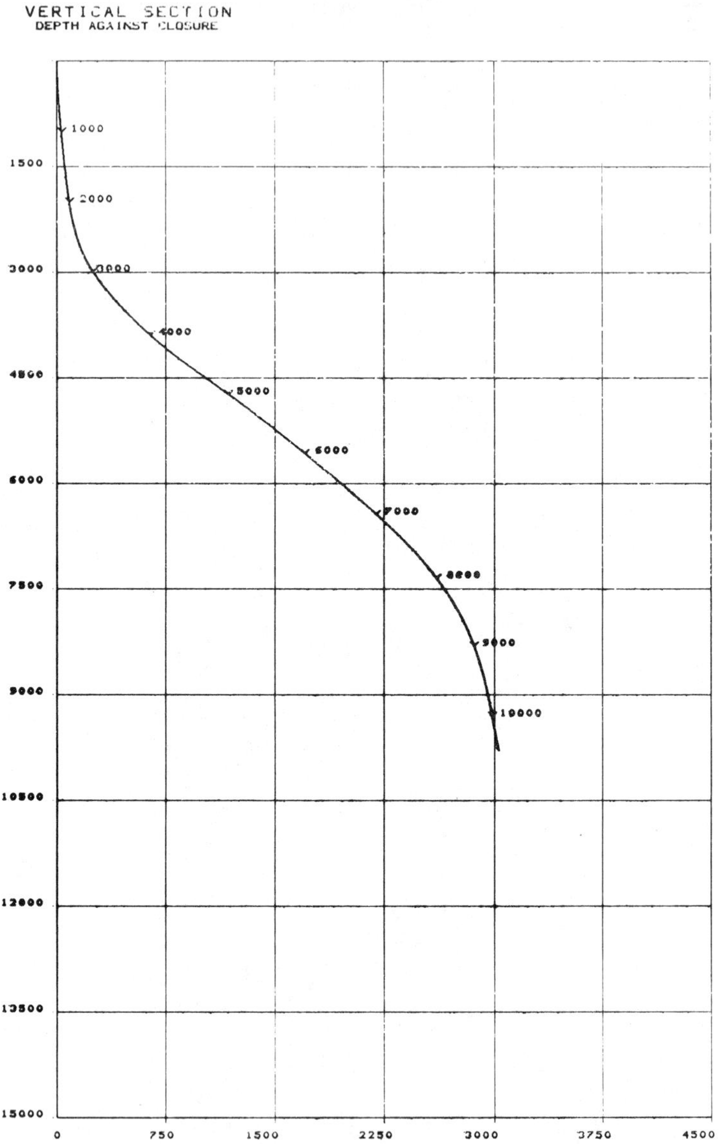

FIGURE 5.10 *Deviation Survey Depth vs. Closure. Courtesy Vizilog, Inc.*

MEASURED DEPTH	VERTICAL DEPTH	HOLE DIRN	DEVIATION ANGLE	NORTH DRIFT	EAST DRIFT	CLOSURE
50.00	50.00	95	0.04	-0.00	0.02	0.02
100.00	100.00	95	0.08	-0.01	0.07	0.07
150.00	150.00	95	0.11	-0.01	0.15	0.15
200.00	200.00	95	0.15	-0.02	0.26	0.26
250.00	250.00	95	0.19	-0.04	0.41	0.41
300.00	300.00	95	0.23	-0.05	0.59	0.59
350.00	350.00	95	0.26	-0.07	0.81	0.81
400.00	400.00	95	0.30	-0.09	1.05	1.06
450.00	450.00	95	0.34	-0.12	1.33	1.34
500.00	500.00	95	0.38	-0.14	1.64	1.65
550.00	550.00	95	0.41	-0.17	1.99	1.99
600.00	599.99	95	0.45	-0.21	2.36	2.37
650.00	649.99	95	0.49	-0.24	2.77	2.78
	699.99	95	0.53	-0.28	3.22	3.23
		95	0.57	-0.32	3.69	3.71
			0.60	-0.37	4.20	4.22
7350.00				-0.41	4.74	4.76
7400.00	7383.58				5.32	5.34
7450.00	7432.08	175				5.95
7500.00	7480.45	173	14.96			
7550.00	7528.61	178	16.29	-130.4[illegible]		
7600.00	7576.40	182	17.95	-145.16	339.4[illegible]	
7650.00	7623.62	177	20.37	-161.57	339.44	375.9[illegible]
7700.00	7670.12	174	22.56	-179.86	340.74	385.30
7750.00	7715.94	176	24.56	-199.74	342.65	396.61
7800.00	7761.23	180	25.44	-220.81	344.67	409.33
7850.00	7806.24	170	26.24	-242.33	347.60	423.73
7900.00	7850.90	169	27.20	-264.26	352.48	440.55
7950.00	7895.22	171	27.92	-286.89	357.27	458.20
8000.00	7939.29	172	28.40	-310.33	360.12	475.39
8050.00	7983.21	171	28.72	-333.64	365.29	494.72
8100.00	8027.23	171	27.57	-357.15	368.36	513.07
8150.00	8071.88	167	26.08	-379.22	372.70	531.71
8200.00	8116.93	178	25.60	-400.69	375.60	549.20
8250.00	8162.01	173	25.81	-422.21	377.41	566.31
8300.00	8206.82	160	26.83	-443.67	382.94	586.08
8350.00	8251.35	141	26.96	-463.48	393.56	608.04
8400.00	8295.95	159	26.72	-483.61	403.74	629.99
8450.00	8340.64	160	26.51	-504.51	411.80	651.23
8500.00	8385.41	152	26.56	-524.29	421.83	672.92
8550.00	8430.01	157	27.15	-545.06	430.72	694.70
8600.00	8474.30	128	28.16	-565.54	441.23	717.30
8650.00	8518.19	96	28.95	-581.39	458.25	740.27
8700.00	8561.61	132	29.52	-594.54	476.27	763.03
8750.00	8605.11	136	30.51	-610.21	497.62	787.39
8800.00	8647.93	136	31.60	-629.09	515.23	813.15
8850.00	8690.25	130	32.72	-647.49	534.45	839.57
8900.00	8732.28	133	32.80	-663.40	556.17	865.69
8950.00	8774.30	130	32.80	-681.53	576.24	892.49
8963.00	8785.23	130	32.80	-686.14	581.57	899.45

FIGURE 5.11 *Deviation Survey Listing. Courtesy Vizilog, Inc.*

The deviation angle of the hole at each station should be checked against the hole angle measured during drilling by the single-shot device. Computed results should be checked for mathematical accuracy and consistency with other information.

CHAPTER SIX

MISCELLANEOUS LOGGING METHODS

FRACTURE IDENTIFICATION

Most fracture-identification systems have evolved from observing undesirable or unusual effects on measurements being used primarily for other information. So-called fracture locating logs were then derived by modification of either tools or data displays to highlight these effects better.

When attempting to locate fractures there are three important points to keep in mind:

1. Almost all logging devices are affected some way by fractures, but not to a degree that can be quantified. At best, relative degrees of fracturing within certain narrow confines can be estimated.
2. Many other common borehole conditions (plus tool malfunctions) will duplicate the same effect on the logging tools as do fractures. Many of these other effects are also impossible to quantify.
3. Because of the "black art" nature of fracture identification, there are many diverse methods, which can usually be grouped by service company, geologic province, formation, or oil company. Most of these methods have intense but sparse advocates. This document will make no effort to recommend, condemn, or order these methods.

Methods Available

Table 6.1 lists methods available to evaluate fractures. The local service company can give assistance in most cases in selecting a method that has a proven track record in a particular geographic area.

SONIC METHODS. Fractures attenuate acoustical energy as it passes through the formation. The energy thus attenuated is either measured or displayed as a log. The most effective devices respond to both the compressional wave and the slower shear wave. A number of conditions affect the accuracy of sonic measurements.

Centering of the Sonic Tool. The tool must be centered very accurately since the detected amplitude will decrease by over 50% if the tool is ¼ in. from the center of the hole. To monitor this tool position, a single-receiver sonic travel-time curve should be recorded. This travel-time curve will follow the normal sonic curve except for the transmitter–receiver spacing and the tool position with respect to the center of the hole. A normal sonic curve is scaled in microseconds per foot of formation, whereas this travel-time curve will include mud time and formation time. If the log is run on a scale of X times the normal scale, where X = transmitter–receiver spacing in feet, the

TABLE 6.1 *Fracture Identification Methods*

Method	Spacing or App. Vertical Resolution (inches)	App. Depth of Investigation (inches)	Percent of Circum. of 8¾ in. Borehole Surveyed (percent)	Effect of Vugs Associated with Fractures	Can Be Recorded in Gas- or Air-Filled Holes	Can Be Recorded in Cased Holes
Spontaneous potential	12	0	100	None	No	No
Calipers						
3-arm bow spring						
Recorded with:						
Induction Electrolog	18	0	25	None	Yes	No
BHC acoustilog	18	0	25	None	Caliper Only	No
1-arm						
Compensated Densilog	6	0	6	None	Yes	No
Sidewall Epithermal Neutron	6	0	6	None	Caliper Only	No
2-arm						
Proximity-Minilog	12	0	36	None	Caliper Only	No
Micro-Laterolog	12	0	36	None	Caliper Only	No
4-arm						
4-arm dual caliper	1	0	4	None	Yes	No
High-Resolution 4-Arm Diplog	12	0	50	None	Caliper Only	No
Minilog						
Micro Inverse 1 × 1 in.	2	1.5	7	Improves	No	No
Micro Normal 2 in.	4	4	7	Improves	No	No

Spectralog	24	6	100	None	Yes	Yes
Correction Curve on Compensated Density	8	(1.5%)	10	Improves	Yes	Yes
Resistivity						
Dual Induction Focused Log:						
Shallow Focused	14	30	100	None	No	No
Medium Induction	60	70	100	None	Yes	No
Deep Induction	72	120	100	None	Yes	No
Proximity Log	12	10	7	Improves	No	No
Dual Laterolog:						
Shallow Laterolog	24	30	100	None	No	No
Deep Laterolog	24	120	100	None	No	No
Micro Laterolog	8	6	7	Improves	No	No
Comparison of Porosity Measurements:						
Sidewall Acoustilog	6	0 to very	4	None	No	No
BHC Acoustilog	24	shallow	100	None	No	No
Compensated Densilog	12	6	12	Improves	Yes	Yes
Compensated Neutron	18	8	30	Improves	No	Yes
High-Resolution 4-Arm Diplog	½	1.5	11	Improves	No	No
Acoustical Variable Density Frac Log						
BHC Acoustilog	36-84	very shallow	100	None	No	Yes
Sidewall Acoustilog	9 or 15	very shallow	4	None	No	No

Source: Adapted from Dresser Atlas.

single-receiver travel time will track the regular sonic travel time with a constant offset. This offset will be constant as long as the mud time is constant. The mud time will vary only when the hole size varies or the tool becomes eccentered. This travel-time curve will allow you to monitor the centering of the tool. By selecting the proper scale, this monitoring can be done easily by overlaying the regular sonic curve with the monitor curve.

Homogenous Mud System. The drilling mud cannot be gas cut or heavily treated with lost circulation material. Either of these will severely attenuate the signal. Heavy concentrations of oil in a water-base mud and invermul muds may also give problems.

Round and Smooth Wellbore. Out-of-round wellbores will both attenuate the signal and make accurate centering more difficult. Rugose wellbores will further attenuate the signal.

Environmental Tool Calibrator or Response. There must be some method of determining exactly what the individual tool being run does in both a high-attenuation and a low-attenuation environment. A low-attenuation environment may be uncemented surface casing, a low-porosity formation without fractures, or an evaporite. A high-attenuation environment may be well-cemented surface casing, washed-out and rugose hole, high-porosity gas sand with little invasion, or a known highly fractured formation. Each log run should include as many of each case as is available under the conditions. Repeats are particularly valuable. A scale should be selected that resolves both high and low attentuation.

SONIC PRESENTATIONS. In addition to a single-receiver travel-time curve that monitors tool centering, the amplitude of the sonic arrival must also be displayed. This can be done with (1) a sonic-amplitude curve, (2) a wave-train display, or (3) an amplitude-ratio curve.

Sonic-Amplitude Curve. The amplitude of the sonic signal arriving at one or more transmitters is displayed by a sonic-amplitude curve. Different service companies take amplitude measurements from different portions of the arriving signal. Thus, for example, some measure compressional arrival amplitude, some shear arrival amplitude, and some integration of arrival amplitudes throughout a time gate. Integration normally includes both compressional and shear arrivals. Figure 6.1 illustrates a sonic-amplitude and a microseismogram presentation across a fracture.

Wave-Train Display. The wave-train display is an analog display of the sonic-wave amplitude at a receiver displayed against time. The X-axis is scaled in time (normally milliseconds) and the Y-axis is scaled in amplitude (normally millivolts). This is sometimes called a *wiggle-trace.* A variation on this display is an X-axis–Z-axis display where the Y-axis excursions are modulated into bands of varying density. Figure 6.2 illustrates this type of display in a fractured reservoir. The circled areas suggest fractured intervals.

Amplitude-Ratio Curve. A ratio of the sonic amplitude arriving at two separate receivers gives an amplitude-ratio curve. This ratio is

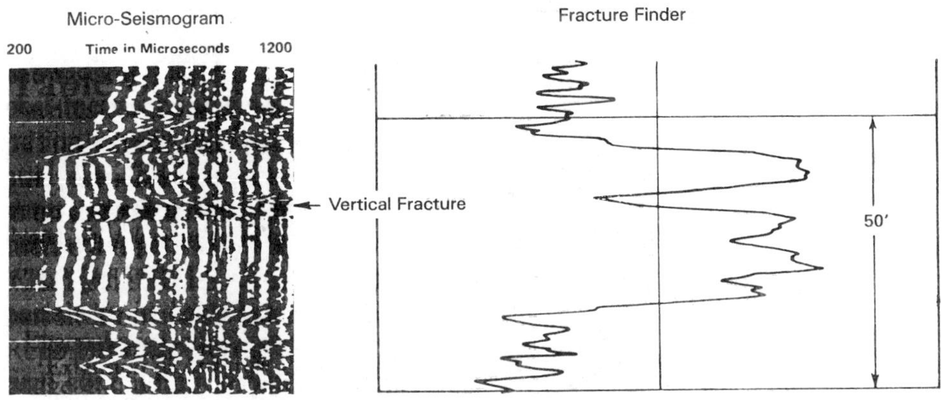

FIGURE 6.1 *Sonic-Amplitude and Microseismogram Presentation. Courtesy Welex, a Halliburton Company.*

normally the far receiver amplitude divided by the near receiver amplitude. The ratio is less than one in all conditions but reduces sharply when fractures are between the two receivers. This approach tends to decrease adverse effects such as eccentering and hole rugosity by presuming that these effects are similar on the two receivers.

RESISTIVITY METHODS. When open fractures are filled with mud or mud filtrate, they provide a low-resistance current path that can provide anomalous responses on resistivity devices.

Dual Induction Focused Log. Horizontal fractures will appreciably affect only the two induction curves. The shallow induction will be more affected and will read lower than the deep induction. Both curves may exhibit spiking in both horizontal and vertical fracturing. The shallow focused log is more affected by vertical fractures and will read less than either or both induction curves (see fig. 6.3). All curves must be corrected for environmental effects before interpretation.

Dual Laterolog. The shallow laterolog will read less than the deep laterolog (see fig. 6.4). This will also be true when $R_{xo} < R_t$ in permeable zones. Both curves may show spikes. Horizontal fractures have limited response on both curves.

Microresistivity Devices. Because microresistivity devices are finely focused, they will try to "see" each fracture as the tool passes the fracture. This will give rapidly changing resistivity values, which will likely not repeat exactly if the tool changes orientation on the repeat pass.

Four-Arm Dipmeter. A special case of a microresistivity device is the four-arm dipmeter. When using this tool to identify fractures, it is assumed that the fractures are vertical and pass through a plane that includes the borehole. Therefore the tool can "see" the fractures only when looking along the fracture plane. When looking perpendicular to this plane, the tool does not "see" the fracture. By having four

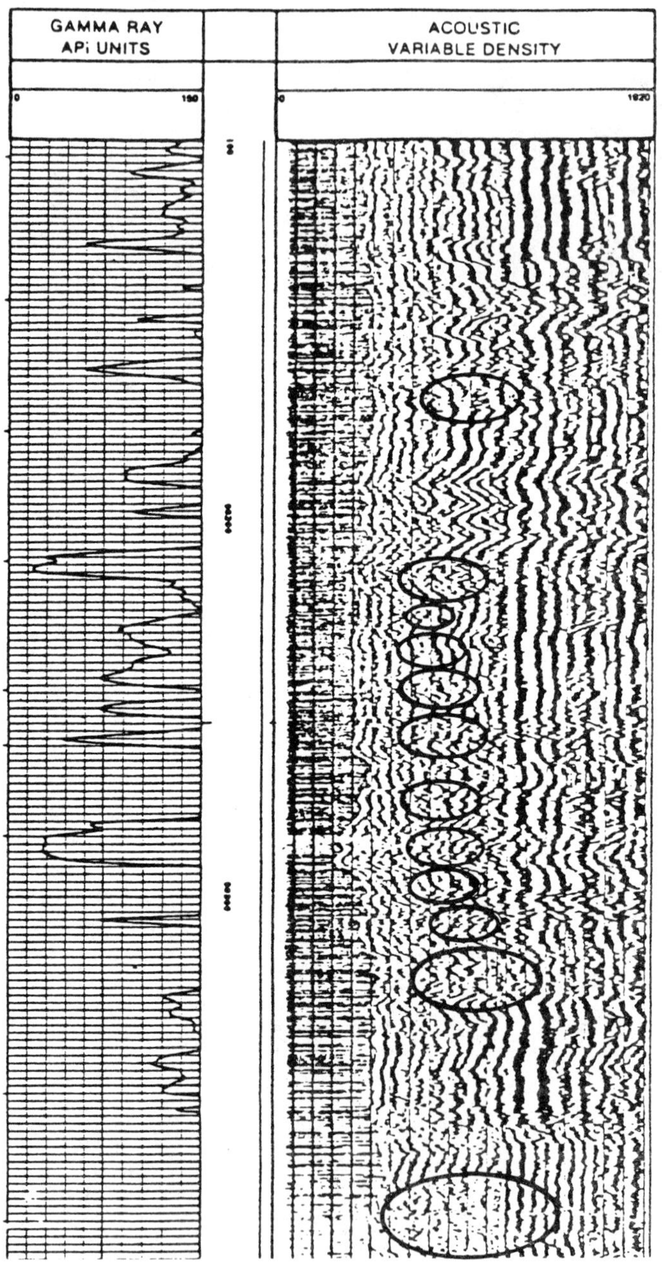

FIGURE 6.2 *Variable Density Display in Fractured Reservoir (Possible Fractures Circled).*

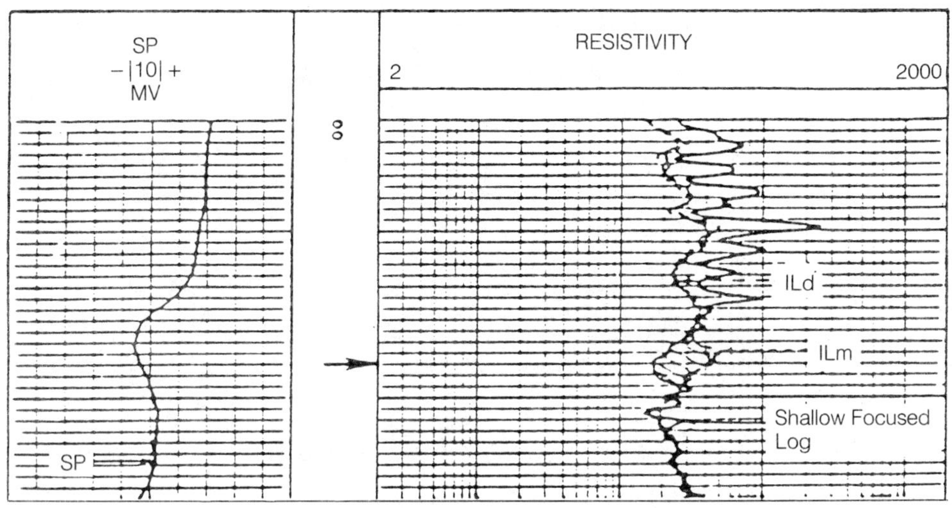

FIGURE 6.3 *Dual Induction Fracture Indications.*

separate measurements oriented at 90° to each other around the wellbore, it is possible to compare two measurements being affected by the fractures directly against two measurements not affected by the fractures. This display can be made on the original recording and the differences or separation between the measurements can then be used to indicate fractures. In addition, this method allows the orientation of the fracture planes as indicated by the azimuth of the lower reading arms. An example of this type of display is given in figure 6.5.

Porosity Devices. Fractures can be indicated by comparing the responses of various porosity logs. This is usually done by comparing porosity values from a total-porosity device such as a Compensated Density or a Compensated Neutron Log to a porosity derived from the Compensated Sonic Log, which is a primary-porosity device. This comparison treats fractures as secondary porosity and indicates fractures in zones where total porosity exceeds primary porosity. Vugs will cause this same response and therefore must be considered when looking for fractures (see fig. 6.6).

The Compensated Density Log may show excessive correction when being affected by fractures (see fig. 6.7). This excess compensation is due to the rugosity sometimes caused by fractures being drilled with a percussion bit.

Caliper Devices. Calipers may indicate fractures in three ways:

1. They may show rugosity associated with the effect of the drilling operation on a fractured formation.
2. They may show mudcake on an otherwise impermeable formation.

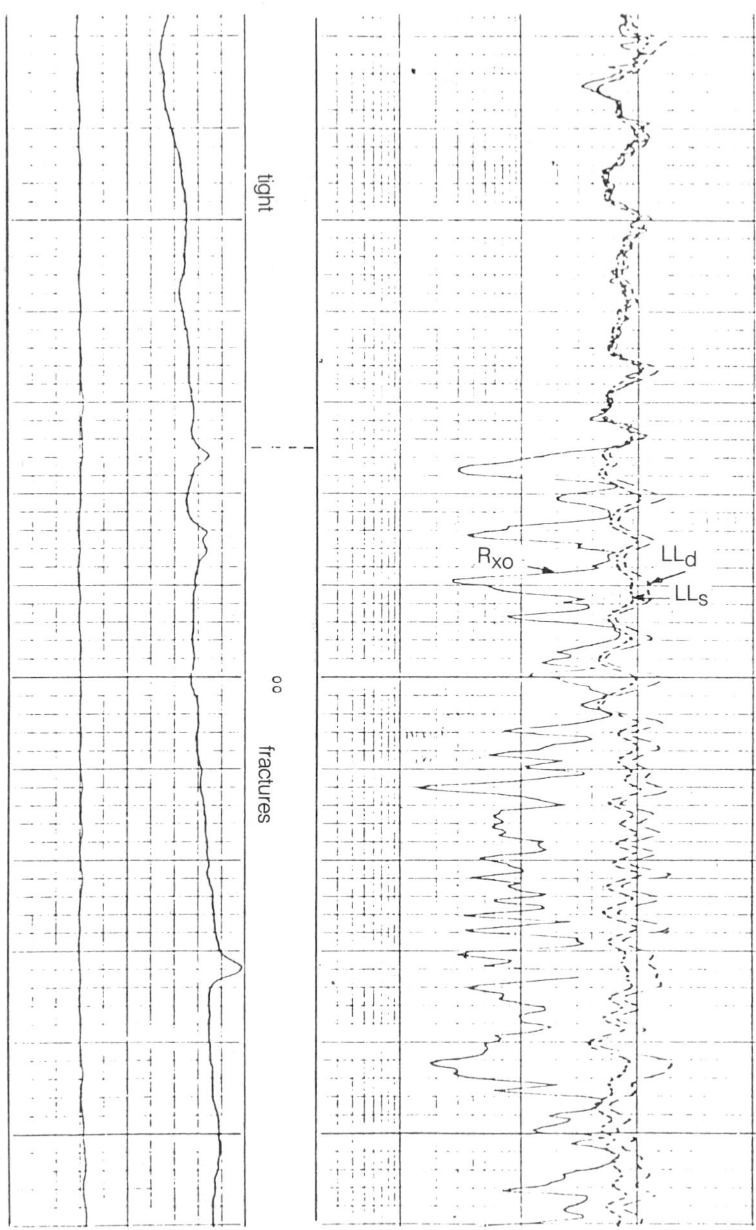

FIGURE 6.4 *Dual Laterolog Fracture Indications.*

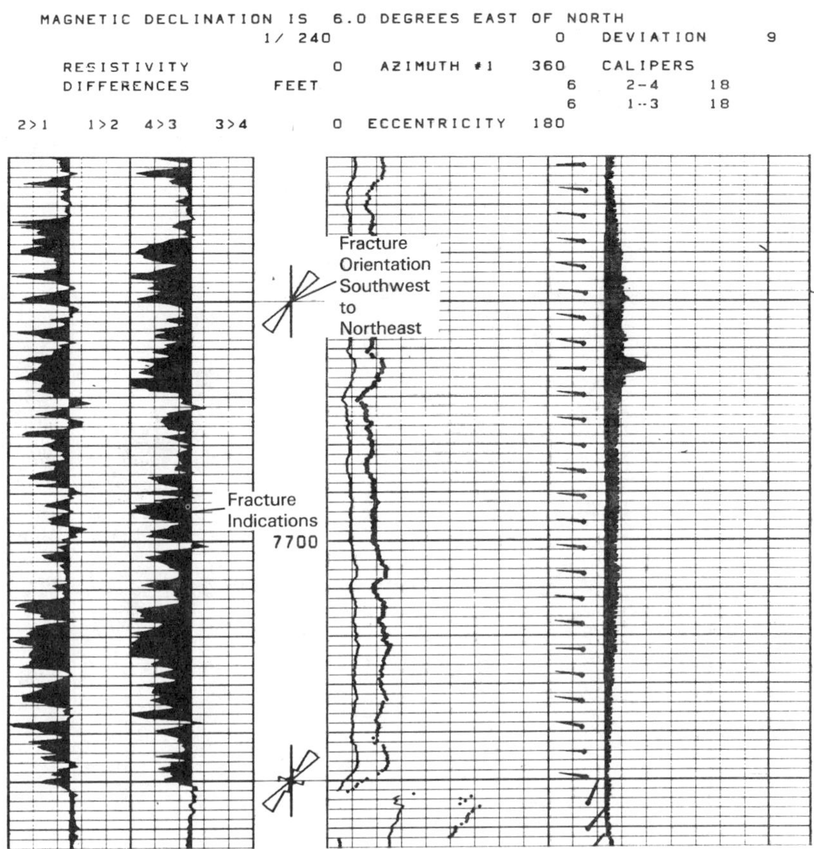

FIGURE 6.5 *Dipmeter Fracture Indications. Courtesy Vizilog, Inc.*

3. In the case of multiple articulated calipers, they may show an elongated or egg-shaped hole. This may be due to the hole eroding along an axis parallel to the fracture planes.

Most devices that use calipers can be used for one or more of these indications.

NATURAL GAMMA RAY SPECTRALOG. Assuming that circulating ground fluids have deposited uranium salts in the fractures, the natural gamma ray spectralog can be used to identify the fractures. The shaded areas on figure 6.8 indicate fractures in a chalk well.

BOREHOLE TELEVIEWER. A sonar-type device is used to produce an image of the formation at the wellbore. The image on a borehole televiewer shows discontinuities and will indicate the fracture or mudcake across the fracture. This is a sonic device, and is subject to the

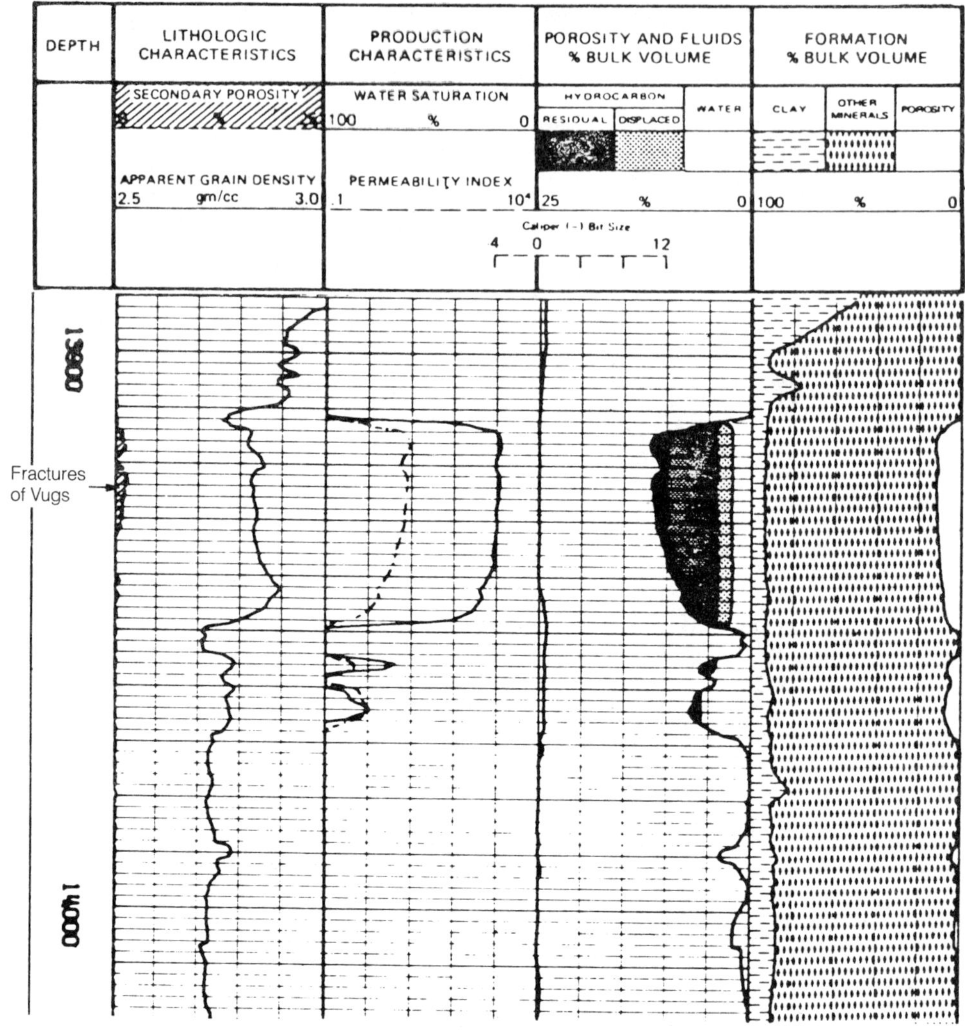

FIGURE 6.6 *Secondary-Porosity Fracture Indications. Courtesy Dresser Atlas.*

same borehole restrictions as other sonic devices. Because of the relatively slow logging speed, the borehole televiewer is best suited to detailed examinations of selected intervals. For more details on the BHTV, refer to chapter 3.

Field Example

A compromise of several logs can be used to pinpoint fractured zones, as shown in figure 6.9. Note that not all of the possible indicators necessarily agree.

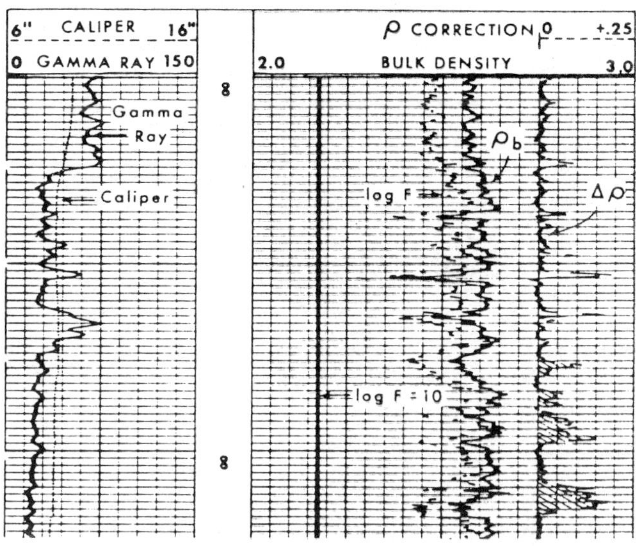

FIGURE 6.7 *Density Log Fracture Indications.*

Summary

Although many devices indicate fractures, in most cases these indications can be ambiguous. Therefore, the most successful identification of fractures will result from using a combination of several *unrelated* systems. Local experience and recommendations from service companies should be given serious consideration in every case.

NUCLEAR MAGNETIC RESONANCE*

Principle of Measurement

The Nuclear Magnetism Log (NML) (mark of Schlumberger) detects the electromagnetic field caused by the magnetic character of hydrogen nuclei. A fraction of the hydrogen nuclei are aligned by and parallel to a magnetic field produced by the tool. The tool then measures the signal generated by those hydrogen nuclei as they rotate (precess) about the Earth's magnetic field after the field that aligned them is shut off. The NML measures how many hydrogen nuclei stay aligned long enough to be measured and how long it takes to align them. This measurement reflects all the hydrogen nuclei except:

those in water in intimate contact with surfaces (The NML does not see the fluid in a shale and does not see the irreducible water in a sand.

*This section on nuclear magnetic logging is based on R. J. S. Brown and Charles H. Neuman: "The Nuclear Magnetism Log—A Guide for Field Use," *The Log Analyst* (Sept.-Oct. 1982) **23**, 4–9, and is included here with their kind permission.

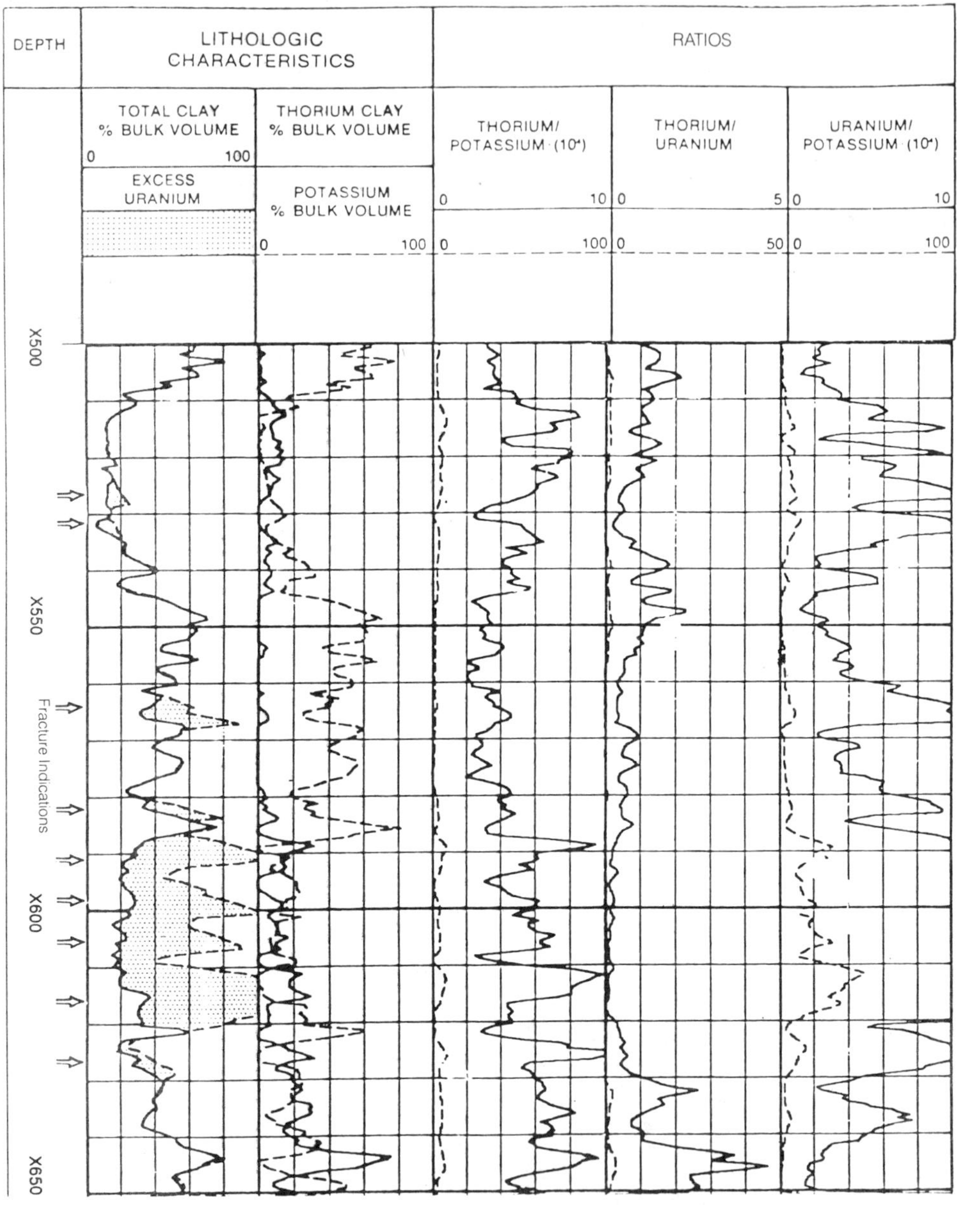

FIGURE 6.8 *Spectralog Fracture Indications. Courtesy Dresser Atlas.*

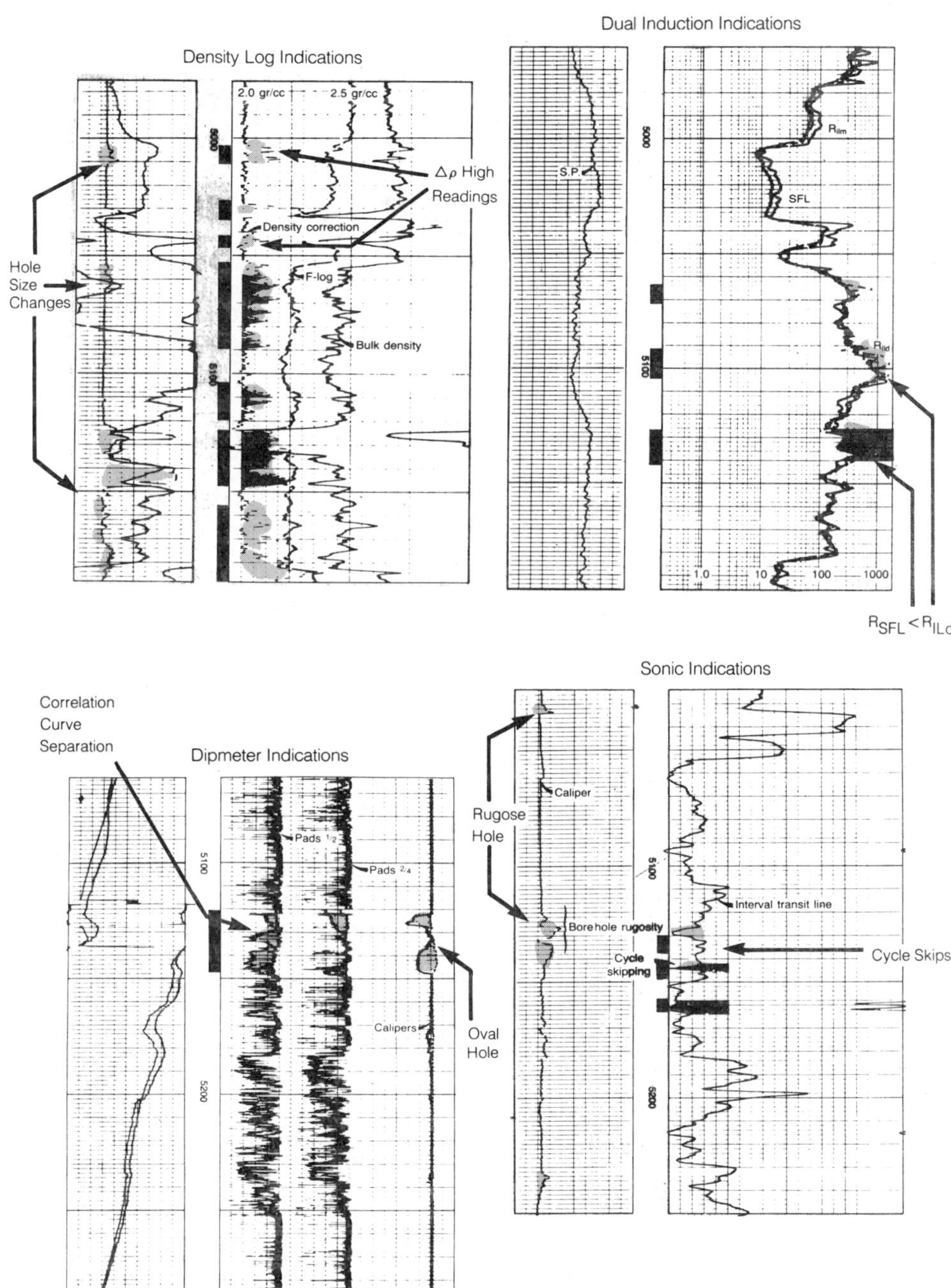

FIGURE 6.9 *Fracture Identification from Several Sources.*

Thus, the fluid it does see is called *free fluid.* In a clean carbonate, even very fine grained, the NML sees all the fluid.)

those in oil more viscous than about 500 cp at reservoir temperature (Oil heavier than 14 to 18° API is usually not seen by NML except at high temperatures.)

those in mud filtrate that has been purposefully doped with chemicals that cause rapid signal decay

those where the magnetic field is significantly changed by the short-range influence of magnetic grains such as those intentionally dispersed in drilling mud or those infrequently found to occur naturally in rocks

Uses of the NML

The nuclear magnetism log can be used to:

1. identify permeable formations—

   The free fluid index (FFI) presented on the log represents the portion of total pore fluids free to flow. FFI is thus zero except where fluids in pores will flow in response to a pressure gradient.
2. reflect permeability differences—

   NML measurements enable prediction of sandstone permeability. Several empirical relations have been shown to reflect how permeability increases with increasing porosity FFI, and time alignment of hydrogen nuclei (T1). Each permeability representation depends on parameters determined from comparisons with core-measured permeability.
3. recognize which zones with heavy oil contain moveable water—

   The signal from very viscous oil decays so rapidly that the NML does not detect it. Thus, the NML shows moveable water only and can be used to predict the response to injected steam.
4. measure residual oil—

   Chemicals can be added to the mud which cause rapid decay of the signal from mud filtrate. An NML recorded after invasion of such mud filtrate measures accurately the residual oil target for tertiary recovery.
5. recognize gas—

   The NML can be used to lessen the ambiguity due to clay minerals in identification of gas-bearing zones using density and neutron logs.
6. measure carbonate porosity—

   The NML records total porosity in clean carbonates independent of whether they are limestone or dolomite.
7. simplify log interpretation in lithologies where other logs are ambiguous—

   The NML has potential to simplify log interpretation in diatomites, chalks, and other special lithologies.

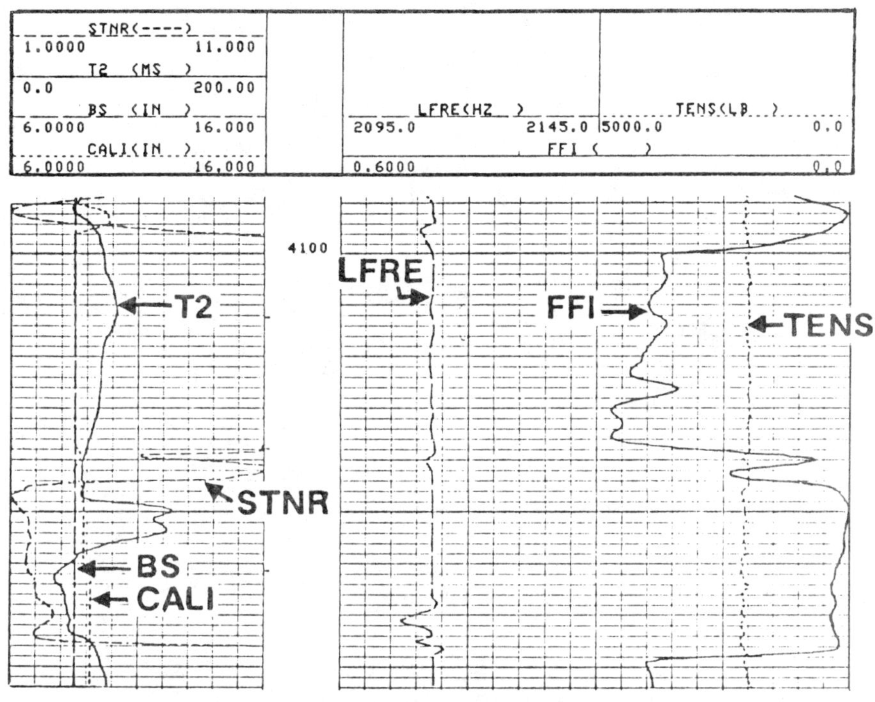

FIGURE 6.10 *NML Presentation (Normal Mode).*

Modes of Operations

(The meaning of the letter labels used for trace curves and parameters of nuclear magnetism logs are given later in this section, under *Traces Presented on Logs* and *Setup Parameters.*)

FFI MODE. In the FFI mode, a single signal is produced every half foot, with the polarizing time (NPP: the time polarizing current is applied to the NML coil), usually 2000 ms. The FFI curve is computed by extrapolating the decaying signal through the deadtime to the time of cutoff of the polarizing field (time of beginning of precession of hydrogen nuclei in the Earth's magnetic field) and applying appropriate calibration and allowances for environmental factors. The curves plotted in this mode are FFI, cable tension (TENS), signal frequency (LFRE), T2 (signal decay time constant), caliper (CALI), bit size (BS), and signal-to-noise ratio (STNR), as illustrated in figure 6.10.

T1-CONTINUOUS MODE. In the FFI mode, the 2000-ms polarizing time is usually long enough to get maximum signal. In the T1-continuous mode, three signals are measured every half foot with three different polarizing times (CBPP times one, two, and four). The number of

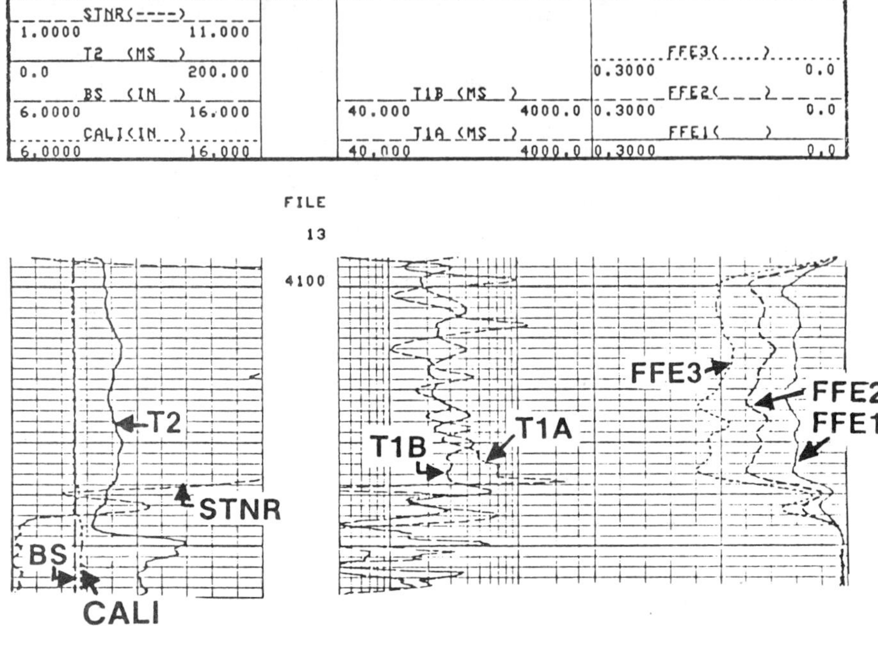

FIGURE 6.11 *Presentation (T1 Continuous Mode).*

curves plotted depends on the choice between amount of computed information desired and amount of clutter acceptable on the log. In simplest form, the three curves FFE1, FFE2, and FFE3 are plotted in Track 3 (see figure 6.11). These are similar to FFI curves, but with short instead of "infinite" polarizing times. If the curves are well separated, the relaxation times are long compared to CBPP. If there is little separation, relaxation times are short compared to CBPP. A fourth curve (FFIB) may be plotted in the same track, giving an extrapolated FFI computed from the two signals with longest polarizing times. FFIA, the FFI calculated from the shortest two polarizing times, can also be displayed, but it is not usually worth the extra clutter.

In Track 2, T1A and T1B are the relaxation times computed from FFE1 and FFE2 and from FFE2 and FFE3. These are usually different because of uncertainty from noisy signals and because the relaxation process is not simply exponential. The relaxation or buildup process usually requires a sum of two exponential terms to represent it.

As in FFI mode, Track 1 displays caliper, bit size, T2, and signal-to-noise ratio.

T1-STATIONARY MODE. The most detailed NML information at a given depth is provided by the T1-stationary mode. The tool is stopped in the hole for about three minutes, and NSPN signals are observed at

TABLE 6.2 *NML Listing for Stationary Measurements*

NMTC STATIONARY LOG 81 /8 /23

STATION DEPTH: 4204.1 FT

STATION PARAMETERS

SBPP (STATIONARY BASE POLARIZATION PERIOD):	100 MS
SPDN (STATIONARY PULSE DURATION NUMBER):	6
NSPN (STATIONARY PULSE NUMBER):	8
NSDP (STATIONARY DEAD PERIOD):	3000 MS

STATION MEASUREMENTS

NT1: 560.377 MS

PPER(MS)	AE
112.500	0.0221077
212.875	0.0266066
411.000	0.0268001
811.375	0.0321215
1610.25	0.0356529
3211.37	0.0350724

T2: 32.0451 MS

FFI: 0.0779406

each of the polarizing times (SPD), starting at SBPP and increasing by factors of two. Ordinarily eight signals are stacked for signal-to-noise ratio improvement at the six polarizing times, 100, 200, 400, 800, 1600, and 3200 ms. Because of the large number of signals averaged, good relaxation data and accurate FFI values usually result from this measurement even when FFI is only a few porosity units.

Results are given as a list of polarizing times (PPER) and relative amplitude estimates (AE) as illustrated in table 6.2. A one- or two-component relaxation curve can be computed from these numbers. Schlumberger, who manufactures the tool, does not at present give the exact relationship between the AEs and FFI. However, they do give a computed FFI and T2 as well as computed T1 (NT1) for a single-component exponential fit.

CHOICES OF MODES. Ordinarily the FFI (NMOD-NORM) curve would be run over intervals of tentative interest, locating zones having sufficient free fluid to warrant taking T1 data. FFI cutoff depends on the objectives of the formation evaluation. Relaxation information from either T1-continuous (NMOD-CONT) or T1-stationary modes can give additional permeability information. If this information is desired over a substantial interval, the T1-continuous should be run. Operation charge for each T1-stationary measurement is about as much as for 140 ft of T1-continuous logging. Usually, the base time CBPP for T1-continuous logging should be 100 ms (giving data for polarizing times of 100, 200, and 400 ms), because information in this time range is usually most relevant to permeability correlations. T1-stationary runs should be made where the maximum information is needed. The stationary data are usually good enough to give meaningful two-component relaxation curves.

If there is interest in good data in low-FFI zones, the repeated signals obtained in the T1-stationary mode give not only T1 values but also T2 and FFI values more reliable than those from the NORMal mode.

Planning NML Runs

HOLE SIZE. The coil section of the NML sonde has a diameter of 5.9 in. and operates in uncased hole only. Minimum hole size for the NML depends on hole condition, but the tool is seldom run in holes smaller than 7 in. (178 mm). Schlumberger has calibrated the NML for hole diameters only up to 10 in. The NML has been run successfully in a 12.5-in. hole, but there is some uncertainty in calibration. In normal operation, the tool is centralized. In a very large hole, stronger signals can be obtained by running the tool at the side of the hole (free in the hole) or on short rubber standoffs.

The NML coil system consists of two coils each about 1½ ft long and connected in opposite directions (known as *bucking*) to cancel induced noise. The coils can be connected in the same direction (known as *aiding*) with a small improvement in signal strength in an 8½-in. hole, and a much larger improvement when the tool is run centralized in larger holes. The noise bucking is presumably not needed for logging under most conditions, where the earth shields out cultural noise, but Schlumberger does need the noise bucking to calibrate the tool in their surface facility. Hence, they do not have FFI calibration for the tool in the aiding mode, although an estimate can be made. Relative (uncalibrated) values of FFI suffice for many large-hole applications.

HOLE CONDITIONS. The 5.9-in.-diameter coil section may warrant special attention to hole conditions. If hole conditions are expected to deteriorate substantially with time, it may be desirable to run the NML early or to make a pipe trip to condition the hole. So far, the centralized NML has given less sticking trouble than wall-contact tools such as the FDC.

DEVIATED HOLES. Accurate centralization is probably not possible in deviated holes. Extra centralizers and rubber standoffs can be used. Calibration is less certain.

OUT-OF-ROUND HOLES. The NML signal does not vary with rotation of the tool when the tool is centralized in a cylindrical hole. If the hole is out-of-round, the signal strength does vary somewhat with tool rotation. The effect is least with the tool centralized.

REGION OF INVESTIGATION. The tool is approximately uniformly sensitive over a depth interval of just under three feet. A little over three-fourths of the signal comes from within the first borehole radius into the formation. Thus, the NML samples several cubic feet of rock.

LOGGING SPEED. At present, the logging speed in the FFI mode (NORM) is about 750 ft/hr. In T1-continuous mode (CONT) the data can be taken at about 1000 ft/hr without real-time data processing. The processed log can be produced later in a computer service unit (CSU), if one is available for this purpose. With real-time processing, the logging speed is presently about 350 ft/hr, although Schlumberger hopes to improve it. The T1-stationary mode now requires about three minutes per station for the usual number of repeated signals observed. If FFI is reasonably large and T2 not too short, adequate signal quality can be obtained in substantially less than three minutes.

BACKUP TOOL. A backup tool is desirable if available. Tool failures with the NML have been more frequent than with conventional logs. Since there are few of the currently commercial tools available, it is sometimes difficult to obtain a backup tool.

Mud Treatment

The drilling mud, which contains hydrogen, is closer to the tool than is the formation, and has nearly 100% porosity. Thus, untreated mud would give a large NML signal. Mud treatment is based on distortion of the Earth's magnetic field by magnetic grains that cause the mud signal to decay so fast it cannot be seen at the end of the 20-ms deadtime of the NML. In some cases, enough steel powder is worn off the drill-string to cause the NML signal to decay in the mud. Usually, a liquid slurry containing magnetite powder is added to the mud. The slurry, manufactured by Baroid, is supplied by Schlumberger. Schlumberger encourages companies that are committed to a number of NML runs to take advantage of the cost saving from obtaining magnetite slurry directly from Baroid. The normal concentration used is 15 gal of slurry per 100 bbl of md. Other mud properties are not changed significantly by the magnetite. The magnetite suspension should be stirred, shaken, or otherwise mixed before being blended into the mud because the solids settle out of some batches of the slurry. Check that the containers are empty after use. Circulate the mud at least twice before logging to assure mixing.

Mud can also be doped by using dry, powdered magnetite at the rate of 15 to 20 lb/100 bbl of mud. The powder, which is dusty and very black, should be very uniformly strung into the mud over one circulation time.

Schlumberger recommends that logging be done within 48 hours of the last mud circulation to avoid the possibility of magnetite settling out. It is not clear, however, that this is always necessary.

In coarse-grained, vugular, fractured, or lost-circulation zones, there is the possibility that whole mud may go into the formation, magnetic grains included. When this is a concern, the magnetite should be added after a mudcake is formed. The mudcake should be disturbed as little as possible. In these cases, it is advisable to add the magnetite while circulating after drilling to TD. When whole-mud invasion is thought to be unlikely, magnetite should be added earlier to minimize rig time.

Log Quality Control

EQUIPMENT DATA. Some NML tools have had frequent failures. Some have higher noise than others. The log heading should specify the equipment used, including the module (NMM), cartridge (NMC), sonde (NMS), caliper (MCD), and digitizing panel (WDM) used to form the complete tool string and surface measurement equipment.

FREQUENCY SETTING. The approximate NML signal frequency (proportional to the strength of the Earth's magnetic field) is often known, either from previous NML runs in the area, or from maps of the strength of the Earth's field. However, it is necessary to go first to a depth where other logs suggest the presence of substantial free fluid and observe NML signals to determine the Larmor frequency using the Playback Larmor Frequency Task (PLFT). Then the default frequency LARM should be set to this value. LARM is the frequency to which the NML coil and filters are tuned and the frequency used in processing. The frequency displayed on the log, LFRE, is determined on individual signals if STNR is over 3.0 and is defaulted to LARM otherwise. LARM must be within ±18 Hz for correct LFRE determination and signal processing.

SHALE SIGNALS. Most boreholes have intervals that can be unambiguously identified from other logs as shale zones having no free fluid. The FFI readings in these zones should be less than one percentage unit, preferably less than 0.5. Larger FFI with a short decay time T2 (perhaps 20 ms) could be from inadequately treated mud. If the mud treatment is marginal and the T2 values for the formation signals are not excessively short, a good log can be obtained by increasing the signal-use delay time EDEL to 35 to 45 ms. If this eliminates the shale signal, the log should be run (or reprocessed) with the increased EDEL. The NML has occasionally produced spurious signals that look much like the actual NML signal but have a different frequency. The newly introduced Schlumberger processing usually suppresses these.

CALIBRATION. The panel calibration procedure should be done before logging and at least once again during logging, preferably near the end of the logging. If logging is several hours, there should be a calibration-tuning check every hour or two. Part of the calibration procedure consists of applying a standard signal to the system and determining a hardware gain and a software gain. The hardware gain involves the signal envelope formed from the NML signal by analog electronic circuitry, and the software gain involves the digitized signal waveform. The gains should not change drastically during logging. The tune word should ordinarily not change at all.

It is probably not necessary to pay attention to the DC levels presented in the calibration summary. They should be, and usually are, much smaller than the noise. They reflect the DC bias of the analog-to-digital converters that digitize oscillating NML signals. Accurate caliper calibration is important because the result is used in the computation of FFI from measured voltages. The computed FFI varies approximately as the square of the caliper reading.

The calibration summary lists the SOFTWARE-DETECTED ENVELOPE AVERAGE NOISE, which is called WALN in the parameter list.

Tool performance sometimes deteriorates during a run, leading to higher noise, calibration changes, bad FFI repeats, etc. If this noise increases substantially or is much over 0.02 μV, an alternate sonde should be used if available.

REPEAT INTERVALS. Repeat FFI values opposite permeable zones should differ by less than 5% of measured FFI plus one porosity unit if T2 is 50 ms or more (averages over a 10-ft interval) and hole size is 8 in. or less. Statistical uncertainty in larger holes increases approximately proportional to the square of the hole diameter. When T2 is less than 30 ms, variations may be 5% of measured FFI plus three or four porosity units. FFI opposite shales should be less than one porosity unit. Noise increases and repeatability decreases with increasing temperature.

The repeatability above reflects fluctuations in NML response due to inevitable electronic noise. Several factors can contribute to larger differences in FFI values on repeat runs:

1. The signal decay time T2 is used to extrapolate measured signal through a deadtime to get FFI. When T2 is short, underestimates have much more effect than overestimates. Underestimates of T2 can lead to unrealistically high FFI. Signal-to-noise ratio, STNR, indicates relative reliability of log results. Values with high STNR should repeat best.
2. Tool position and angle can affect NML sensitivity in an out-of-round borehole.
3. Some changes in tool sensitivity that do not show up in the calibration record give repeat runs differing by a fixed ratio, usually large enough to be easily detected.

CHOICE OF BASE POLARIZING TIMES CBPP AND SBPP (T1 MODES). These parameters are set by the logging engineer with concurrence of the company representative. They should usually be the shortest time value available, which is at present 100 ms. However, if FFE3 (CONT mode) is less than half the previously measured FFI, CBPP should be increased. Values of CBPP over 350 ms require prohibitively slow logging speeds.

CHOICE OF POLARIZING TIME NPP (FFI MODE). NPP should be long compared to T1. It is usually set at 2000 ms (2 seconds), which is longer than T1 in most fields. If T1 is longer than about 500 ms, NPP should be increased. For this reason, it is sometimes advantageous to run a short section in the T1 mode opposite high FFI formation before doing all of the FFI log. NPP should be at least four times the longest T1 measured.

Traces Presented on Logs

The traces presented on nuclear magnetism logs are identified in the following list:

FFI	Free Fluid Index	Free fluid as a fraction of formation volume
CALI	CALIper	Hole diameter in inches (is used in computation)
TENS	TENSion	Cable tension in pounds-weight
BS	Bit Size	Bit diameter in inches
T2		Signal decay time constant in milliseconds
STNR	Signal-To-Noise Ratio	Large values give more reliable results in Hertz
LFRE	Larmor FREquency	NML signal frequency observed
FFE1 FFE2 FFE3	Free Fluid Estimate #1, etc.	Free fluid readings obtained with polarizing times of one, two, and four times CBPP in the T1-continuous mode
FFIA	Free Fluid Index-A	Free fluid index calculated from FFE1 and FFE2, extrapolating signal to that for infinite polarizing time
FFIB	Free Fluid Index-B	Same from FFE2 and FFE3
NT1	Nuclear T1	T1 (longitudinal relaxation time) computed in the T1-stationary mode, in milliseconds
T1A		T1 computed from FFE1 and FFE2
TIB		T1 computed from FEE2 and FEE3 in T1-continuous mode, in milliseconds

Setup Parameters

Most of the parameters of nuclear magnetism logs are fairly standard. At present, the minimum available polarizing time is 100 ms, and this is usually the choice for CBPP and SBPP.

TD	Total Depth	hole depth in feet
SHT	Surface Hole Temperature	degrees F

WALN	Waveform Average Level of Noise (referred to as SOFTWARE-DETECTED ENVELOPE AVERAGE NOISE in the calibration summary)	microvolts
T2LO	T2 LOw	lowest value of the signal decay time constant T2 allowed, usually 20 ms
NSF	Nuclear Signal Factor	a tool calibration factor in reciprocal microvolts
NPP	Nuclear Polarizing Period	polarizing time in FFI mode, usually 2000 ms
NMOD	Nuclear magnetism log MODe	NML mode: NORM for single FFI curve; CONT for T1-continuous mode with curves for one, two, and four times the base polarizing period CBPP
LARM	LARMor frequency	operator-set value of signal frequency used in processing—LARM is set by observing signals in a good free-fluid zone using PLFT
EDEL	E DELay	delay before the waveform-computed envelope is used in determining FFI and T2—EDEL is usually 30 ms, but can be increased to avoid a fast-decaying mud signal if the formation signal is substantially longer
CBPP	Continuous Base Polarizing Period	the shortest of the three polarizing times used in the T1-continuous (CONT) mode—should usually be 100 ms
NFAC	Noise FACtor	fraction of noise power subtracted in signal computation
BS	Bit Size	bit size in inches
BHT	Bottom Hole Temperature	degrees F
PDEL	Processing DELay	delay before waveform data are used in computing software signal envelope—usually 18 ms
WSCA	Waveform SCAle factor	number of binary digits shifted during scaled portion of recorded waveform

WSD	Waveform Scale Delay	usually 100 samples at 0.1-ms intervals (should not be more)
T2HI	T2HIgh	highest value of the signal-decay-time constant allowed, usually 200 ms
NALN	Nuclear Average Level of Noise	noise determined from hardware (analog signal detection) envelope, as opposed to WALN determined from the wave forms—microvolts
MINC	Magnetic INClination	angle between the vertical and the Earth's magnetic field
DEVI	DEVIation	angle between borehole axis and vertical
AZIM	AZIMuth	angle between *magnetic* north and vertical plane containing deviated borehole
CCUR	Cable CURrent	polarizing current in milliamps
SBPP	Stationary Base Polarizing Period	shortest time in T1-stationary mode—usually 100 ms
SPDN	Stationary Pulse Duration Number	number of different polarizing times (pulse durations) in T1-stationary mode. The times vary by factors of two. SPDN is usually 6. If SBPP is 100 ms, the polarizing times are 100, 200, 400, 800, 1600, and 3200 ms
NSPN	N Stationary Pulse Number	number of signals stacked at each polarizing time in T1-stationary mode—usually 8
NSDP	N Stationary Dead Period	time allowed for polarization to die away after each signal in T1-stationary mode, about 3000 ms
PPER	Polarizing PERiod	polarizing times in milliseconds actually measured in T1-stationary mode (as opposed to the round-numbered 100, 200, 400, etc., ms nominal times)
ESTI	ESTImator	type of signal estimate computation—usually least squares (LS)
SPEE	SPEEd	logging speed in feet per hour
SFW	Search Frequency Width	frequency range, centered at LARM, searched in PLFT—usu-

		ally 200 Hertz. LARM may then be reset to the frequency determined.
BAND	BANDwidth	used to control the software narrow band filter for waveform processing—default value is 50 Hz, often set to 30 Hz
WDC	Waveform Depth Constant	each signal is averaged with the WDC preceding ones and the WDC following ones—the usual value is two, giving an average over five signals
PREM	Pulse REMoval	for removal of occasional spikes on NML signals—usually ALLOwed
ELIM	ELIMination	elimination of wild signals in waveform depth averaging—usually ALLOwed
PLFT	Playback Larmor Frequency Task	ALLOwed for determining Larmor frequency in order to set LARM: DISAllowed in normal logging
RLFC	Real-time Larmor Frequency Computation	ALLOwed for computation of individual signal Larmor frequency LFRE

THE BOREHOLE GRAVIMETER (BHGM)

Principle of Measurement

By measuring the acceleration due to gravity g at two different stations in a well the density of the slab of rock between these stations can be calculated. Since the variations in g due to rock density are very small, a very sensitive device is required. The nominal value of g at the surface of the Earth is 980 cm/s^2 or 980 gals. To be of practical use, a BHGM tool needs to measure microgals. Given that such measurements can be made in an accurate and repeatable fashion, the average density of a layer of rock between two points in a well can be calculated. Figure 6.12 illustrates the principle.

The further apart the two measurements are made the greater the accuracy of the calculated result. For example, if the difference in g between two stations, Δg, is measured to an accuracy of 7 microgals, then the corresponding accuracy for the calculated density of the layer of rock encompassed between those two stations will be 0.028 g/cc, if the spacing is 10 ft, but 0.014 g/cc if the spacing is 20 ft. This interplay of tool accuracy (sensitivity), station spacing, and detectable density variation is illustrated in figure 6.13.

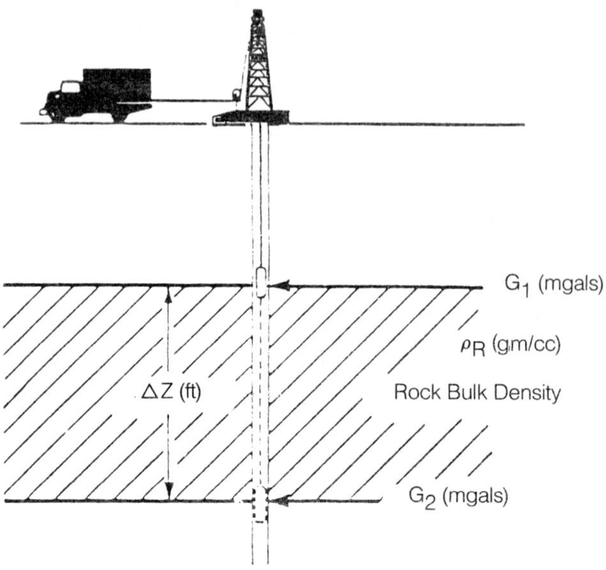

FIGURE 6.12 *BHGM Principle. Courtesy Edcon.*

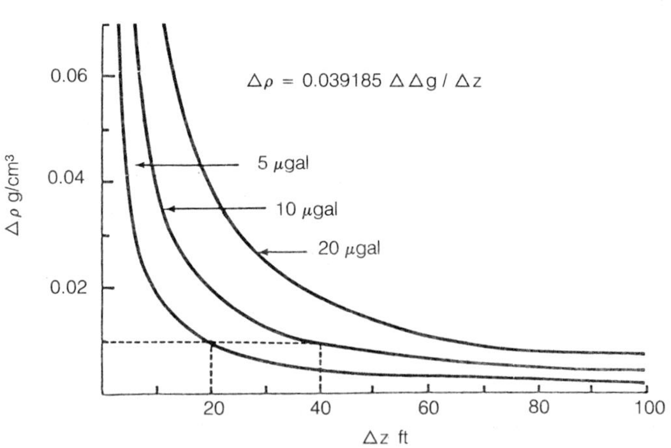

FIGURE 6.13 ***Relationship Between Δz, Δg, and Accuracy of Density Estimate. Courtesy Edcon.***

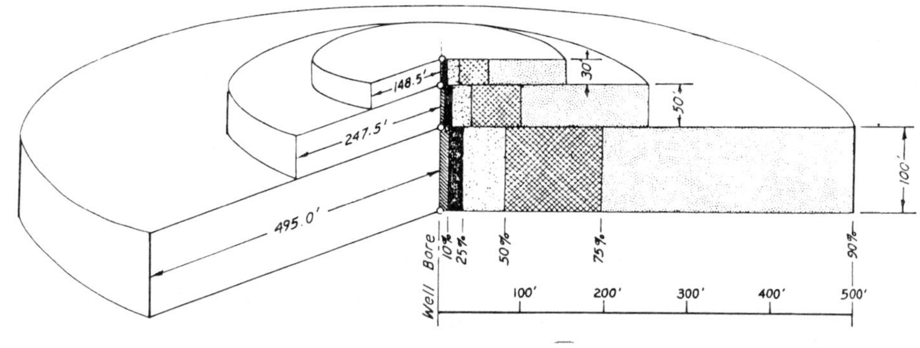

FIGURE 6.14 *Zones of Cumulative Gravity Effect for the BHGM. From McCulloh et al. 1968.*

The volume of rock investigated by BHGM surveys is a function of the spacing between stations. Short spacing measurements investigate small rock volumes and vice versa (see fig. 6.14). If measurements are made at the top and bottom of a slab of formation 100 ft thick then 90% of the measured Δg will come from within an annulus round the borehole of radius 500 ft. For a 30-ft station difference, the 90% response is from within a radius of 150 ft. A rough rule of thumb is that the BHGM reads out to 5 times the spacing between stations.

At all events, the tool investigates a very large volume of rock compared to a conventional formation density tool, which reads a few inches at most into the formation.

Applications

There are currently two main applications for the BHGM: (1) obtaining formation density in old wells completed but unlogged by a modern logging suite, and (2) detecting lithology, porosity, and fluid changes in the formation some distance from the borehole.

Examples of type (1) applications would be detection of gas zones in an old well that only has an electric log. At the time these wells were logged and completed, gas production was not an economic proposition. Now that it is, the question remains of how to distinguish gas-bearing zones seen on the old ES log as high-resistivity zones from low-porosity tight zones that have the same high resistivity. A BHGM survey can determine formation density over 10- to 20-ft intervals, and gas-bearing zones are likely to show densities closer to 2.0 g/cc than the 2.5 g/cc or more shown in tight zones.

Examples of type (2) applications would be the detection of better porosity or gas some distance from an otherwise dry well. The BHGM has been particularly successful in the Niagara Reef plays in Michigan. If the density distant from the borehole is calculated to be less than the density indicated by the conventional density log, then the well may be *fraced* (artificially fractured by pumping in frac fluid under

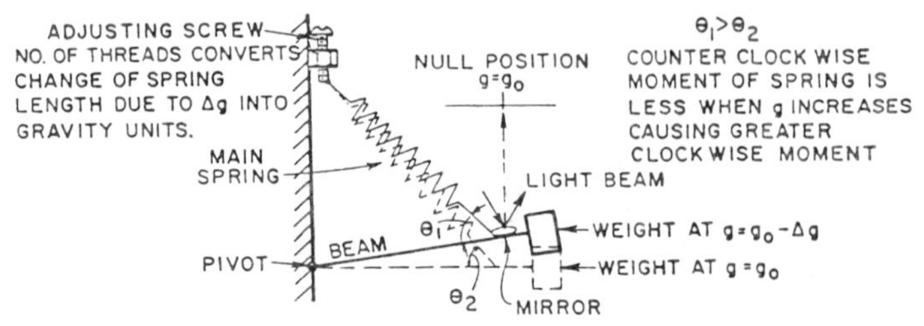

FIGURE 6.15 *The LaCoste Romberg/BHGM Device. Courtesy Edcon.*

high pressure) over the more attractive gas-bearing or higher-porosity zones.

Tools Available

There are two types of commercially available BHGM tools: (1) the vibrating-string type as pioneered by Texaco, and (2) the LaCoste-Romberg zero-length-spring type developed by the USGS and a six-company consortium.

The LaCoste-Romberg device is the one used most frequently due to its superior accuracy, repeatability, and temperature rating. The remainder of this section will refer to this device, illustrated in figure 6.15.

Special BHGM Requirements

Before undertaking a BHGM survey, several important considerations need to be taken into account:

1. Tool size is 5½ in. OD. Do not attempt a survey in a hole less than 7⅞ in. ID or in casing smaller than 6⅝ in. OD.
2. Maximum temperature allowable is 250°F. The tool is held at constant temperature. If that temperature is exceeded the tool cuts out and no further measurements can be made.
3. The logging truck should be equipped with no more than 18,000 ft of Teflon insulated cable, and each conductor must be checked for insulation at 5000 MΩ or better.
4. Maximum hole deviation allowed is 7°.
5. Give as much advance notice as possible when calling for BHGM service since the tool requires special handling and experienced operators.
6. Plan on a dummy run to TD with the empty tool housing, as a precaution against tight spots and tool sticking. If any are found, recondition the well with a wiper trip.
7. Plan on a slow descent to TD at an even speed and a 45-minute stabilization period. Thereafter plan on 10 minutes/station.
8. If high accuracy is required, each measurement must be repeated.

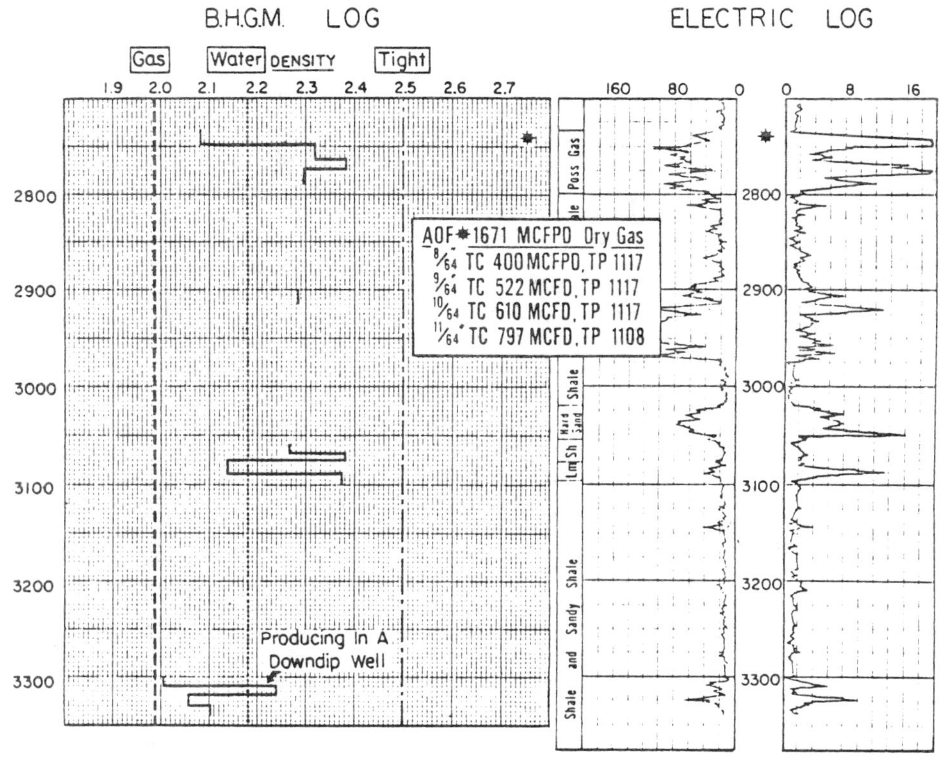

FIGURE 6.16 *Gas Detection by BHGM in Old Well.*

Errors in depth then become significant and special steps have to be taken to precisely pinpoint each station depth by marking the cable.

9. Postsurvey calculations can be long and involved and may require corrections for tide, drift, borehole washout (if severe), terrain, hole deviation, etc.

Field Examples

Figure 6.16 illustrates both an electric log and a BHGM survey in a Gulf Coast Miocene sand/shale series. Of interest are the two high-resistivity kicks seen at 2735 to 2750 ft and 2765 to 2780 ft. Either could be hydrocarbon bearing *or* tight. The BHGM survey successfully predicted gas production from the upper sand on the basis of its calculated density of 2.08 in contrast to the 2.38 density in the lower sand. The well was perforated in the upper sand for an AOF of 1.7 MMCFD.

Figure 6.17 illustrates a carbonate well in which a featureless zone (6732 to 6750 ft) on the FDC log was successfully completed for 1.5 MMCFD due to the disparity between the BHGM density of 2.58 and the FDC density of 2.72.

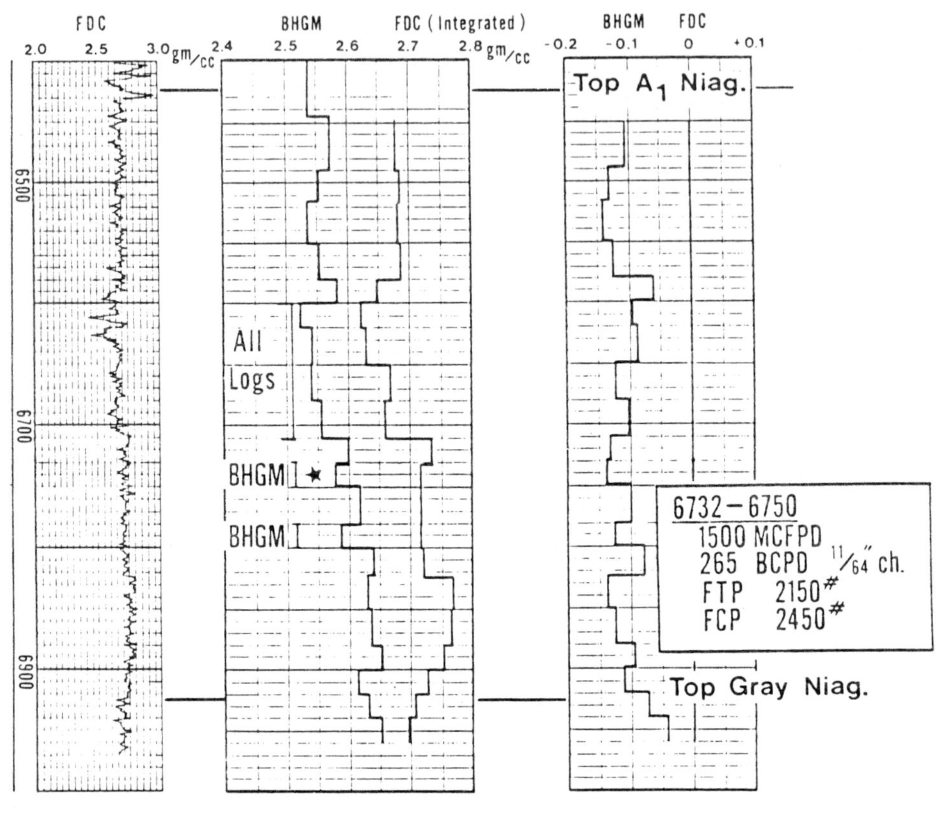

FIGURE 6.17 *Successful Completion in a Carbonate Reservoir.*

SIDEWALL CORING

Sample taking involves bringing a physical sample of a formation to the surface for examination. Some earlier means of sample taking, such as the sidewall core slicer, are no longer commercially available. The most commonly used mechanism for taking samples today is the sidewall core gun, which carries explosive-propelled cylinders (bullets) that are fired into the formation and then retrieved by means of cables connected to the gun. At the surface, the formation samples are pressed from the cylinders into small jars, which are then sealed and labeled by depth. Another device, the core plugger, is available for hard-rock sampling. Note that conventional sidewall cores from core guns are of necessity damaged by the process. In general, porosity is overestimated from these samples, and permeability underestimated, although grain density is correct. The core plugger, however, retrieves samples without damaging them and thus core porosities and permeabilities are much closer to the true values.

TABLE 6.3 *Sidewall Coring Tools*

	Gearhart	Schlumberger	Dresser Atlas	Welex
Core guns				
Min. length (ft)	6.67	9	10.3	10
Max. length (ft)	17	24	19.1	21
Diameter (in.)	4.0	4⅜-5¼	4.0	4⅞-5
Cores per trip	52	72	60	60
Core diameter (in.)	1⅛	⅝ to 1⅛	1 1/16 to 1⅛	¾ to 1
Max. temperature (°F)	350	430	350	350
Max. pressure (psi)	20,000	20,000	20,000	20,000
Core plugger				
Length (ft)	19.5			
Max. OD (in.)	5			
Max. pressure (psi)	10,000			
Max. temperature (°F)	350			
Core dimensions (in.)	⅝ × 2¼			
Cores per trip	3 for 60° angle cut 10 for horizontal cut			

When to Take Samples

Sidewall core sample taking should be considered for the following reasons:

1. Oil or gas shows in the formation can be determined from sidewall core samples.
2. Under most soft-rock conditions, an analysis similar to regular core analysis can reveal porosity, permeability, and fluid saturation subject to the proviso mentioned above regarding damage.
3. Fluid compatibility and solubility can be determined for use in stimulation design for the formation.
4. Detailed laboratory studies can be made when a good sample is obtained.

In most hard-rock areas, sample taking is not successful. Some soft formations require special core bullets or powder charges. When sample taking is being ordered, estimated porosity, type of rock, depth, and hole size must be stipulated to the service company. When sample recovery is particularly critical, extra shots should be requested so that "double shooting" will be possible in those critical zones. The service company should be made aware of the critical zones and charged with the responsibility of using their best techniques to effect recovery.

Tools Available

CONVENTIONAL SIDEWALL CORE GUNS. The commonly available sidewall coring tools and their manufacturers are given in table 6.3. A typical core gun is illustrated in figure 6.18. Note that the SP electrode for

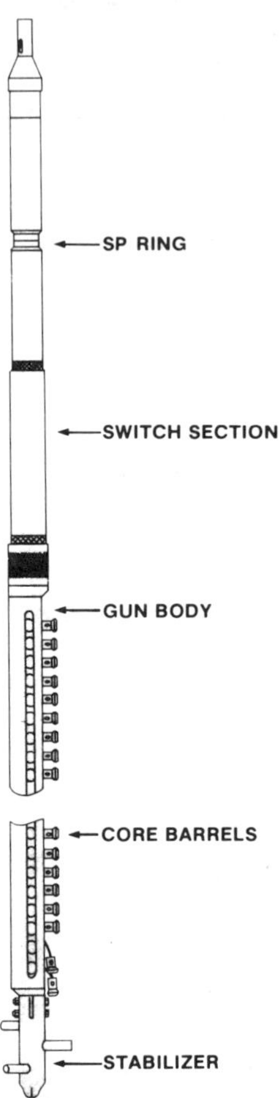

FIGURE 6.18 *Typical Sidewall Core Gun. Courtesy Welex, a Halliburton Company.*

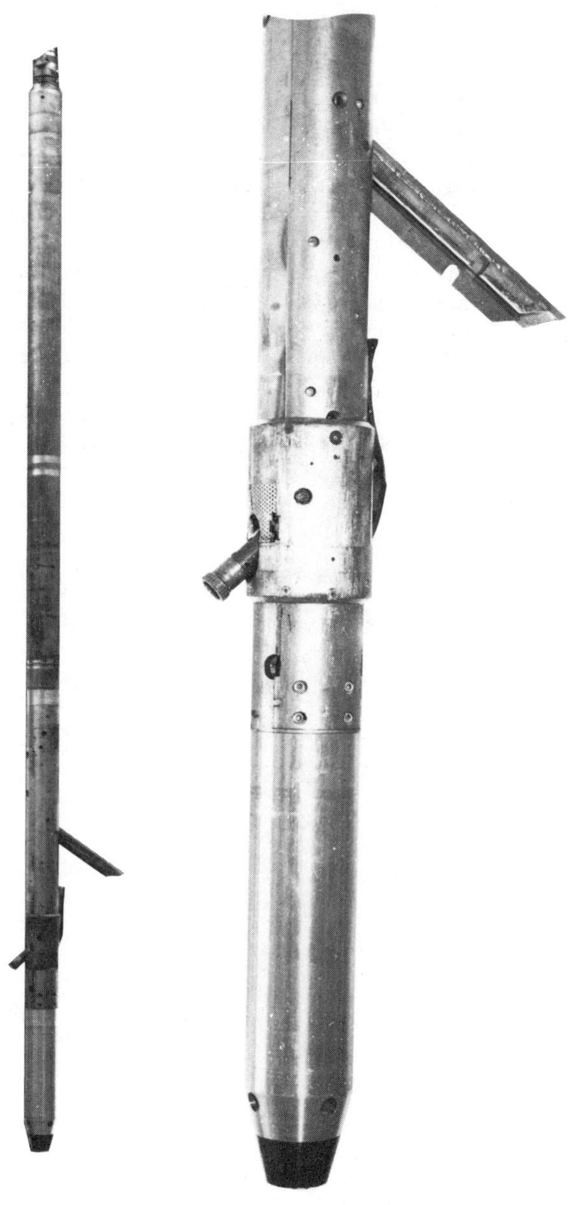

FIGURE 6.19 *Core Plugger.*

depth control may be on the gun itself or may be positioned on the logging-cable bridle.

THE CORE PLUGGER. Figure 6.19 illustrates a core plugger, a device that uses a motorized bit to cut cores one at a time and carry them back to the surface for analysis. This tool, engineered by Amoco, is now available from Gearhart. Pertinent details are given in table 6.3.

Quality Control

All samples should be pressed from the core bullets into a container and sealed as soon as possible. Care must be exercised to insure that each sample is marked with the correct depth and lease name.

Sometimes the steel sample bullets are lost in the formation or pulled off in the hole. Notes should be kept of all missing bullets and the depth at which they were lost. This information may be of value if the bullets interfere with subsequent well operations.

CHAPTER SEVEN

COMPUTED LOG PRODUCTS

PROPRIETARY COMPUTED LOGS

This section is intentionally left blank so that the user may file here any material that is pertinent to proprietary computed logs products that the user may have access to.

INDUSTRY COMPUTED LOG PRODUCTS

Overview

Log analysis performed by a computer is available to the log user at three levels:

1. Real-time, "quick-look," products run at the same time the log is being run include R_{wa} curves, "F" logs, etc.
2. Wellsite analysis is made after the logging is completed by playing back taped logs and using an appropriate log analysis program, such as a shaly-sand analysis or a dipmeter computation.
3. Computing-center products are provided well after the logging is completed (days or weeks later), and are generally more comprehensive than either of the wellsite products.

In general, quality control for all computed logs has the following requirements:

a. Check that all curves are "on depth" with each other before any computation is performed.
b. Check that all input parameters (such as R_w, shale picks, etc.) are reasonable.
c. Check that the interpretation model used fits the geology of the formations logged.

Wellsite "Quick-Look" Logs

Real-time logs generate presentations that aid the log analyst, the driller, the completion engineer, and the geologist in making decisions on future actions with the well. The products generally available are summarized in table 7.1. Three of these presentations—"F" log, R_{xo}/R_t vs. SP, and borehole volume integration—are illustrated in figures 7.1, 7.2, and 7.3 respectively.

Wellsite Log Analysis

Table 7.2 summarizes the products generally available at the wellsite for log analysis in replay time. These computations, if correctly made using an appropriate model and accurately picked parameters, largely reduce the need for any further detailed log analysis.

THE CROSSPLOT LOG.* The crossplot (X-Plot) log combines data from all tools logged, displays their responses on a five-track log, performs all the tedious crossplotting of information using complex mathematical formulas, and prints a display of data used for computation (see fig. 7.4). Computed curves can include crossplotted porosity (ϕ_{xp}), apparent delta travel time through the matrix (Δt_{map}), apparent water resistivity (R_{wa}), apparent matrix density (ρ_{ma}), formation resistivity (R_w)

*The following information on the crossplot log, the well evaluation log, and Cyberdip logs has been provided, with permission, by Gearhart Industries, Inc.

TABLE 7.1 *Wellsite Analysis Available in Real Time*

Generic Name	Derivation	Log Input Required	Presentation	Gearhart	Dresser	Schlumberger	Welex
R_{wa}	Assumes all fms. contain 100% water. Computes apparent R_w. $R_{wa} = R_t/_F$	Simultaneous resistivity and porosity. Usually sonic and induction	Single curve in Track 1 on logarithmic scale	R_{wa}	R_{wa}	R_{wa}	R_{wa}
R_{xo}/R_t	Uses deep resistivity and shallow focused resistivity to estimate R_{xo}/R_t ratio	Simultaneous deep resistivity and shallow focused resistivity	Single curve in Track 1 compatibly scaled as a pseudo SP curve	R_{xo}/R_t	R_{xo}/R_t	R_{xo}/R_t	R_{xo}/R_t
"F" overlay or R_o overlay	Derives F from a porosity curve which is played onto logarithmic resistivity as R_o	Deep resistivity and porosity	Single dashed curve in Tracks 2 and 3 on logarithmic scale	R_o	"F" curve	R_o	"F" curve
Compatible porosity scales	Any combination of porosity logs using same lithology assumptions to compute porosity	Simultaneous porosity logs	Coded curves in Tracks 2 and 3 with Gamma ray and caliper in Track 1	Compatible porosity scales	Compatible porosity scales	Compatible porosity scales	Compatible porosity scales
Hole volume	Uses caliper logs to compute hole volume for cement calculation	Caliper curve—preferably 2 curves at 90° such as 4-arm dipmeter	Pips or tic marks in depth column at every 10 ft^3 and 100 ft^3	Borehole volume	Borehole volume	Borehole volume	Borehole volume
Fracture-locating log	Uses differences in adjacent pad readings from 4-arm dipmeter to infer fractures	4-arm dipmeter	Adjacent pad readings are superimposed and any separation is coded	Fracture detector log	Fracture location log	Fracture identification log	Dipmeter fracture log

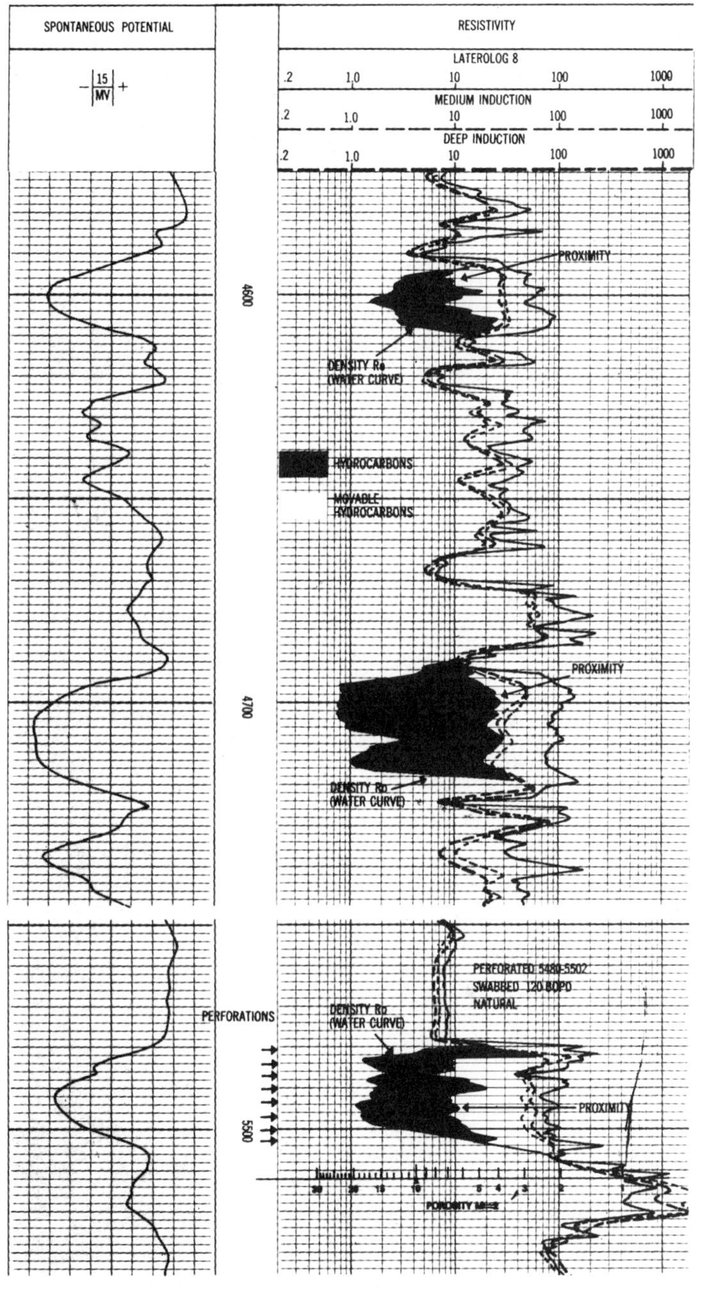

FIGURE 7.1 *"F" Log Presentation. Courtesy Schlumberger Well Services.*

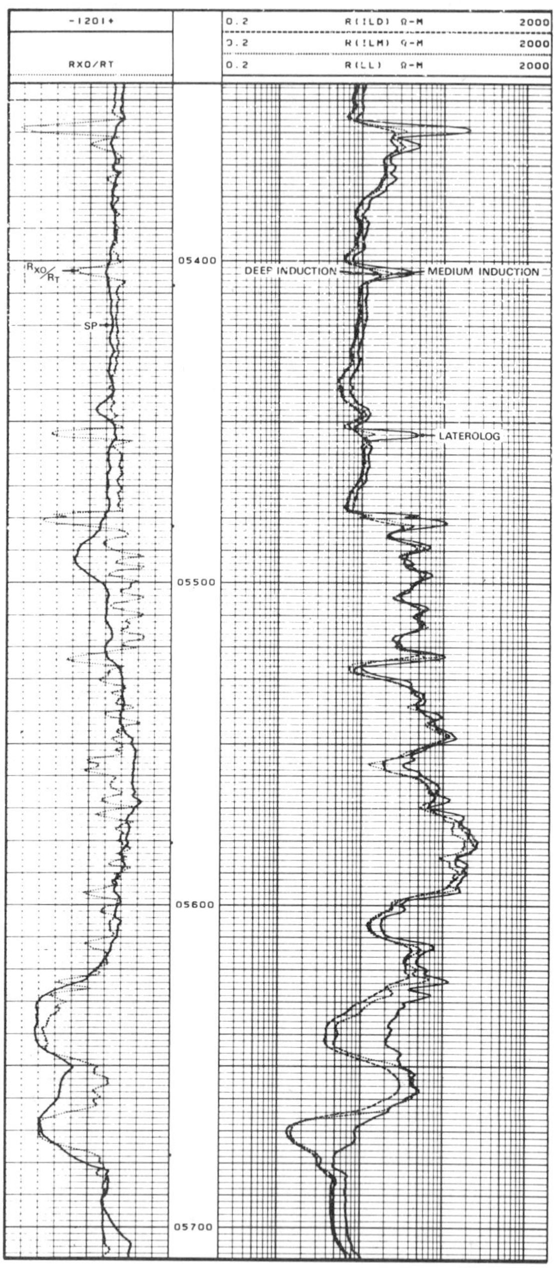

FIGURE 7.2 ***R_{xo}/R_t vs. SP Presentation. Courtesy Gearhart Industries, Inc.***

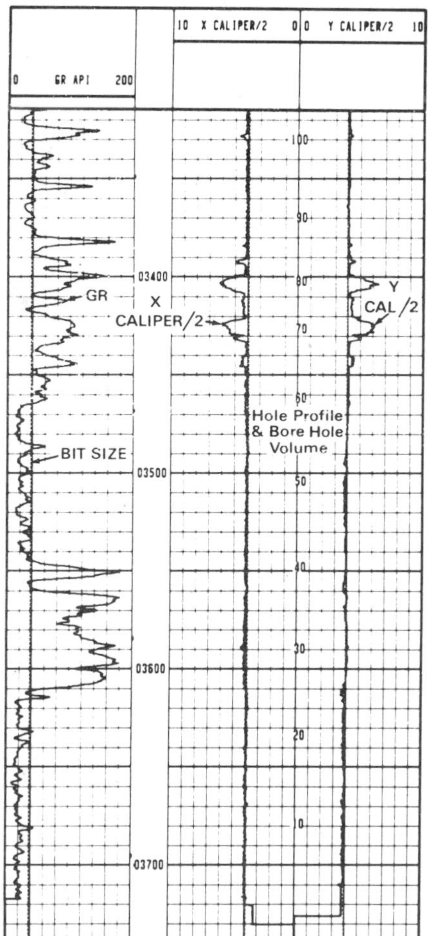

FIGURE 7.3 *Borehole Volume Integration. Courtesy Gearhart Industries, Inc.*

from the SP, etc. (Since there was no sonic tool run in the well from which the example log in figure 7.4 was taken, no Δt_{map} is displayed.) The X-Plot also provides instantaneous verification that all tool responses were recorded on depth as well as a check of log quality.

THE WELL EVALUATION LOG. The well evaluation log (WEL) gives a data display and a log that uses information from individual porosity and resistivity tools and/or crossplot data integrated, with additional refinements, into the computer computations. The displayed log is a simply read visual interpretation of the well's potential (see fig. 7.5).

Track 1 shows formation characteristics with a graphic display of crossplotted porosity as well as the matrix/shale percentages. The

TABLE 7.2 *Wellsite Analysis Available in Replay Time*[a]

Generic Name	Derivation	Log Input Required	Presentation	Gearhart	Dresser	Schlumberger	Welex
Merged and depth-shifted data	Replays all logs, shifts depths and makes simple calculations such as R_{wa}, R_{xo}, R_{xo}/R_t, compatible porosity scales, or crossplot porosities	Any logs run on tape	Usually 3 to 5 tracks (varies by service company), displays only log data	CrossPlot (X-plot)	Prolog	Cyber look Pass 1	
Wellsite evaluation log	Uses all logs to provide a first-order computer analysis	Resistivity and porosity	Usually 3 or 4 tracks, has reservoir data derived from log data	Well-Evaluation Log	Prolog	Cyber look	CAL
Formation dip computations	Computes formation dip from 4-arm dipmeter	4-arm dipmeter	Formation dip, hole deviation, calipers	FED DDL	Pro-Dip	Cyberdip	
True vertical depth log	Computes TVD of any point from dipmeter orientation data	Continuous dipmeter plus any log to be converted to TVD	Replay of any log on TVD depth scale	TVD	TVD	TVD	

[a]All logs available in real time are also available in replay time if recorded on magnetic tape.

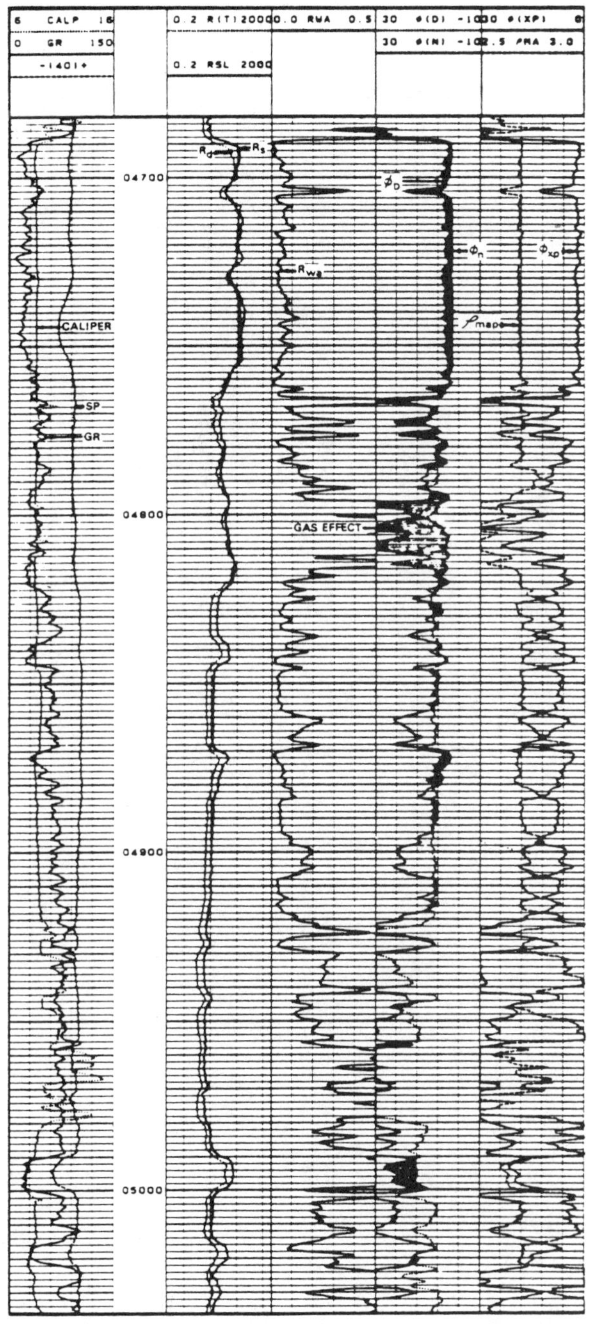

FIGURE 7.4 *"Cross-Plot" Analysis. Courtesy Gearhart Industries, Inc.*

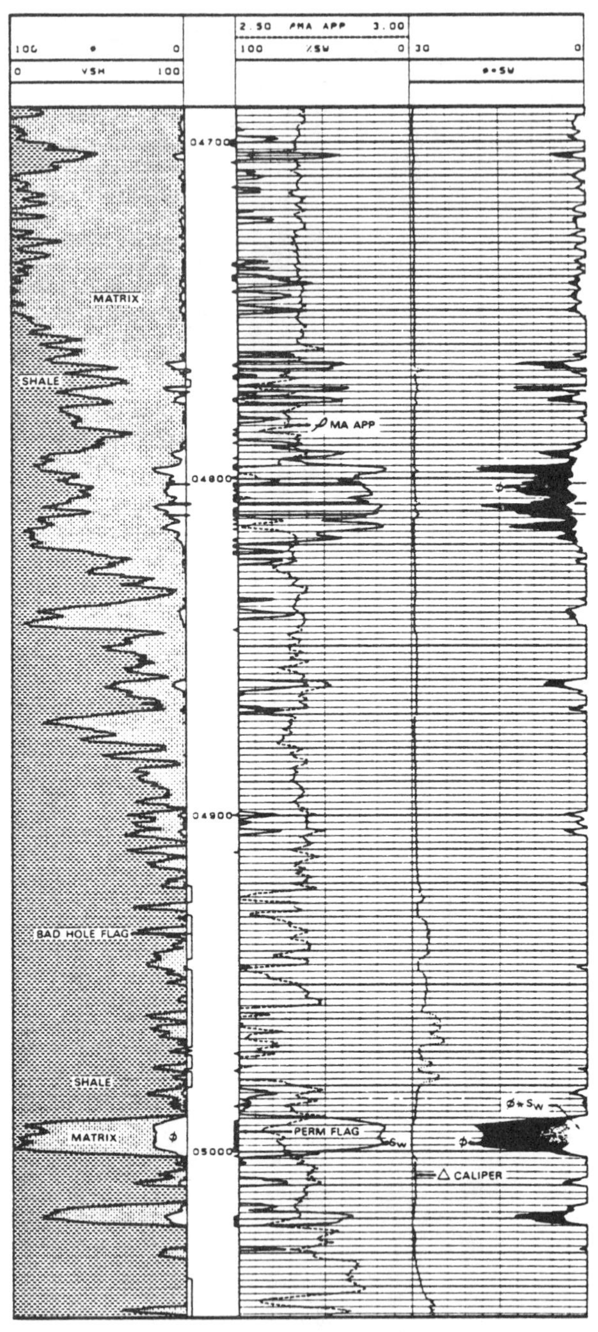

FIGURE 7.5 *Well Evaluation Log. Courtesy Gearhart Industries, Inc.*

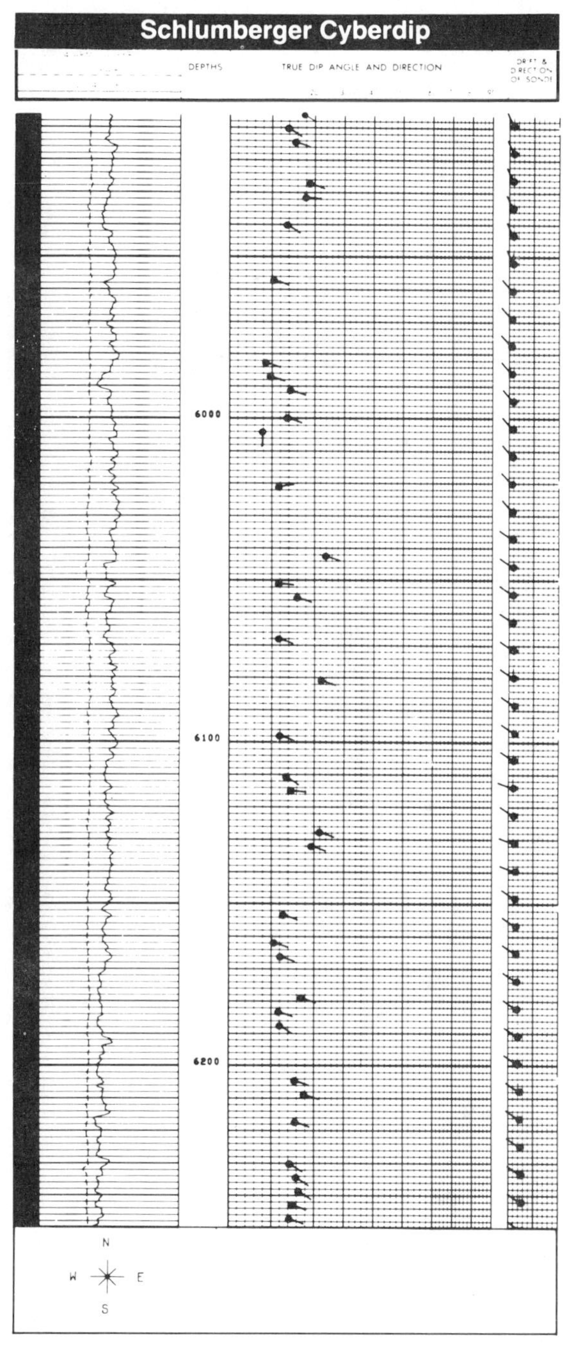

FIGURE 7.6 *Cyberdip. Courtesy Schlumberger Well Services.*

TABLE 7.3 *Computing Center Log Analysis Products*[a]

Generic Name	Derivation	Log Input Required	Presentation	Dresser	Schlumberger
Advanced sandstone analysis	Uses most sophisticated analytical and statistical methods to correct and compute logs in sandstones and shaly sandstones	Resistivity, density, neutron, gamma ray with sonic desirable	Usually 4 tracks; presentation of lithology, saturation, porosity, and bulk volume	Epilog (sandstone analysis)	SARABAND VOLAN
Advanced carbonate analysis	Uses most sophisticated analytical and statistical methods to correct and compute logs in carbonate and lithologically complex reservoirs	Resistivity, density, neutron, gamma ray with sonic and microresistivity desirable	Usually 4 tracks; presentation of lithology, saturation, porosity, and bulk volume	Epilog (complex reservoir analysis)	CORIBAND
Advanced dipmeter computations	Uses most advanced correlation logic to compute dips followed by a statistical sorting to retain the most reliable data	4-arm dipmeter	Arrow plot with caliper, correlation curve and hole deviation. Also available: azimuth frequency diagrams, modified Schmidt plots, stickplots, histograms, and listings	Dresser Computed Dipmeter	CLUSTER (structural) and GEODIP (stratigraphic)

[a]Log analysis is available from company computing centers.

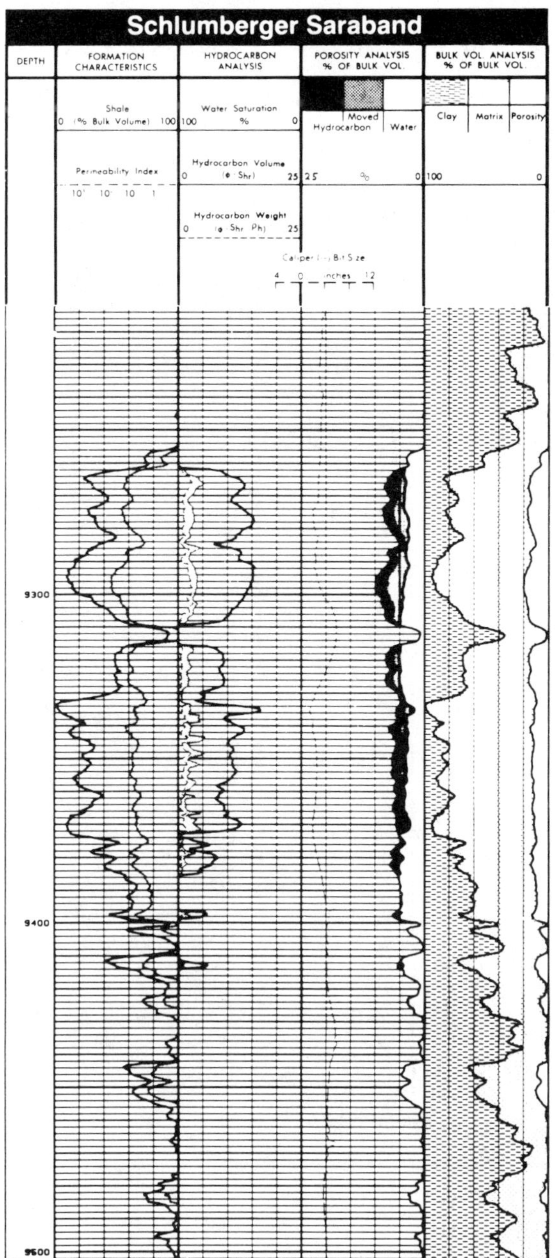

FIGURE 7.7 *Shaly Sand Analysis. Courtesy Schlumberger Well Services.*

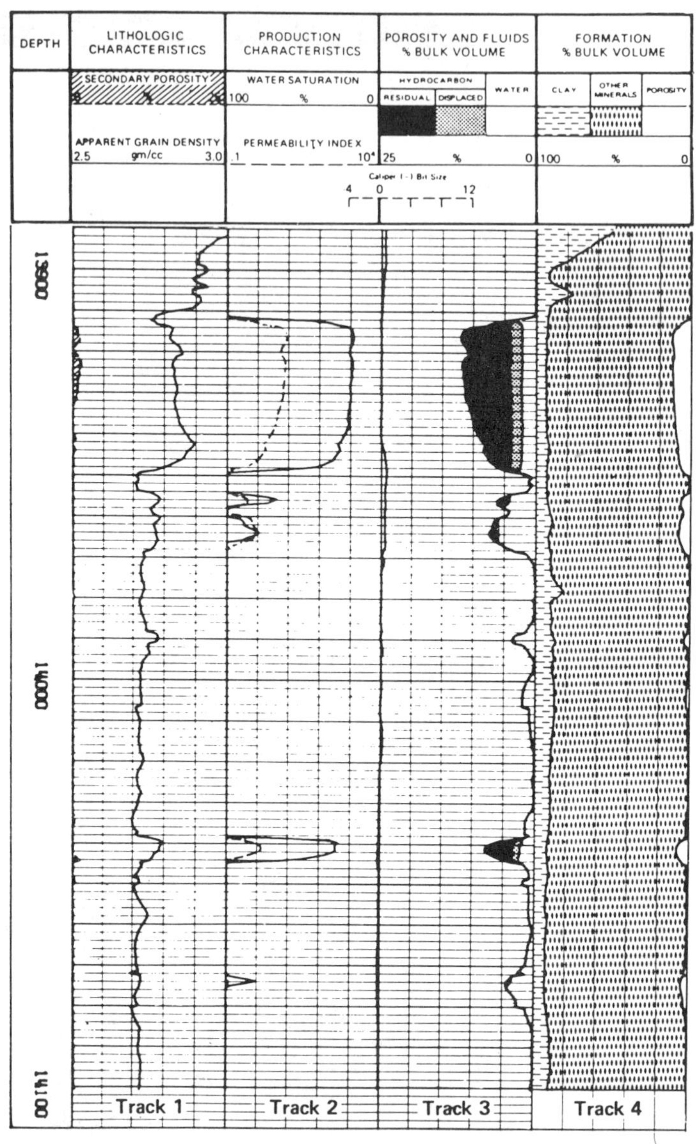

FIGURE 7.8 *Complex Lithology Analysis. Courtesy Dresser Atlas.*

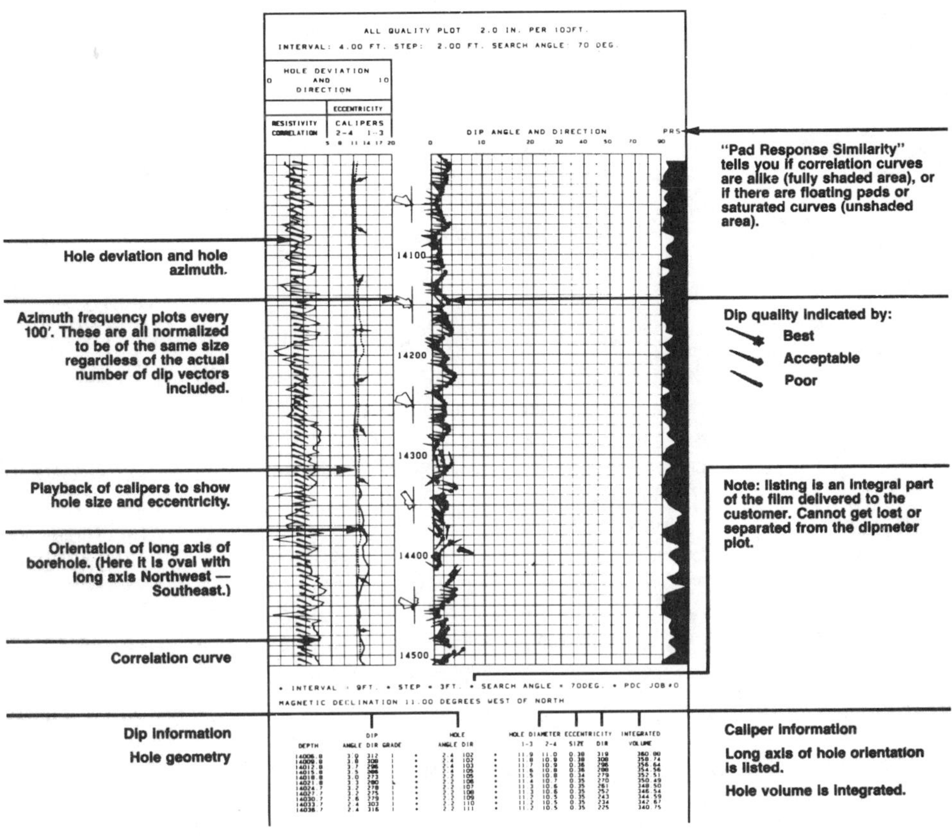

FIGURE 7.9 *Dipmeter Analysis. Courtesy Vizilog, Inc.*

presence of coal, if any, can be shown on the log as black streaks across Track 1 plotted vs. depth. Flags along the left side of the depth column indicate bad hole conditions, which preclude presentation of otherwise erroneous information. Flags on the right side of the depth column show permeable zones from microlog data. Track 2 is for fluid analysis and shows the percentage of water saturation (S_w) as computed from input data. Track 3 gives bulk volume analysis of the pore space by displaying total crossplot porosity, with the hydrocarbon and water content, graphically. A differential caliper curve is traced between Tracks 2 and 3.

Different parameters can be entered for each lithological zone to insure that accurate information is displayed throughout the entire logged interval. All input data is printed on the log for future reference.

CYBERDIP LOGS (MARK OF SCHLUMBERGER). Dipmeter computation can be requested wherever CSU equipment is in service. It makes use of the same logic that has been used for many years in computer-center processing of dipmeters, though the Cyberdip program is necessarily simpler and less flexible.

Like other quick-look logs, Cyberdip processing is geared to the need for wellsite decisions. With reasonable hole conditions and deviations, the Cyberdip results produce reliable definition of structural dip and major geological features.

The Cyberdip log is output as an optical film and a magnetic tape. The graphic log is in the familiar arrow-plot configuration (see fig. 7.6).

Computing Center Analysis

More complex analysis is available from service-company computing centers when time is not critical and if the formation evaluation problem merits the additional cost. In general, these analyses fall into three categories: (1) shaly-sand analysis, (2) complex lithology analysis, and (3) dipmeter processing.

The most widely used computer-center products are summarized in table 7.3. Other less frequently used products such as tar-sand analysis, mechanical properties, etc., are not included in the table; details may be obtained directly from the service companies. Figures 7.7, 7.8, and 7.9 illustrate, respectively, examples of a shaly-sand analysis, a complex lithology analysis, and a dipmeter analysis.

CHAPTER EIGHT

PRODUCTION LOGGING

PROFILE LOGGING

Application

Profile logging may be used (1) to monitor injection rates in injection wells; (2) to monitor production rates in producing wells; and (3) in either of the above, to detect casing, tubing and/or packer leaks, channeling behind pipe in poorly cemented zones, etc. Although some tools can handle both environments, some can be used only for injection profiling.

In general, profiles may be obtained without disturbing dynamic well behavior by using the proper pressure control equipment and operating techniques. That is, logs can, and should, be run through tubing without having to kill the well or pull the tubing.

Before attempting to obtain a profile log, plan the operation in advance with the service company, paying particular attention to the following:

expected flow rate
casing and tubing size, type, and weight
expected wellhead pressure
type of christmas tree connections
tubing restrictions
corrosive or poisonous production fluids
completion records
openhole logs

Tools Available

Profiling tools used for measurement of fluid flow rates fall into three major categories: (1) continuous flowmeters (including fullbore flowmeters), (2) packer flowmeters, and (3) radioactive tracers (velocity and tracer modes). Some of the available tools are described in table 8.1 with their respective limitations. Three types of flowmeters are illustrated in figure 8.1. A radioactive tracer tool and its associated survey are shown in figure 8.2.

Limitations

Various factors influence the accuracy of fluid flow rate measurements:

the number of commingled phases
the well deviation

TABLE 8.1 *Profiling Tools Available*

	Flowmeters					
	Gearhart	Dresser	Schlumberger			Welex
	Continuous Flowmeter	Continuous Flowmeter	Continuous Flowmeter	Packer Flowmeter	Full Bore Flowmeter	Continuous Flowmeter
Tool diameter (in.)	$1\frac{11}{16}$ and $1\frac{7}{16}$	$2\frac{1}{8}$ and $1\frac{11}{16}$	2 and $1\frac{11}{16}$	$2\frac{1}{8}$ and $1\frac{11}{16}$	Variable to casing ID	1 and $\frac{11}{16}$
Max. temperature (°F)	300	300	350	285	350	300
Max. pressure (psi)	15,000	18,000	15,000	10,000	15,000	15,000
Min. flow rates (B/D)	150	200	300	10	50	200
Max. flow rates (B/D)	60,000	60,000	60,000	1,900	50,000	60,000
	Radioactive Tracer Tools					
Tool diameter (in.)		$1\frac{1}{2}$		$1\frac{11}{16}$		
Max. temperature (°F)		350		275		
Tracer capacity (cc)				20		
Number of ejections				150–200		

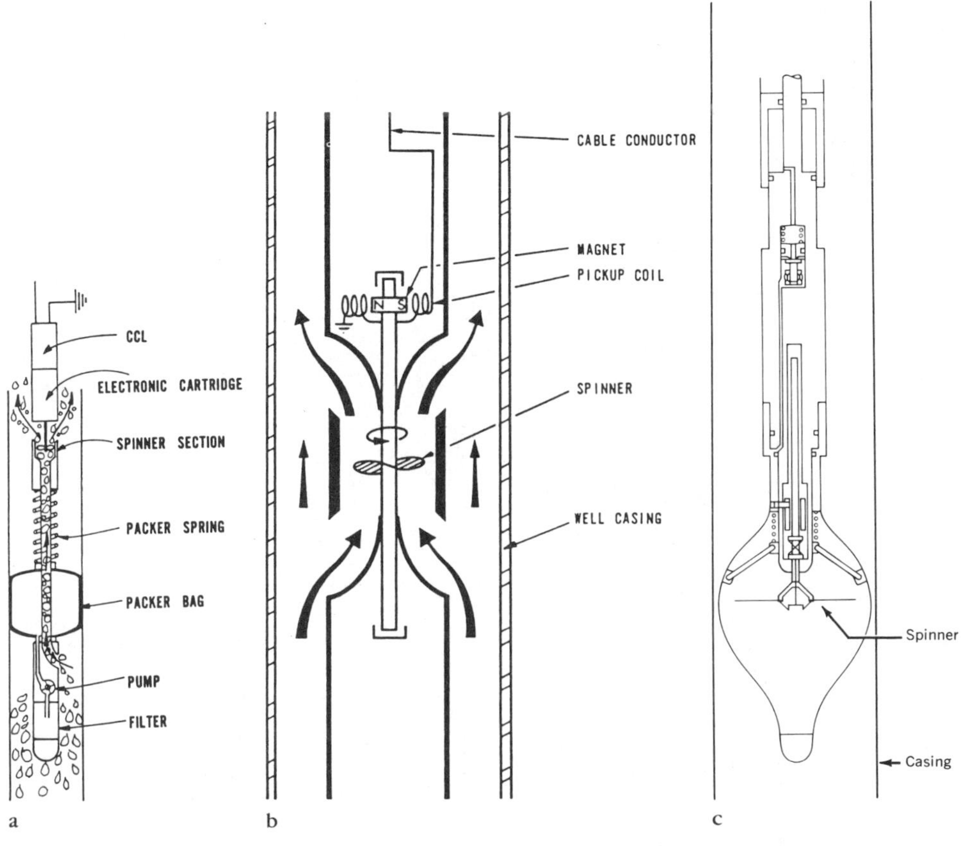

FIGURE 8.1 *Flowmeters: (*a*) Packer, (*b*) Continuous, and (*c*) Fullbore. Courtesy Schlumberger Well Services.*

the type of tool and the way it is run
hole diameter variations
production/injection-rate variations

Greater confidence in results can be expected when there is only one phase flowing (oil or water or gas), when the well is vertical, and when the appropriate tool is used for the particular well conditions. A lesser degree of confidence can be placed in results in deviated wells, under conditions producing froth or slug flow, in wells that are heading, and where the design limitations of the tools are exceeded (e.g., continuous flowmeters in low flow rate wells). For safety reasons, radioactive tracer surveys should be run only in injection wells.

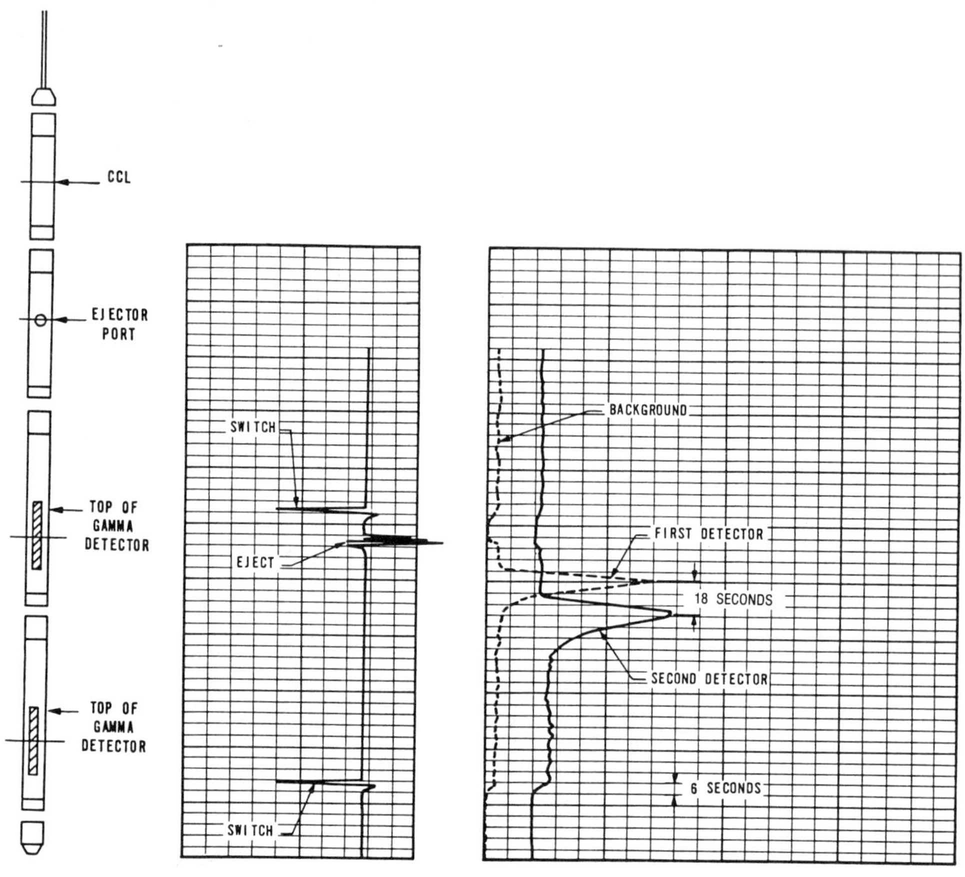

FIGURE 8.2 *Radioactive Tracer Tool and Survey. Courtesy Schlumberger Well Services.*

Log Quality Control

PACKER FLOWMETERS

a. At each station check for proper packer inflation by slacking off a few feet of cable and observing cable tension dropoff.
b. Repeat each station measurement.
c. Be sure that the spinner OD and pitch are known so that the correct interpretation chart can be used.
d. Make at least one station above all perforations where flow rate is known or can be calculated from surface flow rate and PVT data.

CONTINUOUS FLOWMETERS

a. Make several passes both logging up and logging down at different cable speeds.
b. Keep cable speeds constant on each pass.
c. If logging in an open hole completion, an accurate caliper measurement will be needed for interpretation.

RADIOACTIVE TRACERS

a. Observe all safety procedures in handling radioactive materials.
b. Double check ejector-to-detector distances.

Field Examples

A fullbore flowmeter survey run at three different cable speeds is shown in figure 8.3. A production profile made from a flowmeter survey is shown in figure 8.4, and figure 8.5 illustrates a radioactive tracer survey made in time-lapse mode. (Note the final destination of the released tracer material in fig. 8.5.)

FLUID IDENTIFICATION

Applications

Production logging tools that can differentiate between oil, gas, and water in a producing well allow diagnosis of a number of completion problems, better understanding of reservoir performance, and monitoring of secondary and tertiary recovery projects. In particular they help (1) to pinpoint gas, oil, and water entries into, and exits from, the production string and, (2) to determine, in combination with flow measurements, how much of which fluid is produced from which horizon.

Tools Available

The many tools available for distinguishing one type of fluid from another may be classified into four groups (see table 8.2):

1. fluid density measurements
2. fluid dielectric constant measurements
3. recovery of a fluid sample at the flowing pressure of the well
4. frequency spectrum of noise generated by fluid flow

Examples

FLUID DENSITY. Two commonly used devices for measuring fluid density are (1) the Gradiomanometer (mark of Schlumberger) (see fig. 8.6), which measures the pressure difference in the wellbore between

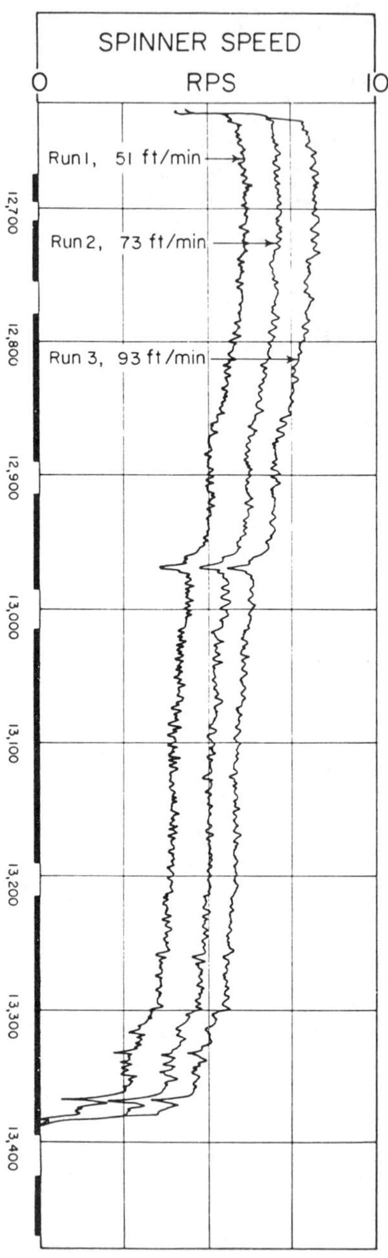

FIGURE 8.3 *Fullbore Flowmeter Survey. Courtesy Schlumberger Well Services.*

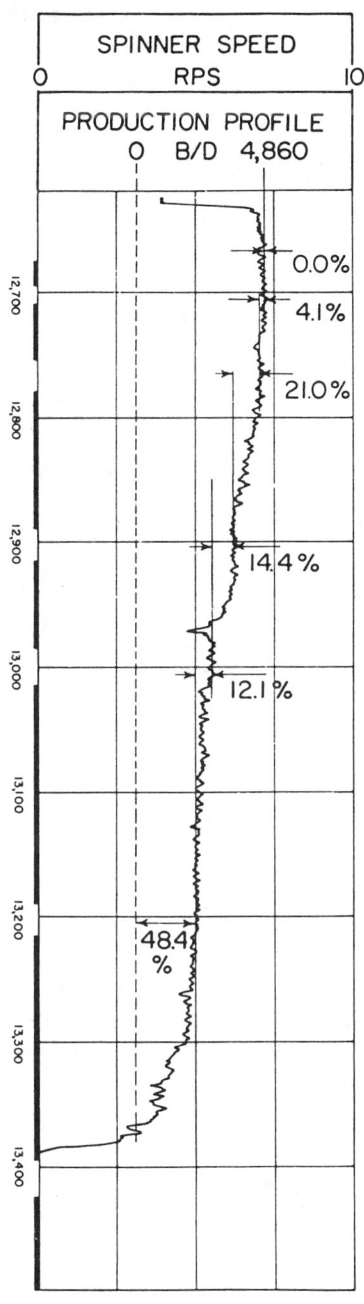

FIGURE 8.4 *Production Profile. Courtesy Schlumberger Well Services.*

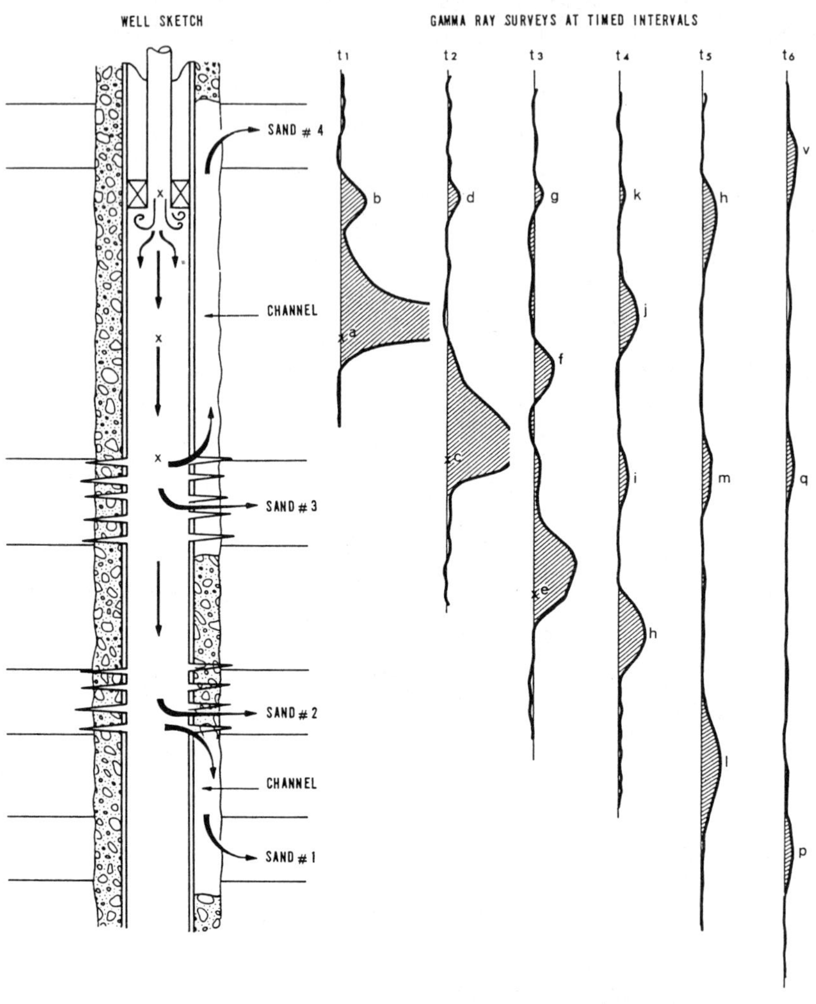

FIGURE 8.5 *Time Lapse Radioactive Tracer Survey. Courtesy Schlumberger Well Services.*

two pressure sensors a fixed distance apart; and (2) the fluid density log (see fig. 8.7), which measures the absorption of gamma rays by the fluid between a gamma ray source and a detector.

DIELECTRIC CONSTANT. The Hydro Log (see fig. 8.8) measures the dielectric constant of the fluid flowing in the wellbore. Due to the difference between oil and water the holdup of the flowing mixture can be estimated.

SAMPLE RECOVERY. A downhole fluid sampler is illustrated in figure 8.9.

TABLE 8.2 *Fluid Identification Tools*

Tool Type and Specifications		Gearhart	Dresser	Schlumberger	NL McCullough
Fluid density	Tool	Fluid Density Logging (Scattered GR)	Fluid Density Log (Scattered GR)	Gradiomanometer (Δ Pressure)	
	Tool diameter (in.)	1 7/16	1 3/4 and 1 7/10	1 11/16	
	Max. temp./press. (°F/psi)	375/15,000	400/15,000	350/15,000	
Dielectric log	Tool	Hydro Log			
	Tool diameter (in.)	1 7/16			
	Max. temp./press. (°F/psi)	300/15,000			
	Tool		Tru-tubing Borehole Fluid Sampler	Fluid Sampler	
PVT sampler	Tool diameter (in.)		1 11/16	1 11/16 and 2 1/2	
	Max. temp./press. (°F/psi)		300/10,000	350/10,000	
Noise log	Tool	Borehole Audio Tracer Survey (BATS)	Sonan Log		Noise Logging Service
	Tool diameter (in.)	1 and 1 11/16	1 11/16		1 1/2
	Max. temp./press. (°F/psi)	300/15,000	350/17,000		350/20,000

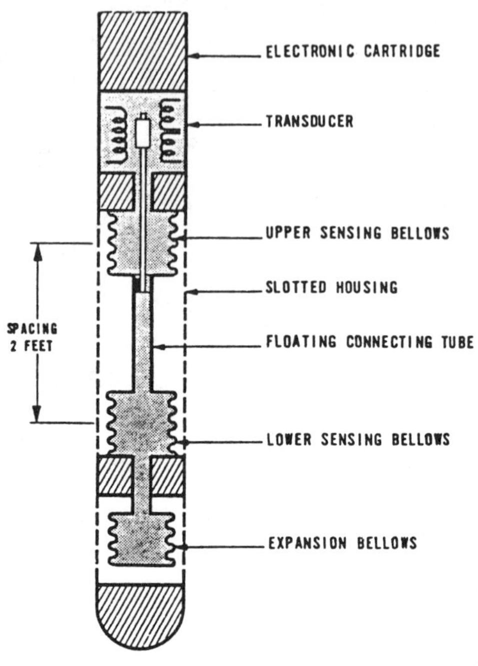

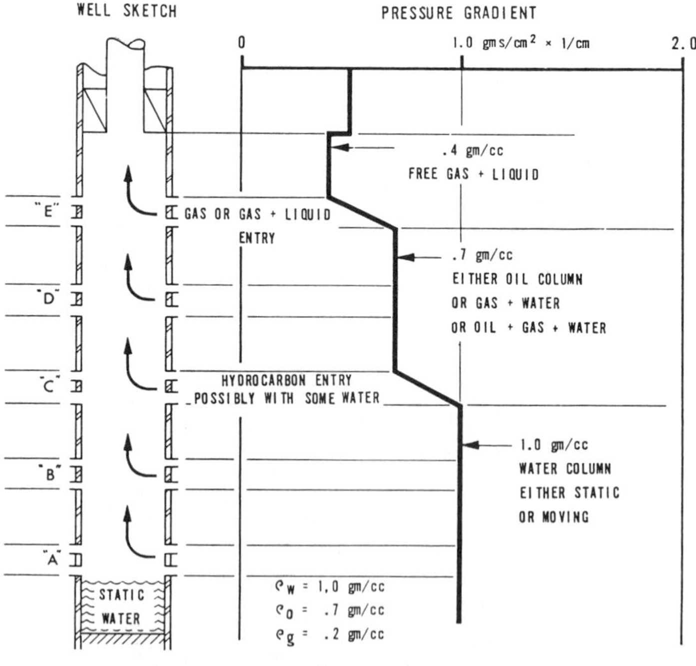

FIGURE 8.6 *Gradiomanometer. Courtesy Schlumberger Well Services.*

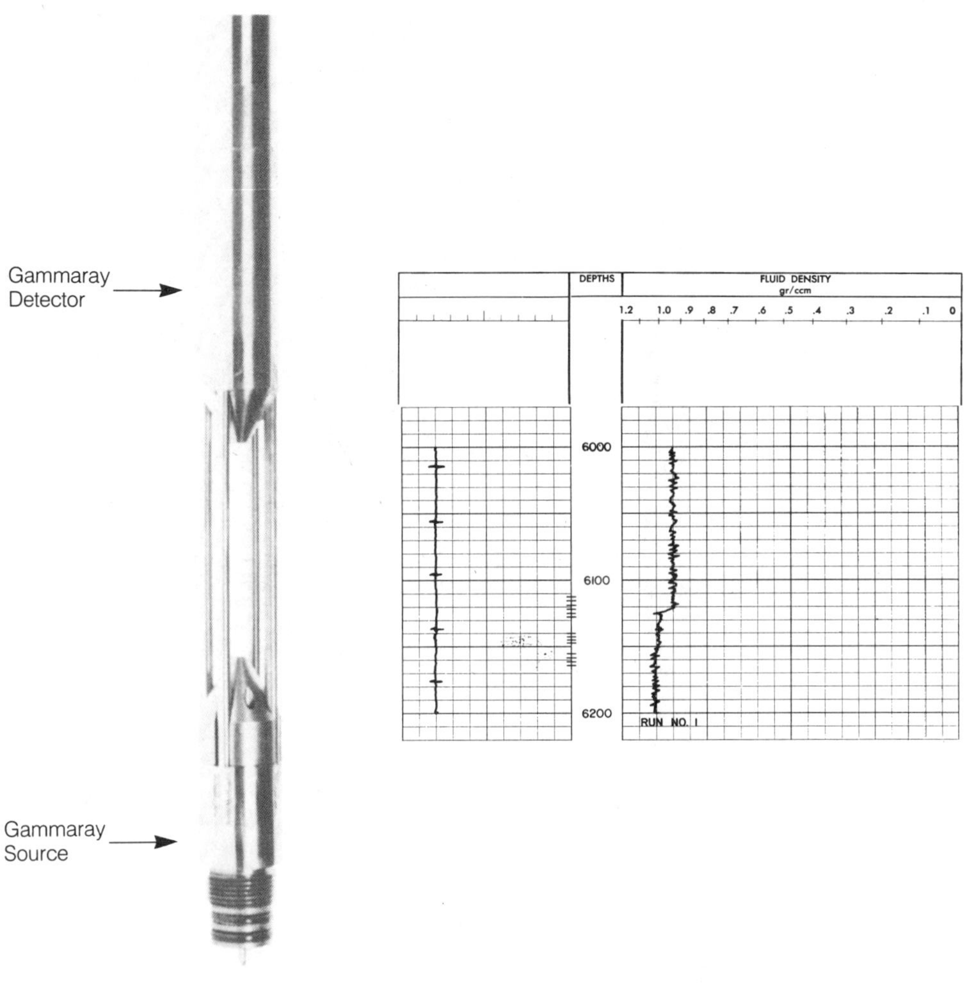

FIGURE 8.7 *Fluid Density Log. Courtesy Gearhart Industries, Inc.*

NOISE LOGS. Turbulent fluid movement generates noise. Both the amplitude and frequency of this noise vary with the quantity and type of fluid and the medium through which the fluid is flowing. Measurements of these characteristic sounds can be interpreted to indicate the type of fluid flowing and its location. In the case of gas, it is possible to calculate the approximate rate of flow. Figure 8.10 shows a borehole audio tracer survey (BATS) log.

Log Quality Control

Gradiomanometer logs measure the pressure difference between two points. With the tool vertical in zero flow, this pressure difference can

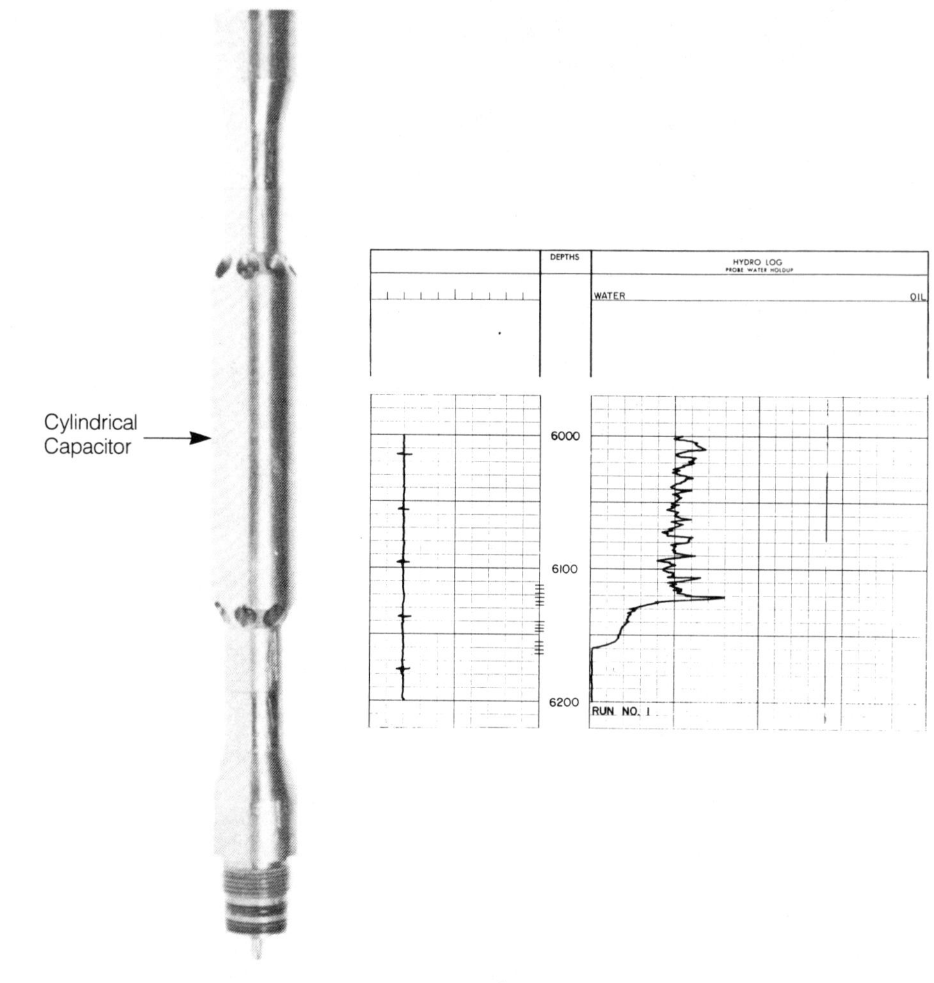

FIGURE 8.8 *The Hydro Log. Courtesy Gearhart Industries, Inc.*

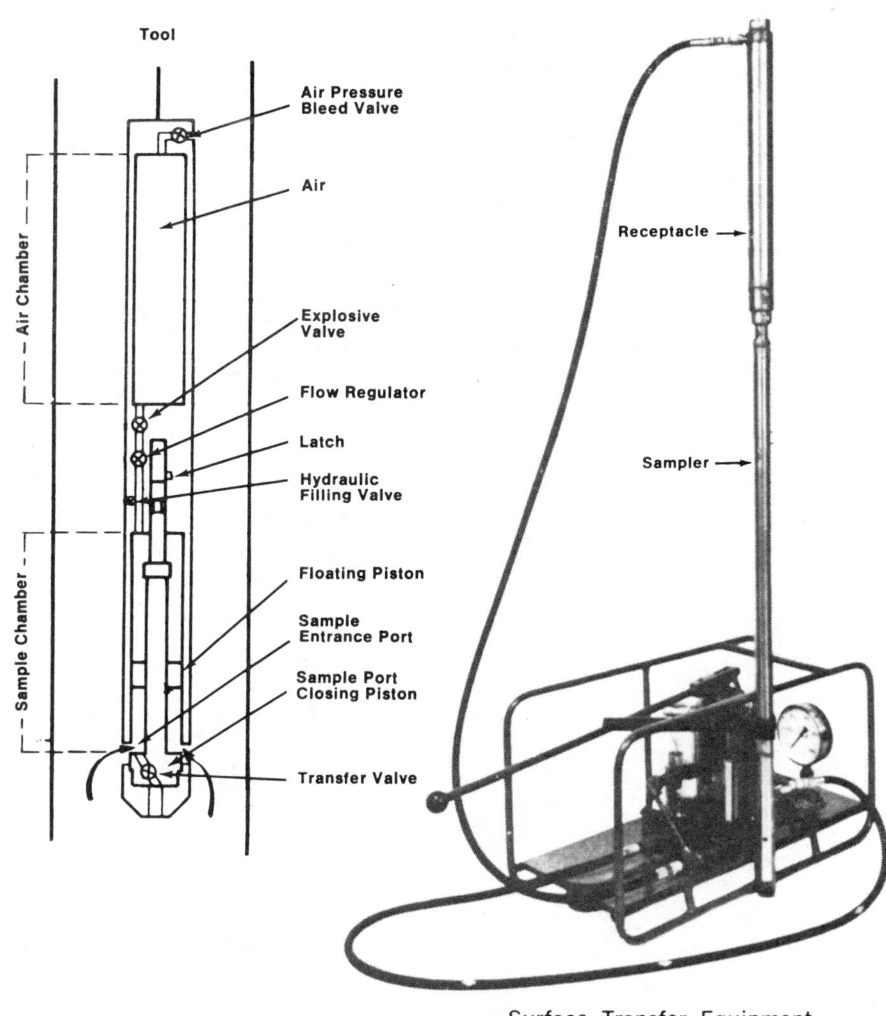

FIGURE 8.9 *Fluid Sampler. Courtesy Schlumberger Well Services.*

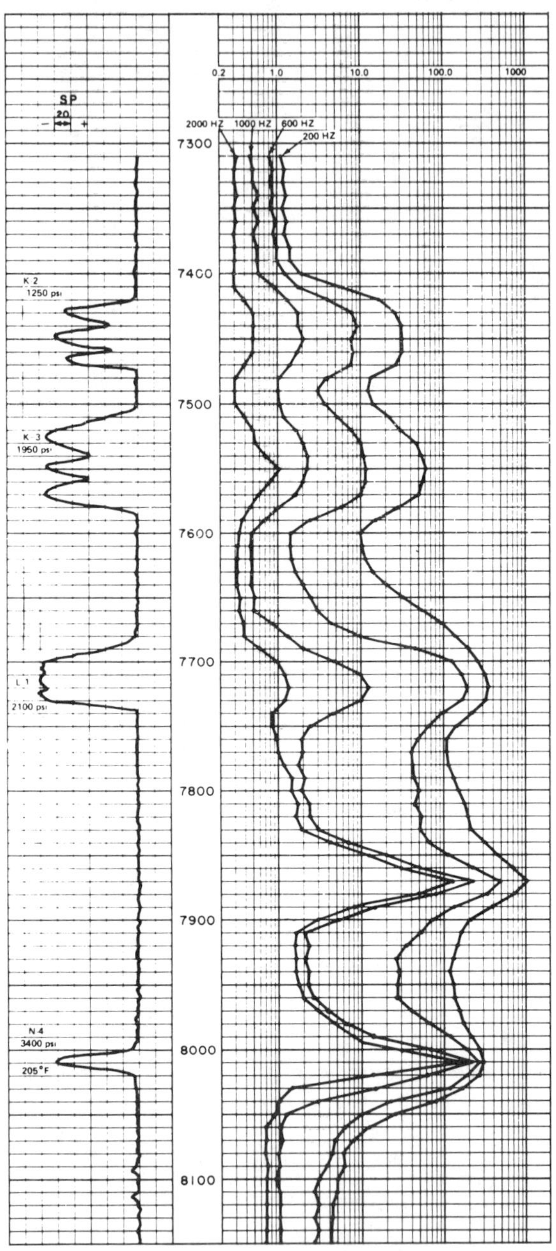

FIGURE 8.10 *BATS (Noise Log). Courtesy Gearhart Industries, Inc.*

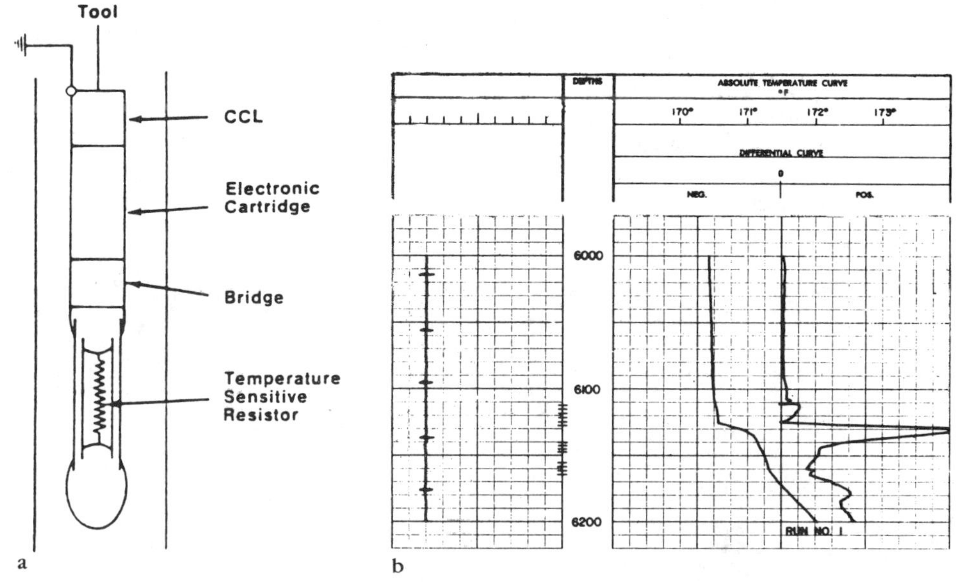

FIGURE 8.11 *Temperature Logging. (*a*) Courtesy Schlumberger Well Services and (*b*) Courtesy Gearhart Industries, Inc.*

be converted to a fluid density. However, in conditions of high-flow rate in small-diameter pipe there will be an additional friction gradient; and in deviated wells a deviation correction is required.

All other fluid-typing measurements should be checked by making measurements in parts of the well where flow rate and fluid type are known or can be inferred, for example, below the tubing packer and above all perforations, or in a static water column at the bottom of the well.

TEMPERATURE LOGGING

Applications

Temperature logs may be used to monitor fluid flow in production or injection wells and have an added advantage that they can detect fluid flow outside the completion string in tubing/casing annulus or casing/formation annulus. They are particularly useful for (1) finding gas entries to, or exits from, the wellbore; (2) finding channels in poorly cemented sections; (3) finding lost circulation zones in openhole; and (4) finding the cement top in a recently cemented well.

Tools Available

Three types of temperature measurements are commonly available: (1) the conventional temperature survey (see fig. 8.11); (2) the differential temperature survey; and (3) the radial differential temperature survey (see fig. 8.12).

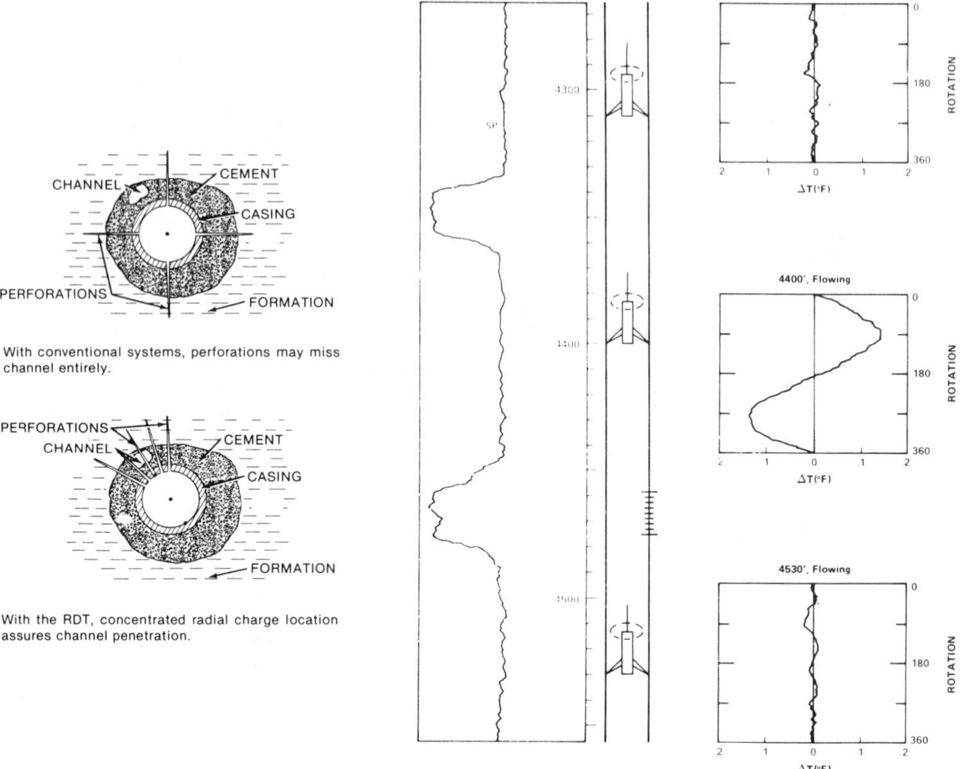

FIGURE 8.12 ***Radial Differential Thermometer. Courtesy Gearhart Industries, Inc.***

Log Quality Control

When making temperature logs, keep the following quality control points in mind:

1. Choose an appropriate scale so that there are no excessive scale changes over the zone of interest.
2. Log going down where possible so that the presence of the tool and cable in the wellbore does not influence the measurement being made.
3. Remember that temperature-measuring devices are normally quite sensitive to temperature changes but not very accurate in absolute terms.

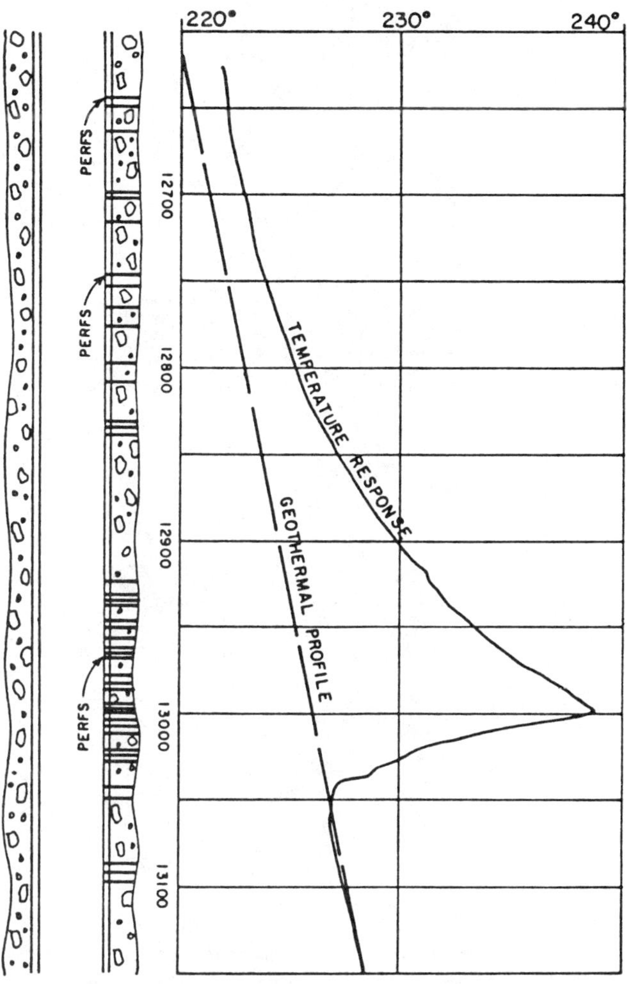

FIGURE 8.13 *Temperature Log: Oil Production. Courtesy Schlumberger Well Services.*

Field Examples

A temperature log showing oil production through a perforated interval is illustrated in figure 8.13.

FORMATION TESTERS

Wireline formation testing can be used to obtain information concerning the formation: (1) the produced fluids; (2) formation pressure; and (3) formation permeability.

FIGURE 8.14 ***Selective Formation Tester. Courtesy Gearhart Industries, Inc.***

The latest models of wireline testers allow multiple-tool settings on each trip into the well, so that virtually any number of pressure recordings may be made and two separate fluid samples may be taken. Samples may be segregated samples from a single test or taken from two separate tests. The sample size varies with the service company.

Figure 8.14 shows the packer section of a selective formation tester and figure 8.15 shows a schematic diagram of a repeat formation tester.

Applications

Wireline testing is used, for example:

a. to investigate a zone of interest under conditions where conventional tests are not feasible—such as too far above TD; no good intervals to

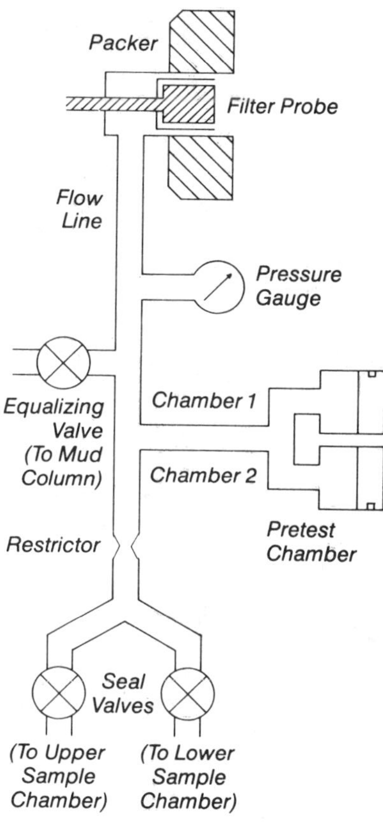

FIGURE 8.15 *Schematic of Repeat Formation Tester. Courtesy Schlumberger Well Services.*

TABLE 8.3 *Repeat Formation Tester Specifications*

	Gearhart	Schlumberger	Dresser Atlas	Welex
Pressure rating (psi)	15,000	20,000	15,000	20,000
Temperature rating (°F)	350	350	350	350
Minimum hole size (in.)	6¼	6	6¼	7
Maximum hole size (in.)	12¾	14¾	12¼	13
Basic make-up length (ft)	16	33	35½	?
Formation pressure readings per trip in hole	Any number	Any number	Any number	Any number
Sample chamber sizes (gal)	2¼, 5	1, 2¾, 6, & 12	2¼, 5	1, 2½, 5, & 7

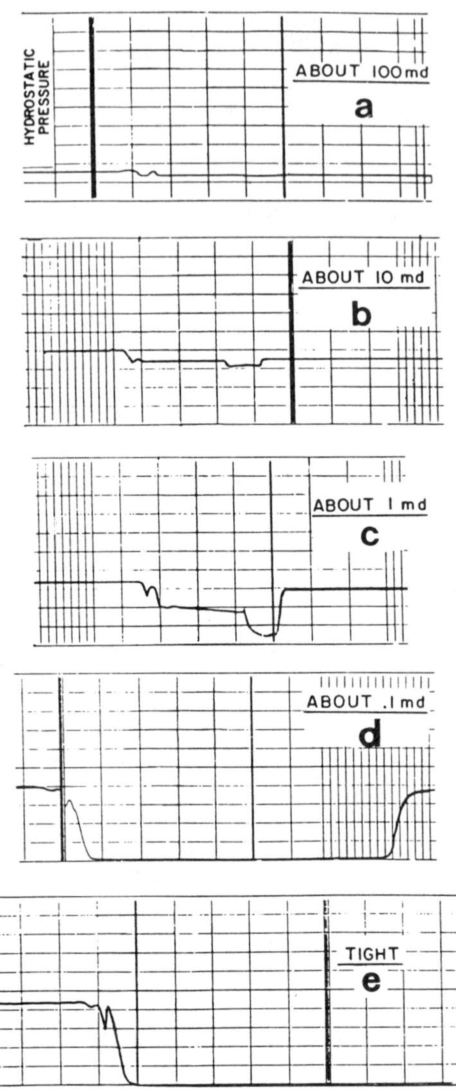

FIGURE 8.16 *Permeability Estimates from Pretest Pressure Records. Courtesy Schlumberger Well Services.*

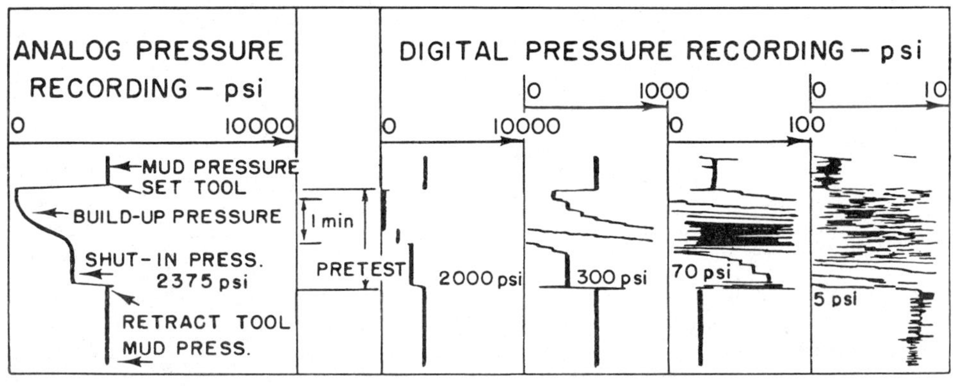

FIGURE 8.17 *Pressure Test. From Schultz et al. 1974. ©1974 SPE-AIME.*

set straddle packers; or a very short interval, where depth control is critical

b. to pin down water-oil, gas-oil, or gas-water contacts
c. where rig time is critical
d. where pressure control is critical due to time of day or rig location

Tool Availability

All service companies should be given maximum notice when wireline testing is being considered for well evaluation. In some areas, several days or even weeks will be required to insure that equipment and trained personnel are available. Be sure to consider the various options or conditions that will need to be accommodated. The equipment needed to deal with variables such as sample size, packer hardness, choke size, and water cushions may not be universally available. If a sample of recovered gas is needed for lab analysis, a special pressure cylinder should be requested from the service company. The sizes and ratings of various service-company testers are given in table 8.3.

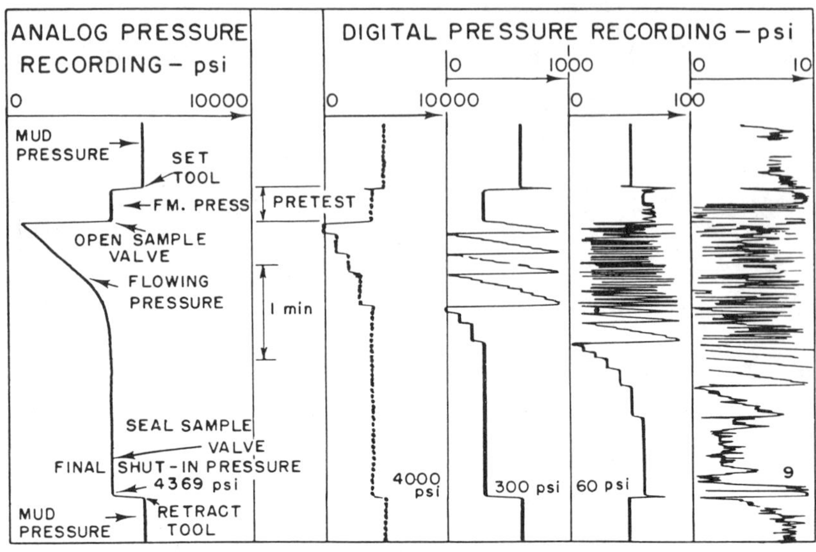

FIGURE 8.18 *Sample Test. From Schultz et al. 1974. ©1974 SPE-AIME.*

Quality Control

Each test attempted should show the SP or GR recording used for depth tie-in. This recording is the proof that depth control was accurate. The test sequence is recorded on a film driven by a time-drive mechanism. This makes it possible to monitor all test activities and pressures vs. time. This recording should include the entire test sequence, with each significant step marked on the film before field prints are made. All fluid recoveries should be separated and their volumes recorded. A representative sample of each fluid should be kept in a sealed container for possible later analysis. A sealed sample of drilling mud should be kept with the recovered fluid samples for later chemical comparison if needed.

Hydrostatic pressures recorded during testing should be verified, within limits of accuracy, with a calculation involving mud density and depth of test.

A valid test is one that recovers significant quantities of fluid and records formation pressures and hydrostatic pressure. A single dry test is indeterminate. In the case of a dry test, the tool should be respositioned several times to determine if:

a. the formation is impermeable, in which case all tests will be dry; or
b. the tool was set in a tight or shale streak, in which case repeated efforts should result in a valid test.

A lost packer seal is also indeterminate. In this case, the tool should be repositioned several times in an effort to effect a packer seat.

Close examination of the openhole logs may help resolve indeterminate tests. Microresistivity logs are particularly helpful in resolving a dry test. Caliper logs, particularly with four-arm calipers, can help resolve packer-seal problems.

Typical pretest pressure recordings in a number of formations of differing permeability are illustrated in figure 8.16.

Field Examples

A typical pressure test is presented in figure 8.17. A sample test is shown in figure 8.18.

CHAPTER NINE

DIGITAL LOG DATA TAPE FORMATS

Increasingly, log data is being recorded directly on magnetic tape in digital form. This chapter gives a guide to the most common formats in which log data is recorded. More exhaustive treatment of the subject is available directly from the service companies themselves.

It is essential that the tape delivered to the wellsite geologist or engineer be clearly labeled with the following:

a. the well name and number
b. the type of tape (i.e., 7-track, 9-track, or cartridge)
c. the recording density (800, 1600, or 6250 BPI, i.e., bits per inch)
d. the logs recorded (e.g., DIL/Sonic Run 2, 5762′ to 10,987′)
e. the sample increment (e.g., 2 samples/ft or 5 samples/m)
f. the service-company format used

Obtaining this data at the wellsite saves time and trouble later. Many service companies will convert their digital log data tapes into other formats.

Before using digital log data, it is strongly recommended that a playback be made and the machine-produced log compared with the original analog print to check for:

a. correct inscription
b. data busts (missing data, excessively high or low values, etc.)
c. correct names used for the curves (e.g., watch for a deep induction curve labeled as a deep laterolog curve)

Optical logs that have been digitized are particularly prone to labeling, scaling, and depth-shifting errors. See, for example, figures 9.1 and 9.2. Figure 9.1 shows a short section of an optical log that was sent out to be digitized. Figure 9.2 shows a playback of the digital log tape. Note that the shallow resistivity curve has been wrongly digitized. It has been swapped left for right and flipped to read resistivity values that are too high. Such a digital tape is bad and the log should be redigitized.

SCHLUMBERGER LIS

The LIS format was designed primarily for use with magnetic tape; it may, however, also be used with other types of storage media. A hierarchy is used for storing and retrieving data; this hierarchy separates physical representation from logical information interpretation.

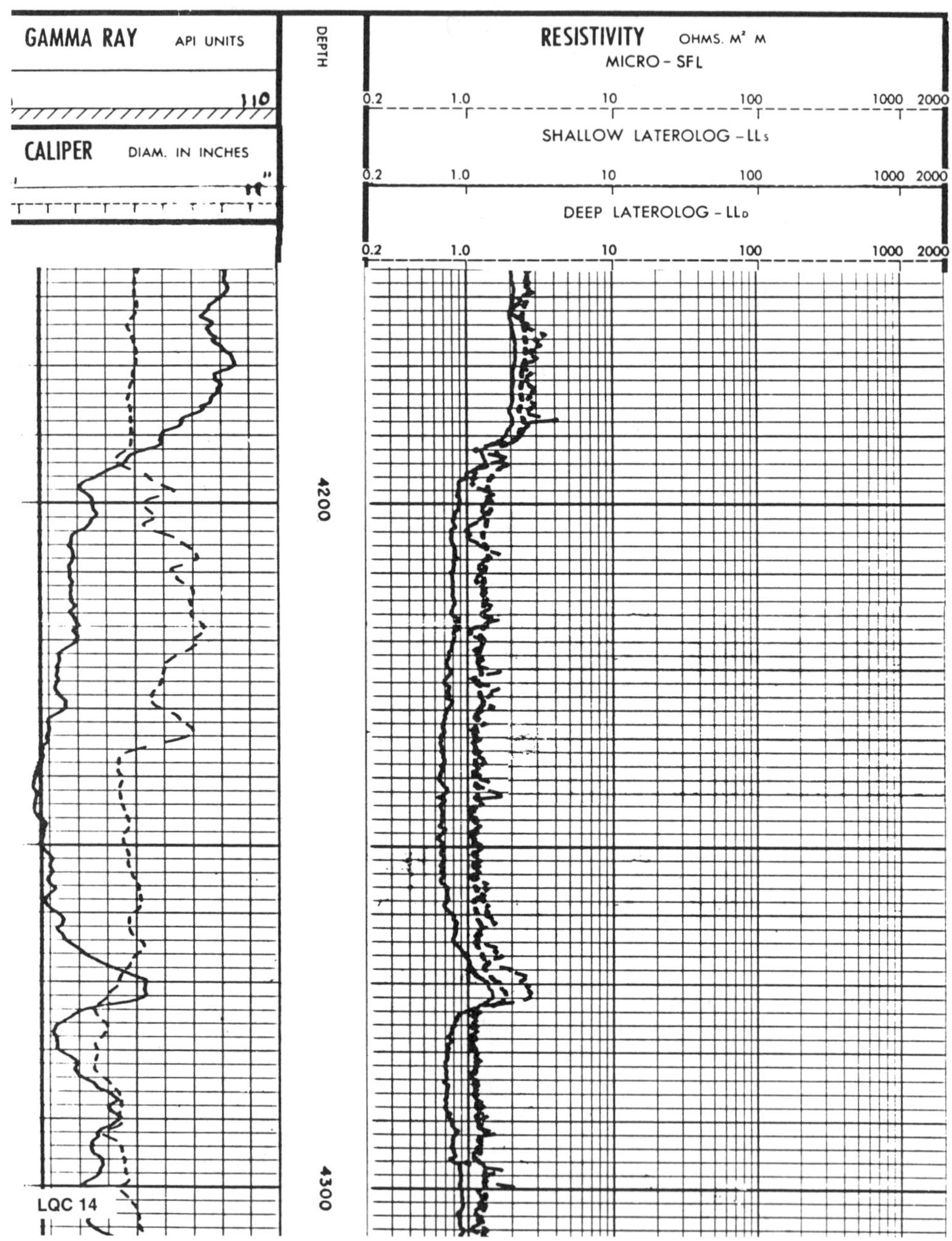

FIGURE 9.1 *Original Log.*

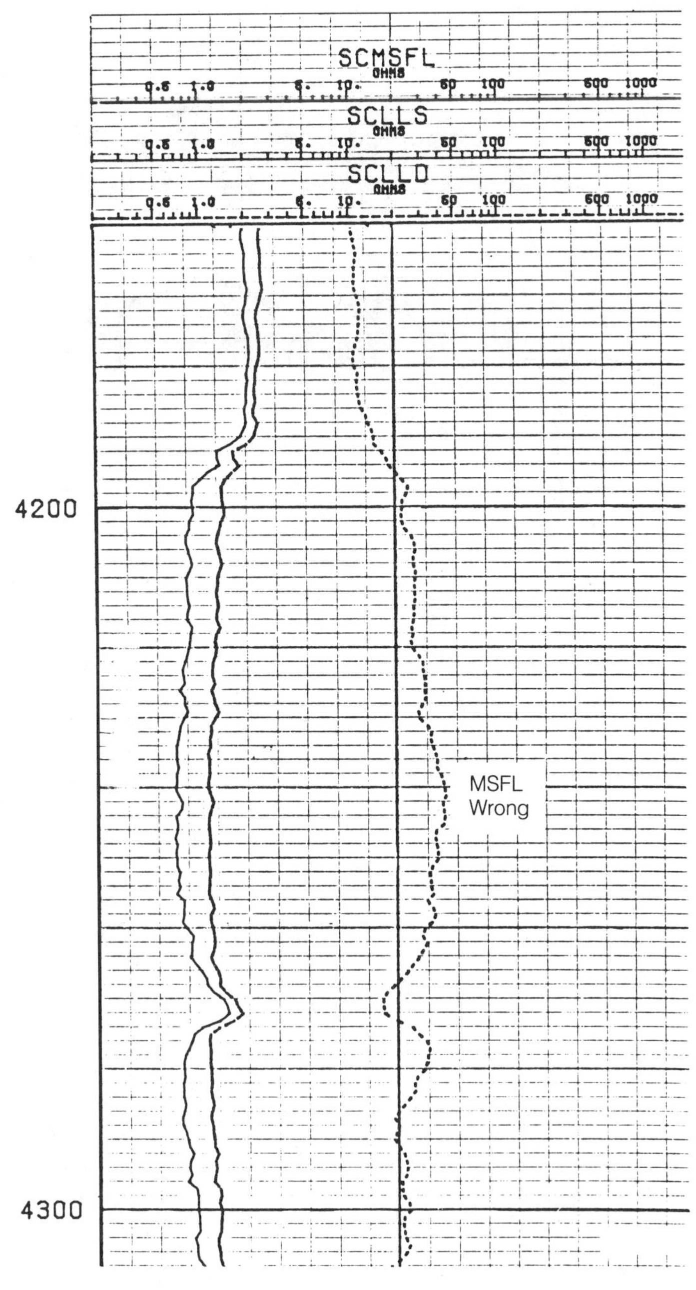

FIGURE 9.2 *Playback of Digital Tape.*

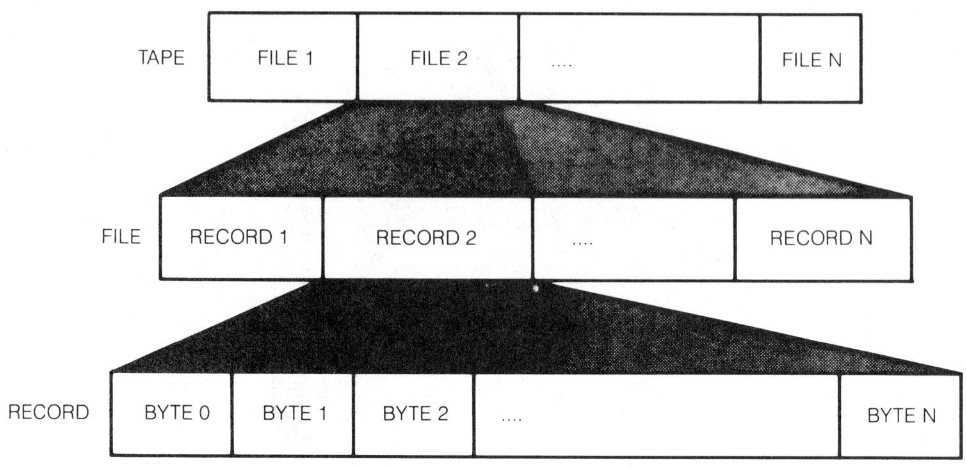

FIGURE 9.3 *Logical Hierarchy. Courtesy Schlumberger Well Services.*

Each magnetic tape is kept logically complete by recording on each tape the information required to interpret the data on that tape.

The LIS format permits adaptation to new concepts while maintaining compatibility with previously written tapes with no significant effect on processing software. This format has two structures: logical and physical. Logical structure refers to the type of information and the organization of the information; physical structure refers to the physical dimensions of the information (e.g., how much space it occupies).

Logical Structure

From the logical aspect, the LIS format is divided into three levels:

logical tape: made up of logical files
logical files: made up of logical records
logical records: made up of bytes

The relationship of these levels is illustrated in figure 9.3.

A *logical tape* is a group of logical files. Each *logical file* is made up of a group of related logical records treated as a unit. Each *logical record* is a collection of related data treated as a unit. Each logical record is composed of *bytes,* or data characters.

LIS defines specific ways to label logical tapes, logical files, and logical records. There is no strict definition of the type of log data a logical tape or a logical file may contain. Sometimes a logical tape may contain all the data from a single well; but this is not always true.

LIS does, however, specifically define many types of logical records. Each logical record has a two-byte header at the beginning of the record. This header contains a code identifying the *record type.* This record type is used to tell the program how to interpret the record.

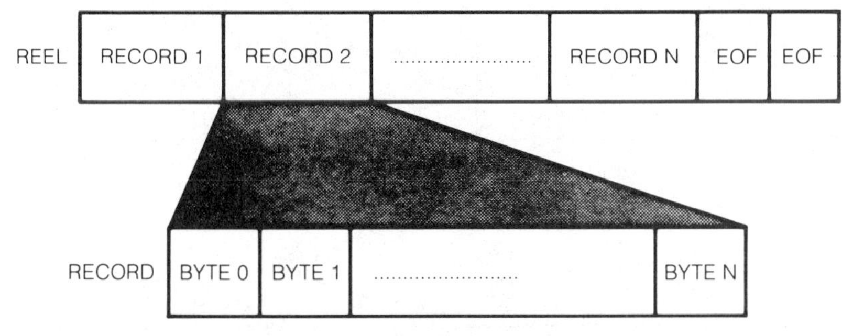

FIGURE 9.4 *Physical Hierarchy of LIS. Courtesy Schlumberger Well Services.*

Unlike many other systems, relative position in the sequence of records is not the key to understanding the contents of a record. Computer programs that read LIS do not have to be able to interpret every type of record they encounter. Normally, unknown types can be ignored with safety.

Files and tapes have both *header* and *trailer records* at their beginnings and ends. These header or trailer logical records contain information about the logical files or logical tapes that they begin or end.

Logical tapes and logical files are not strictly defined with respect to the detailed identification of log data. In some instances, a logical tape corresponds to the data from a single well. This correlation does not hold in all cases, however, and programs that read LIS tapes should not assume such a restriction.

Physical Structure

Physical structure in the LIS format describes the way that logical tapes, files, and records are represented on a physical reel of tape. A physical reel of tape is composed of a group of physical records arranged in a specific order. A *physical record* is a group of eight-bit bytes, also arranged in a designated order. Only at the byte level are logical and physical structures equal. There are no physical files in the LIS format, so the relationship within the physical structure is:

physical reel: made up of physical records
physical records: made up of bytes

The relationship is shown in figure 9.4.

A special type of physical record, called a *file mark* or an EOF, is used to identify the end of a logical file. The purpose is to allow special EOF-recognition hardware to be used for high-speed tape searches. Two consecutive EOFs or an end-of-tape reflective marker indicate the end of a physical reel of tape.

A type of logical record called a *reel header* is used to identify a physical reel at its beginning. As with logical files and logical tapes, there is also a corresponding type of logical record called a *reel trailer*.

BYTE	0	1	2	3
BITS	0–7	8–15	16–23	24–31

PHYSICAL RECORD

PHYSICAL RECORD HEADER 4 BYTES	ALL/PART OF A LOGICAL RECORD	(OPTIONAL) PHYSICAL RECORD TRAILER T BYTES

FIGURE 9.5 ***Physical Record Format. Courtesy Schlumberger Well Services.***

Record Structure

An LIS tape is generated on a 9-track magnetic tape encoded at 800 or 1600 BPI (bits per inch) written in odd parity. The information on the tape is further divided into files and finally into individual records. Record length is defined in bytes; a *byte* is eight bits.

Every physical record on the reel of magnetic tape uses the first four bytes as a *physical record header.* The first two bytes of the header contain the length of the record (the number of bytes), including the physical record header. The third and fourth bytes of the record specify attributes of the physical record. Bits of the attribute bytes are used to identify some properties of the physical record. A physical record can be terminated with a *physical record trailer*; however, the trailer is optional. Figure 9.5 shows the physical record header and the general construction of a physical record.

The length of a logical record, or logical record fragment, is given by:

$$\text{logical record length} = (\text{physical record length} - 4 - T)\ \text{bytes}$$

The value of T may vary from 0 to 6 bytes.

PHYSICAL RECORD HEADER. The physical record header is 32 bits in length. Bit 0 is high order; bit 31 is low order. If bits 19, 21, and 22 are set, each bit reserves two bytes in the physical record trailer. Listed as a bit string, the contents of the physical record header are as follows:

Bit	*Definition*
0–15	physical record length
16	undefined (reserved)
17	physical record type
18	checksum type bit 1
19	checksum type bit 2
20	undefined (reserved)
21	file number presence
22	record number presence
23	undefined
24	undefined (reserved)
25	parity error
26	checksum error
27	undefined
28	undefined (reserved)
29	undefined
30	predecessor continuation bit
31	successor continuation bit

Physical record length (bits 0–15) is a 16-bit, unsigned, binary integer which specifies the length, in bytes, of information in the physical record. The length, which includes the physical record header and the trailer (if present), may be smaller than the actual number of bytes written. That is, a physical record may be padded with null characters to guarantee a minimum record size.

Physical record type (bit 17) is the type code for a physical record. Figure 9.5 illustrates a type 0 physical record, which is the only type of physical record defined by LIS at this time.

Checksum type bits 1 and 2 (bits 18 and 19) designate the type of checksum associated with this physical record. If a checksum is present, it is the last entity of the physical record trailer.

Predecessor continuation bit (bit 30), if 0, indicates that this physical record is not associated with the previous physical record; if 1, it indicates that this physical record is associated with the previous physical record.

Successor continuation bit (bit 31), if 0, indicates that this physical record is not associated with the next physical record; if 1, it indicates that this physical record is associated with the next physical record. So, if a logical record were to span four physical records, the values of the predecessor and successor continuation bits for the four records would be:

	Predecessor	*Successor*
Record 1:	0	1
Record 2:	1	1
Record 3:	1	1
Record 4:	1	0

If a logical record spans several physical records, then only the first physical record contains a logical record header. The following physical records contain only the succeeding sections of the logical record.

If the predecessor continuation bit in the physical record header is 0, then the physical record contains a logical record header. If it is 1, the physical record does not contain a logical record header.

The complete logical record may be reconstructed by concatenating the logical record components, in order, from each of the physical records.

PHYSICAL RECORD TRAILER. Physical records may optionally have trailers. The presence, length, and contents of the trailer are defined by three physical record attribute bits, defined under Record Structure. The sequence of information in the trailer is shown below. Any or all of the items may be missing:

Entry	Size	Representation Code	Header Bit Controlling
Record number	5	79	22
File number	2	79	21
Checksum	2	79	19

Bits 19, 21, and 22 in the physical record header indicate the presence or absence of each of these three pieces of information in the physical record trailer.

Logical Records

Logical records form the basic coherent bodies of information in the LIS format. A logical record may be contained within a single physical record, or it may span several physical records.

LOGICAL RECORDER HEADER. Logical records begin with a 2-byte logical record header. The first byte of the logical record header contains the record type. The second byte is reserved.

Figure 9.6 shows how logical and physical records could be organized on a tape. Figure 9.7 shows a typical grouping of the various logical record types.

Reel Trailer (Type 133)

The physical reel is terminated by a *reel trailer* and two hardware EOF records. The *reel trailer record* is the final record of the physical reel. It contains identification of the next physical reel in the set. This is an optional record. No encored information follows the double EOF.

If multiple logical tapes are required, they are stacked sequentially on the reel of tape (fig. 9.8). The trailer of one tape may name the header of the next. If multiple physical reels are required, the reel trailer may name the next physical reel.

The LIS format technically allows physical records of greater than

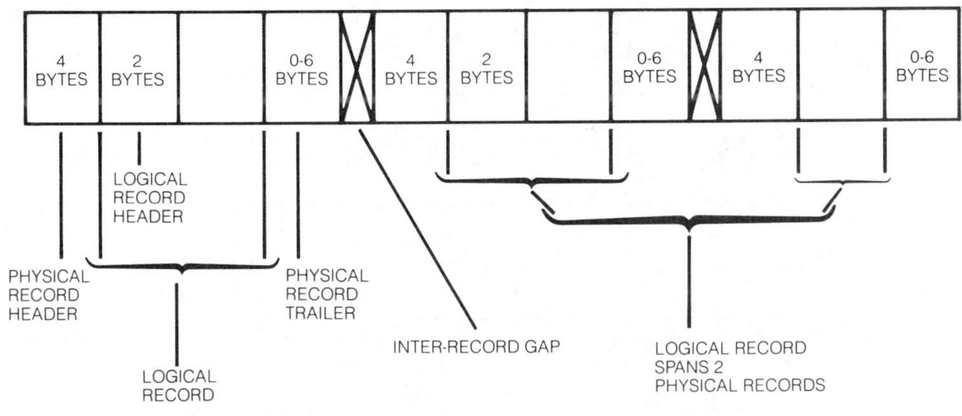

FIGURE 9.6 *Organization of Logical and Physical Records on a Tape. Courtesy Schlumberger Well Services.*

32,000 bytes in length. In practice, some maximum lengths exist, although they are rather arbitrary and could increase at some future time. The current maximum physical record length is 1024 bytes, with the exception of tapes containing waveform data. The maximum physical record length on a waveform data tape is approximately 5000 bytes.

DRESSER ATLAS

Dresser Atlas records data in either the BIT or EBIT formats on 9-track tape at 800 BPI, odd parity. A 1600 BPI recording is optionally available.

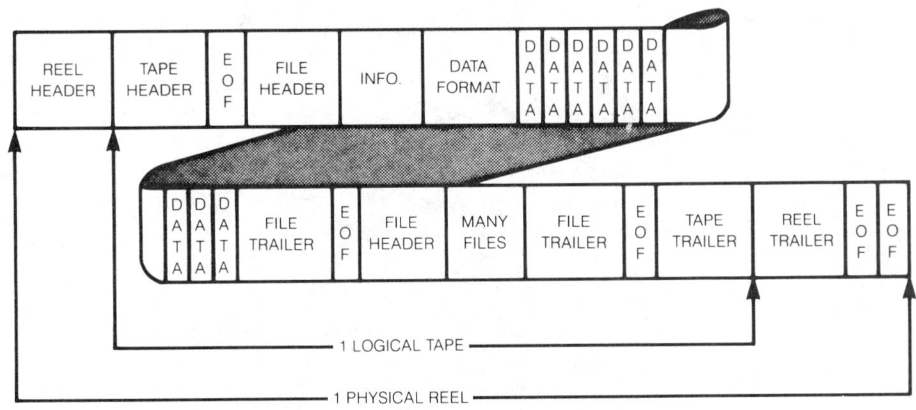

FIGURE 9.7 *Reel of Tape. Courtesy Schlumberger Well Services.*

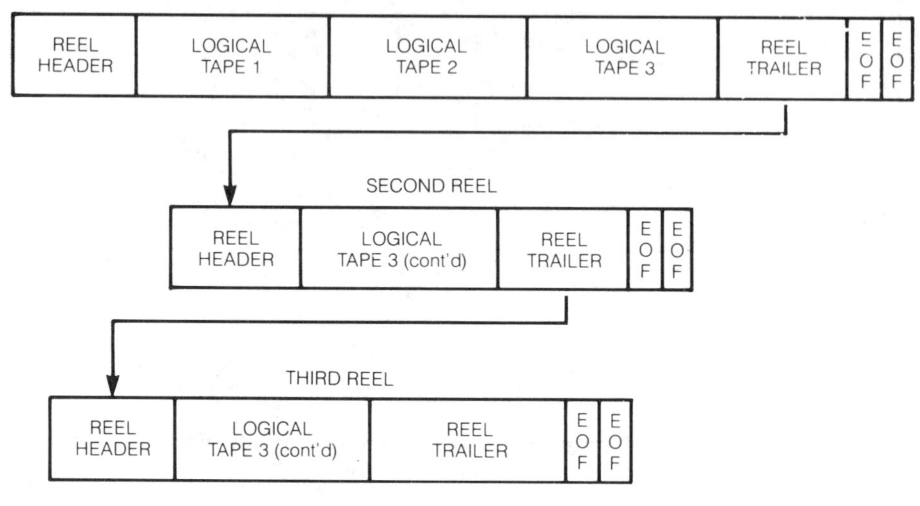

FIGURE 9.8 ***Relationship Between Logical Tapes and Physical Reels. Courtesy Schlumberger Well Services.***

Data is allowed one of three formats:

Alphanumeric	(AN) ASCII character set
Integer	(12) INTEGER-2 16-bit, two's complement
Floating point	(FP) IBM, single precision with 8-bit exponent and 24-bit mantissa

BIT Format

The Dresser Atlas BIT format utilizes the tape structure shown in figure 9.9. The first item on any tape is a *general heading,* which is illustrated more fully in table 9.1. The general header is 276 bytes long. The record structure is as shown in figure 9.10. It can be seen that each data record contains SPCRR samples of each of the NOLOG curves. Within each record, the order of the curves is as given in the general heading. The maximum number of curves possible using this format is 20.

EBIT Format

The EBIT format is more complete than the BIT format (see table 9.2). The *general heading* (of 1024 bytes) is followed by any number of *service heading* records (1024 bytes/record; see table 9.3). Following the service heading records come the data records, which are of the same type as in the BIT format.

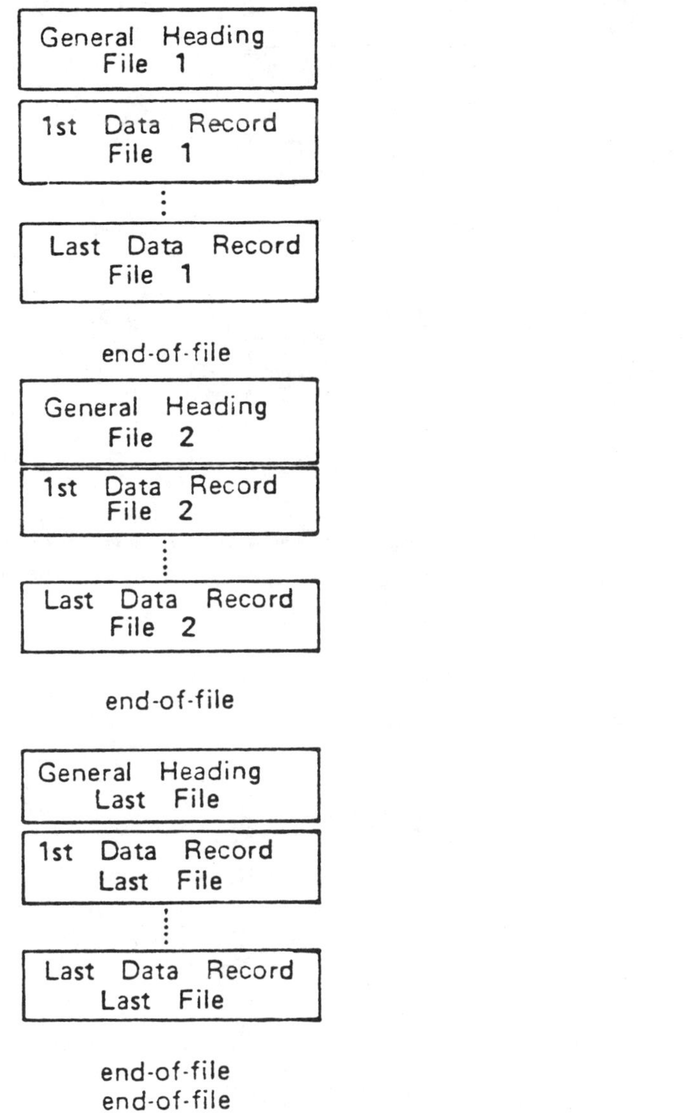

FIGURE 9.9 *BIT Tape Structure. Courtesy Dresser Atlas.*

1st data record

curve 1
curve 2
. . .
curve NOLOG

1st SPCPR samples of curve 1
1st SPCPR samples of curve 2

1st SPCPR samples of NOLOG

1st sample is at depth STDEP
2nd sample is at depth STDEP + RLEV
3rd sample is at depth STDEP + 2 x RLEV

2nd data record

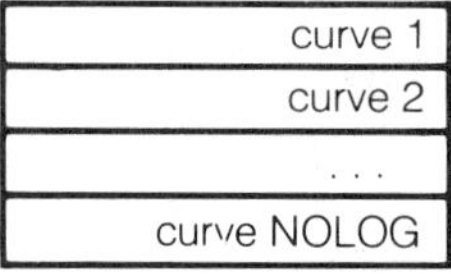

2nd SPCPR samples of curve 1
2nd SPCPR samples of curve 2

2nd SPCPR samples of NOLOG

FIGURE 9.10 *BIT Tape Structure. Courtesy Dresser Atlas.*

TABLE 9.1 *BIT Tape General Heading*

Position			Length			
Byte	Half	Full	(bytes)	Name	Description	Type
1	1	1	4	ECC	Reference no.	FP
5	3	2	80	ICO (40)	Company name	AN
85	43	22	80	IWELL (40)	Well name	AN
165	83	42	2	NOLOG	No. of curves on tape	I2
167	84		2	•IMFG	Metric flag	I2
169	85	43	80	LOG (2,20)	Curve names	AN
249	125	63	4	STDEP	Starting depth	FP
253	127	64	4	ENDEP	Ending depth	FP
257	129	65	4	RLEV	Level spacing	FP
261	131	66	2	•IDC	Currently not used filler	
263	132		2	•IZ2		
265	133	67	4	SPCPR	No. of samples per record	FP
269	135	68	6	•ISERV	Service no.	I2
275	138		2	•IFILE	File no.	I2

•Unique to bit format

TABLE 9.2 *EBIT General Heading*

Position			Length					
Byte	Half	Full	(bytes)	Name	Description	Type	Note	Usage
1	1	1	4	IRLAC (2)	Log analysis center code	AN	1)	R
5	3	2	4	●ECC	Reference number	FP	2)	R
9	5	3	40	●ICO(20)	Company name	AN	3)	R
49	25	13	40	●IWELL(20)	Well name	AN	4)	R
89	45	23	40	IFIELD(20)	Field name	AN	5)	R
129	65	33	30	ICNTY(15)	County name	AN	6)	R
159	80		10	ISTATE(5)	State	AN	7)	R
	85	43	4	●STDEP	Starting depth	FP	8)	R
		44	4	●ENDEP	Ending depth	FP	9)	P
169			4	●SPCPR	Number of Samples/rec.	FP	10)	R
173	87			IDM	Drilling measured from	AN	22)	0
177	89	45			Elevation of PD	FP	23)	0
631	316	158.5	2			FP	24)	0
633	317	159	4	●EPD			25)	0
637	319	160	4	EKB	Elevation of KB			0
641	321	161	4	EDF	Elevation of DF	FP		
645	323	162	4	●EGL	Elevation of GL	FP	26)	
649	325	163	9	FLAGS(2)	Future use	FP		
657	329	165	4	DECLM	Magnetic declination	FP	27)	P
661	331	166	364	FILLER(91)	Future use	FP		
			1024					

●Common to bit format

TABLE 9.3 *EBIT Service Heading*

Position			Length					
Byte	Half	Full	(bytes)	Name	Description	Type	Note	Usage
1	1	1	4	ISERV(2)	Service mnemonic	AN	1)	R
5	3	2	6	ISVCD(3)	Service code	AN	2)	P
11	6		2	IRUN	Run no.	AN	3)	R
13	7	4	8	IDATE(4)	Date	AN	4)	R
21	11	6	6	ISON(3)	Service order no.	AN	5)	P
27	14		2	ILDP	Log measured from	AN	6)	P
29	15	8	4	DP	Depth driller	FP	7)	0
33	17	9	4	DL	Depth logger	FP	8)	0
	19	10	4	BLD	Bottom logged depth	FP	9)	0
37		11	4	TLD	Top logged depth	FP	10)	0
41	21		4	CSZD	Casing size-driller	FP	11)	0
45	23	12		TDEPS	Top depth this service	FP	33)	R
153	77	39	4		Bottom depth this service	FP	34)	R
157	79	40	4	BDEPS		AN	35)	0
161	81	41	96	ITS(4,2,6)	Tool-series serial no.		36)	0
275	129	65	4	TLN	Tool length	FP		P
261	131	66	200	ICVNM(4,25)	Curve names	AN	37)	
461	231	116	100	ITRK(2,25)	Plot track	I2	38)	P
561	281	141	100	SCLL(25)	Left plot scale	FP	39)	P
661	331	166	100	SCLR(25)	Right plot scale	FP	40)	P
761	381	191	100	CHDEP(25)	Scale change depth	FP	41)	P
861	431	216	4	IGRID(2)	Type of grid	AN	42)	P
865	433	217	40	IENG(20)	Engineer	AN	43)	P
905	453	227	80	ICOM(40)	Comments	AN	44)	0
985	493	247	40	FILLB(10)				

WELEX

The Welex Library Tape format conforms to the LIS format with all of the reel, tape, and file headers and trailers. It is not set up to continue reels or tapes, but can have multiple files for separate wells or different runs on the same well.

The tapes are 9 track, 800 or 1600 BPI. Standard bit density is 1600, unless 800 is requested. The merged log data is in DEC10 internal floating-point numbers, but a representation code is available as an option for Data General internal floating point, which is the same as IBM.

Verification copies are sent out with all tapes, with a printout of all the records except the data records, which will generally be the first frame of each data record.

GEARHART

Gearhart Industries DDL Customer Tapes are recorded on 9-track tapes at 800 BPI, although field recordings are made on 4-track cartridges.

The bit pattern array is one or more bytes (maximum 8). The *data word format* is as follows:

Alphanumeric	Raytheon ASCII character set
Integer*2	INTEGER*2 16-bit, two's complement
Integer*1	INTEGER*1 unsigned 8-bit number
Integer*4	INTEGER*4 32-bit (Raytheon type)

The *record structure* is as follows: The first record will be a *library header record* and it is entirely in ASCII code generated by the operator. Following the EOF will be an operator-generated record, the *log descriptor record.* Its purpose is to describe the cartridge image that follows.

The sequence in the preceding paragraph is repeated for each cartridge tape. After the last cartridge image, the operator closes the tape with a *library trailer record.* Following the library trailer record will be:

Type 0 merged data tape descriptor record
1 calibration data record
2 log data record
3 log stop record
4 statistical data record
7 log complete record

The length of record is variable in most cases.

API-D9

The reader is referred to the *API Bulletin D9,* third edition, February 1981, for a complete description of this format. Sections 1 through 7 of the bulletin are reproduced here (without the tables) together with the API D9 tape format organization diagram (see fig. 9.11).

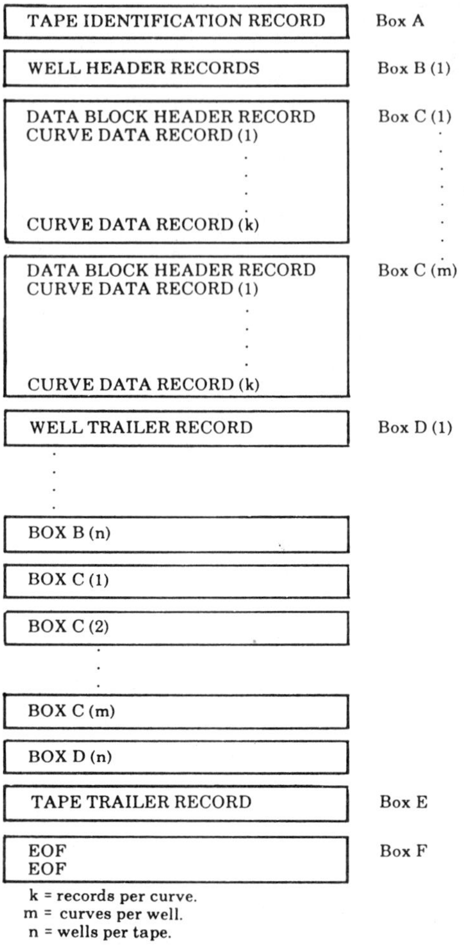

FIGURE 9.11 *API D9 Tape Format Organization. Reprinted by permission of the API, from the* API Bulletin D9, *3rd ed. (1981).*

RECOMMENDED STANDARD FORMAT FOR RECORDING DIGITAL WELL LOG DATA ON MAGNETIC TAPE*

Section 1. Introduction

1.1 *Historical Comments.* The first API Subcommittee on Format for Recording Logging Signals for Data Processing originally initiated work to standardize wellsite digitization of logs. It soon became apparent that such standards could not be practically implemented by well logging contractors. Consequently, that Subcommittee developed format standards applicable to digital log data subsequent to wellsite operations. The resulting *API Bulletin D9* was issued in August 1965 and revised in February 1973. New logging tool developments, computer hardware improvements, and substantial digital log processing experience gained by the industry in recent years warranted a review and update of *Bulletin D9,* Second Edition.

1.2 *Format Purpose and Objectives.* A revised recommended format, designated as *API2,* has been developed and is documented herein. The purpose of this new format is to facilitate the *exchange* of digital log data between various users and organizations. The format is not necessarily intended as a specification for company proprietary log processing, or storage and retrieval systems; however, it may be useful as a guide for developing such inhouse formats. The objective is to provide a format that can be conveniently used by many organizations, particularly those with limited computer hardware and programming resources. Several major innovations have been included in the *API2* format, and a number of options existing in the previous format have been deleted, in an effort to assure long-term use.

1.3 *Format Innovations.* The format specifications are designed to be readable by a computer program written in FORTRAN II. The Metric Système International d'Unites (SI) is used for all units of measure within the format. The *minimum* information required to read a tape in the *API2* format is specified. The numerical code system for describing log type, curve type, and curve class has been expanded and updated. An alphanumeric curve identifier has been added to the format as an alternative or supplement to this numerical curve coding system.

1.4 *Suggested Revisions.* Continual industry development of new tools, technology, and applications will establish needs for revision of the contents of this Bulletin. Suggestions for revisions or additions are welcomed and should be submitted to the Director, API Production Department, 211 N. Ervay, Suite 1700, Dallas TX 75201. All revision suggestions will be periodically called to the attention of responsible industry administrative committees. These committees will render a decision as to the appropriate timing to issue a revised edition or supplement to the existing edition of Bulletin D9.

1.5 *Bulletin Organization.* Section 2, "Data Representation," defines the character sets, logical and physical record lengths, and decimal point notation used in the format. The types, numbers, and organization of records are presented in Section 3, "Format Organization." The specific information contained in each record is summarized in Section 4, "Information Content," and Section 5, "Numerical Codes." Metric (SI) units and conversion factors for most common well logging terms are noted in Section 6, "Metrication and Units." Section 7, "Computer Program Documents," includes a FORTRAN program, which will read an *API2* format tape and a sample tape dump in 80-character format.

Section 2. Data Representation

2.1 *Character Sets and Bit Configuration.* The recording should be EBCDIC (9-track, odd parity, 800 bpi preferred, 1600 bpi acceptable) or BCD (7-track, even parity, 800 bpi preferred, 556 bpi acceptable). An exact description of the bit configuration for both the EBCDIC and BCD character sets is listed in Table 1.

2.2 *Record and Character Lengths.* Logical and physical records are 80 characters long. A character is defined as an 8-bit EBCDIC or 6-bit BCD character.

2.3 *Decimal Point Notation.* Implied decimal point notation is used to represent all numerical information.

a. Decimal point location for curve data is specified as part of the format information; details appear under "Information Content."
b. Decimal point location for all other data is specified in the API format documentation; details appear under "Information Content."

2.4 *Trace Sequential Specification.* All well log curves are formatted trace-sequentially, i.e., serially.

2.5 *Depth Specification.* All well log curve data are referred to wireline, i.e., measured depths.

Section 3. Format Organization

3.1 *Record Types.* The format includes records of the following types:

a. Tape Identification
b. Well Header (10 record types)
c. Data Block Header
d. Curve Data

e. Well Trailer
f. Tape Trailer

[Figure 9.14] shows the record types required and organization for the tape format.

3.2 *Tape Identification Record.* The Tape Identification Record contains the characters "*API2*" as the first four characters. The remainder of the record is free field.

3.3 *Well Header Records.* Well Header Records are of the following types:

a. Well Location .... Type 01
b. Well Name .... Type 02
c. Elevation Reference .... Type 03
d. Logging Data .... Type 04
e. Casing Information .... Type 05
f. Drilling Bit Information .... Type 06
g. Drilling Fluid Information .... Type 07
h. Timing Information .... Type 08
i. Well Comments .... Type 09
j. Log Comments .... Type 10

Data included on these records appear under "Information Content." At least one Well Header Record of each type must be present for each well.

1. Only one each of types 01, 02, and 03 above will be accepted; each has a sequence number of 99.
2. From 1 to 99 of types 04 through 10 will be accepted.
3. Well Header Records must be in increasing numerical order by type.
4. The last (or only) record of each type is designated by a sequence number of 99.

3.4 *Data Block Header Record.* The Data Block Header Record contains information necessary to read the Curve Data Records which immediately follow; data included on this record appear under "Information Content."

a. A record type of 98 is assigned to the Data Block Header Record.

3.5 *Curve Data Record.* Each Curve Data Record contains a 10-character depth field followed by ten 7-character data fields.

a. The first character in each data field will be reserved for alpha notation only.
b. Each set of curve data records will have data fields that are monotonic with depth.
c. Missing data fields will be designated by −99999.

d. The last Curve Data Record in a set is designated by having only a depth field of −99999999 specified.

3.6 *Well Trailer Record.* A Well Trailer Record is designated by a Data Block Header Record having a record type of 98 and a sequence number of 99.

3.7 *Tape Trailer Record.* A Tape Trailer Record is designated by a record having a record type of 99 and a sequence number of 99.

3.8 *Minimum Record Specification.* The minimum records required to read an *API2* format tape are as follows:

a. Tape Identification
b. Well Header Record Types 01–10 (one each)
c. Data Block Header and Curve Data Records (if curve data is available)
d. Well Trailer
e. Tape Trailer

The minimum information required on these records is defined under "Information Content."

Section 4. Information Content

4.1 *General Comments.* The information content specifications for each record type are shown in Tables 2-16. The format and units associated with each information item are defined. Metric units are required throughout the *API2* format; details, including conversion factors, appear under "Metrication and Units." Information defined by numerical codes is identified and references provided; details on codes unique to the *API2* format appear under "Numerical Codes."

4.2 *Tape Identification Record.* Table 2 defines the Tape Identification Record. The record must contain the format identification "*API2*" in the first four characters. The remainder of the record may contain any information in free format.

4.3 *Well Header Records.* Tables 3-12 define Well Header Record Types 01–10, respectively.

4.4 *Data Block Header Record.* Table 13 defines the Data Block Header Record. The record contains all information necessary to read Curve Data Records, which follow it in the format organization.

a. The Sample Increment information must contain a sign indicating downhole (+) or uphole (−) orientation, or progression, of the digitized data. Lack of sign implies (+). A Sample Increment of zero or blank implies that the data records following do not have a uniform sampling interval, but are arranged with one depth and one data value per record. The data may be in any depth sequence.

b. The Implied Decimal Locator information must also contain a sign indicating right (+) or left (−) shift of the decimal point from a zero location at the right edge of the data fields. The decimal point is shifted a number of locations equal to the value of the Implied Decimal Locator. Lack of sign implies (+).
c. The Curve Identification information is available for alphanumeric description of a curve as an alternative or supplement to the Curve Codes.
d. The Curve Identification is at the user's option and definition.

4.5 *Curve Data Record.* Table 14 defines the Curve Data Record.

a. The first character in each 7-character data field is reserved for an alpha (only) notation related to that data value. These alpha notations are undefined in the *API2* format.

4.6 *Well Trailer and Tape Trailer Records.* Tables 15 and 16 define the Well Trailer and Tape Trailer Records, respectively.

4.7 *Minimum Information Content Specification.* The minimum information required for an *API2* format tape is as follows:

a. Unique Well Identification (primary or secondary)
b. Well Name and Number (nonblank)
c. First Sample Depth
d. Sample Increment
e. Implied Decimal Location
f. Log Type Code, Curve Type Code, Curve Class Code, and Curve Modifier Code: or Curve Identification (alphanumeric)

Section 5. Numerical Codes.

5.1 *General Comments.* Numerical codes are required for definition of specific information items on the Well Header Records and Data Block Header Records. These items are generally identified as "codes" under "Information Content." Numerical code specifications unique to the *API2* format are shown in Tables 17-23; other available codes utilized in the format are specified in References 2 and 7 (Section 8).

5.2 *Header Record Codes.* References 2 and 7 specify codes required for the Unique Well Identification (primary) and related geographic information. For convenience, Country Codes (from Reference 7) are reproduced in Table 18. The remaining Header Record codes are shown as follows:

a. Table 17—Company Codes
b. Table 19—Mud Codes
c. Table 20—Log Type Codes

The Company Codes specify Service Contractor, Logging Company, and Digitizing Company information.

5.3 *Data Block Header Record Codes.* The Curve Codes included in the Data Block Header Record are shown as follows:

a. Table 21—Curve Type Codes
b. Table 22—Curve Class Codes
c. Table 23—Curve Modifier Codes

Table 24 illustrates how curve data are identified using the Curve Codes specified in Tables 21-23.

Section 6. Metrication and Units

6.1 *General Comments.* All units of measure within the *API2* format are in metric units of the Système International d'Unites, i.e., the SI System. The objective of this recommendation is to assure long-term utility of the format.

6.2 *Well Header Data Units.* The units of measure associated with information on the Well Header Records appear under "Information Content" and are specified in Tables 3-13. Conversion factors and associated units are shown in Table 25.

6.3 *Curve Data Units.* The units of measure for well log curve data on the Curve Data Records are specified in Table 21. Conversion factors and associated units for the more common log curve data are shown in Table 26. Conversion factors for other well log units of measure are specified in References 3, 4, and 5 (Section 8). Curve data having unspecified units should be represented as a fraction of the deflection between the upper and lower scale limits.

6.4 *Rounding Values.* The procedure for rounding a figure to fewer digits than the total number available is as follows:

a. When the first digit discarded is less than 5, the last digit retained is unchanged (3.46325 becomes 3.463 when rounded to four digits).
b. When the first digit discarded is greater than or equal to 5, the last digit retained is increased by one unit (8.37652 becomes 8.377 when rounded to four digits).

Rounding the specifications are shown in Tables 25 and 26.

Section 7. Computer Program

7.1 *General Comments.* Several computer documents are included as part of this Bulletin. The purpose of these documents is to provide an illustration of the *API2* format specifications contained herein.

7.2 *Program Listing.* Appendix A is a FORTRAN computer program which will read an *API2* format tape and also produce the listing shown in Appendix C. The program is not intended for direct user application; however, it may be used as a guide for development of such programs. Comment statements in the program note the scope and limitations.

7.3 *Tape Dump Example.* Appendix B is an 80-character dump of an *API2* format tape containing data from two wells. The listing shows the organization of data on the tape.

7.4 *Mini-Dump Specification.* The definition of an abbreviated, or mini-dump, listing of an *API2* format tape is included as part of the specifications. The mini-dump includes the following records:

a. Tape Identification
b. Well Header (all record types)
c. Data Block Header
d. Curve Data (first and last data records in each set of curve data records)
e. Well Trailer
f. Tape Trailer

Appendix C illustrates an interpreted mini-dump of that data contained in the 80-character tape dump shown in Appendix C.

DIPMETER FORMATS

CSU/LIS

The following curves are stored in the standard LIS format (at a rate of one reading every 3.2 inches):

DEVN (hole deviation)
RELB (relative bearing of pad #1 to high side of tool)
HAZI (azimuth of hole direction) and/or P1AZ (azimuth pad #1)
C1 (caliper 1)
C2 (caliper 2)
SP (optional, spontaneous potential)

An additional curve—normally called RHDT—at least 80 bytes in length, contains the four resistivity curves, plus the accelerometer, each sampled 16 times. The individual readings are therefore 1 byte in length with a maximum range from 0 to 255. This works out to a sampling for each resistivity curve of 60 to the foot.

TTR Dipmeter Tapes

The TTR Dipmeter format is stored on 7-track tape at 200 BPI, gapless except for file ends. Three of the tracks are used for parity bits, depth, and frame synchronization flags. Only 4 tracks remain to hold data values, thus it takes two successive data recordings to make up 1 byte of data. The byte is the recording mode for all data curves with a range of 0 to 255. As a result of this, all of the data is uncalibrated and must be converted to inches or degrees before computing. The sampling rate is the same as with the CSU dipmeter data—that is, 60 or 3.75 readings per foot depending on the type of curve. Because of the byte format, the depth readings are recorded only every 10 feet where the frame is flagged in nondata tracks, and only the hundred and tens digits are recorded.

Dresser Dipmeter Tapes

This format is stored on 9-track tapes, at 800 BPI. There are two types of record, ID records of 29 bytes and data records of 1920 bytes. The data records are split into 20 sequences each of 96 bytes (48 words integer). Each sequence comprises eight *short sequences*.

A short sequence consists of sampling of each of 5 fast curves—that is, 4 resistivity curves with 1 blank curve, followed by a slow curve. In each sequence, the slow curves are the 5 dipmeter computation curves:

1. pad #1 azimuth
2. relative bearing
3. inclination
4. caliper 1
5. caliper 2

Followed by the depth in short sequences 6 and 7, with a sync word in short sequence 8.

The data recordings are in 12-bit integer, the top 2 bits in each byte not being set. The sampling rate of the resistivity curves is 64 to the foot, and of the other curves 8 to the foot.

Gearhart Dipmeter Tapes

Normally, Gearhart dipmeter tapes are recorded on 800 BPI cassettes; each cassette holds 4 separate recording tracks.

Each record has 1024 words of data. The first 24 words contain the format specification and header. The most important of these are word 18, which is the format specification number determining the order of data curves; word 12, which gives the size of each data sequence; and word 4, which gives the number of sequences in the record. Word 6 gives the number of samples to the foot (60 is standard), and word 1 gives the record type (2 for data). All curves occur on the tape at the same frequency. The depth is recorded only once on each record, in the second and third words, to the nearest 0.10 ft.

APPENDIX A: TOOL DIMENSIONS

Planning a logging suite requires knowledge of service-company tool dimensions and ratings. This Appendix illustrates the various types of tool (e.g., induction) and tabulates, by service company, the pertinent data for each.

Accurate log quality checks also require knowledge of the memorization lengths for each multiple-sensor tool string and the key distance from lowest measure point to the bottom of the tool. Since there are so many possible combinations, the safest procedure is to have the logging engineer either physically measure the respective distances or provide an up-to-date listing from his service catalog.

Figure A shows schematically the important dimensions of a generalized logging tool. These are the dimensions that should be known before a logging tool goes in the hole.

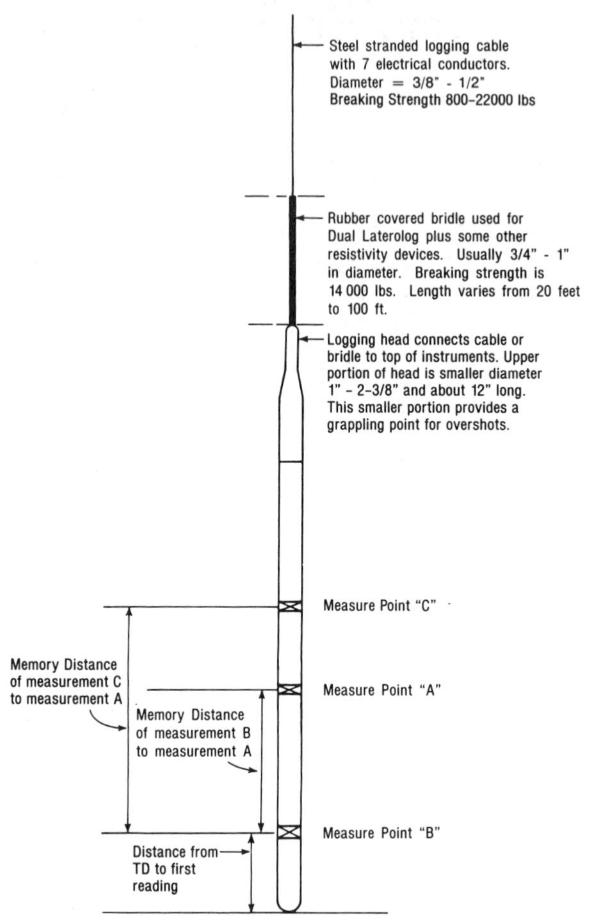

FIGURE A *Key Tool Dimensions.*

Induction Electrical Log

Service Company	Gearhart	Dresser	Schlumberger	Welex
Trademark Name	Induction Electrical Log (IEL)	Induction Electrolog (IEL)	Induction Electric Survey (IES)	Induction Electric Log
Distance from bottom to measure point, Ft	1	5	1	5
Length, Ft	16	20	20	21
Diameter, Ins	3 7/8	3 5/8	3 7/8 (2.75*)	3 3/8
Weight, Lbs	174	225	225	225
Max Temp, °F	350	350 (400*)	350 (500*)	300
Max Press, PSI	18500	20000 (25000*)	20000 (25000*)	20000
Minimum Hole Size, Ins	5	5	5	5
Maximum Hole size, Ins	24	24	24	24
Oil base Mud	Yes	Yes	Yes	Yes
Cased Hole	No	No	No	No
Air or Gas Filled Holes	Yes	Yes	Yes	Yes
Tools that can be combined	Caliper Gamma Ray Micro Electrical Log Sonic Density Neutron	Gamma Ray	Gamma Ray	Caliper Gamma Ray Neutron

* Available by Special Order

Induction Spherically Focused Log

Service Company	Gearhart	Dresser	Schlumberger	Welex
Trademark Name			Induction Spherically Focused Log (ISF)	
Distance from bottom to measure point, Ft			2	
Length, Ft				
Diameter, Ins			3.5	
Weight, Lbs			175	
Max Temp, °F			350	
Max Press, PSI			20000	
Minimum Hole Size, Ins			4.5	
Maximum Hole size, Ins			24	
Oil base Mud			Yes	
Cased Hole			No	
Air or Gas Filled Holes			Yes	
Tools that can be combined			Sonic Density Neutron Gamma Ray Caliper	

Dual Induction

Service Company Trademark Name	Gearhart Dual Induction Laterolog	Dresser Dual Induction Focused Log	Schlumberger Dual Induction Spherically Focused Log	Welex Dual Induction Log
Distance from bottom to measure point, Ft	1.5		5	5
Length, Ft	21.2	24	21	20
Diameter, Ins	4	3 5/8	3 5/8	3 5/8
Weight, Lbs	345	306	300	300
Max Temp, °F	300	350 (400*)	350	325
Max Press, PSI	15000	18000 (25000*)	20000	15000
Minimum Hole Size, Ins	5	5	5	5
Maximum Hole size, Ins	24	24	24	24
Oil base Mud	Yes	Yes	Yes	Yes
Cased Hole	No	No	No	No
Air or Gas Filled Holes	Yes	Yes	Yes	Yes
Tools that can be combined	Sonic Micro Electrical log Density Neutron Caliper Gamma Ray	Sonic Density Neutron Caliper Gamma Ray	Sonic Density Neutron Caliper Gamma Ray	

* Available by Special Order

Dual Laterolog

Service Company	Gearhart	Dresser	Schlumberger	Welex
Trademark Name	Dual Laterolog	Dual Laterolog	Dual Laterolog (DLL)	Dual Guard Log
Distance from bottom to measure point, Ft	8.5			
Length, Ft	17.5	29.5	30	
Diameter, Ins	3.5	3 5/8	3 5/8	3 3/8
Weight, Lbs	275			
Max Temp, °F	350	400	350 (500*)	350
Max Press, PSI	20000	20000	20000 (25000*)	20000
Minimum Hole Size, Ins	5	5	5	5
Maximum Hole size, Ins	16	16	16	16
Oil base Mud	No	No	No	No
Cased Hole	No	No	No	No
Air or Gas Filled Holes	No	No	No	No
Tools that can be combined	Gamma Ray Micro SFL Caliper	Gamma Ray	Gamma Ray Micro SFL Caliper	Gamma Ray Caliper Neutron Density F_o R_{xo}

* Available by Special Order

Microlaterolog

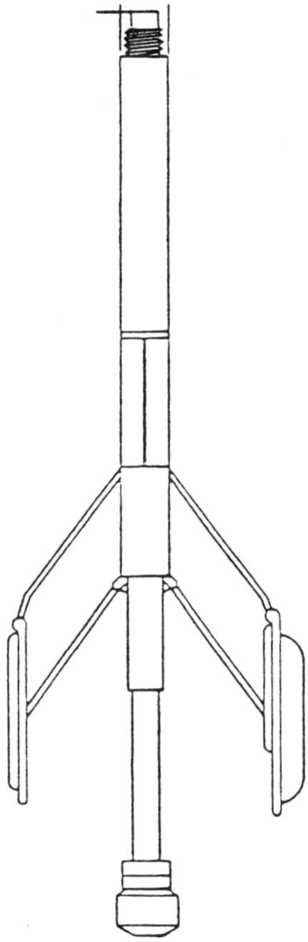

Service Company	Gearhart	Dresser	Schlumberger	Welex
Trademark Name	Microlaterolog	Micro Laterolog	Microlaterolog (MLL)	$F_o R_{xo}$
Distance from bottom to measure point, Ft	1	1	1	1
Length, Ft	6.4	12	12	
Diameter, Ins	3.5	4.5	5 1/16	3 7/8
Weight, Lbs	85		150	
Max Temp, °F	300	300	350	300
Max Press, PSI	15000	20000	20000	20000
Minimum Hole Size, Ins	5	6	6.5	5.5
Maximum Hole size, Ins	16	16	20	16
Oil base Mud	No	No	No	No
Cased Hole	No	No	No	No
Air or Gas Filled Holes	No	No	No	No
Tools that can be combined	Gamma Ray	Gamma Ray	Microlog	Gamma Ray

Microlog

Service Company	Gearhart	Dresser	Schlumberger	Welex
Trademark Name	Micro-Electrical Log	Minilog	Microlog (ML)	Contact Log
Distance from bottom to measure point, Ft	1	1	1	1
Length, Ft	6.4	7	12	
Diameter, Ins	4	3.75	5 1/16	3 7/8
Weight, Lbs	85		150	
Max Temp, °F	300	300	350	300
Max Press, PSI	15000	20000	20000	20000
Minimum Hole Size, Ins	5	5.5	6.5	5.5
Maximum Hole size, Ins	16	16	20	16
Oil base Mud	No	No	No	No
Cased Hole	No	No	No	No
Air or Gas Filled Holes	No	No	No	No
Tools that can be combined	Gamma Ray Induction Electrical Dual Induction		Microlaterolog	Gamma Ray

Micro Spherically Focused Log

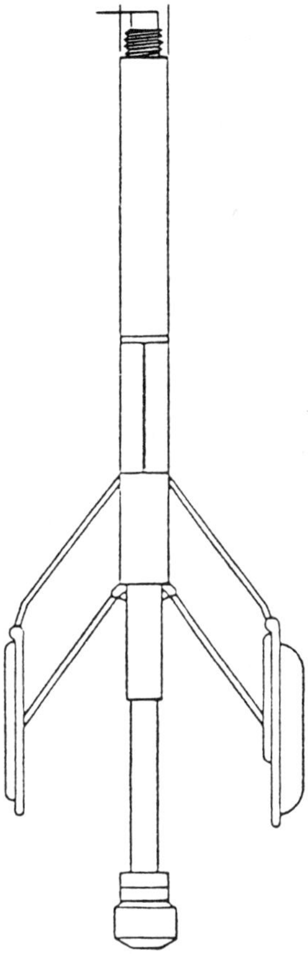

Service Company	Gearhart	Dresser	Schlumberger	Welex
Trademark Name			Micro-SFL (MSFL)	Not Available
Distance from bottom to measure point, Ft			2	
Length, Ft				
Diameter, Ins			5 1/4	
Weight, Lbs				
Max Temp, °F			350	
Max Press, PSI			20000	
Minimum Hole Size, Ins			6.5	
Maximum Hole size, Ins			22	
Oil base Mud			No	
Cased Hole			No	
Air or Gas Filled Holes			No	
Tools that can be combined			Dual Laterolog Induction Spherically Focused Log	

Compensated Neutron

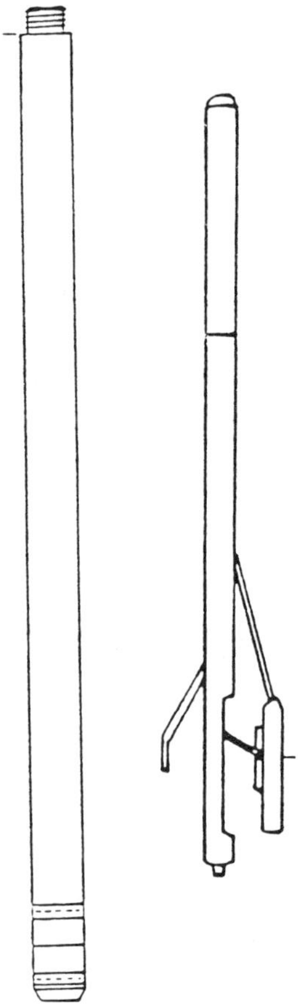

Service Company	Gearhart	Dresser	Schlumberger	Welex
Trademark Name	Compensated Neutron (CNS)	Compensated Neutron Log (CN)	Compensated Neutron Log (CNL)	Neutron Log Dual Spaced (DSN)
Distance from bottom to measure point, Ft	2		2	
Length, Ft	9.75	6 7	6	17.6
Diameter, Ins	4	3 5/8 (2.75*)	3 3/8 (2.75*)	3 3/8
Weight, Lbs	250			
Max Temp, °F	300	300	400 (500*)	400
Max Press, PSI	20000	20000	20000 (25000*)	20000
Minimum Hole Size, Ins	5	4.5	4 1/4	4
Maximum Hole size, Ins	18	14	16	16
Oil base Mud	Yes	Yes	Yes	Yes
Cased Hole	No	Yes	Yes	Yes
Air or Gas Filled Holes	No	No	No	No
Tools that can be combined	Density Dual Induction Induction Electrical Micro Electrical Gamma Ray Caliper	Density Dual Induction Induction Electrical Gamma Ray Caliper	Density Sonic Induction SFL Dual Induction Gamma Ray Caliper	

* Available by Special Order

Sidewall Neutron

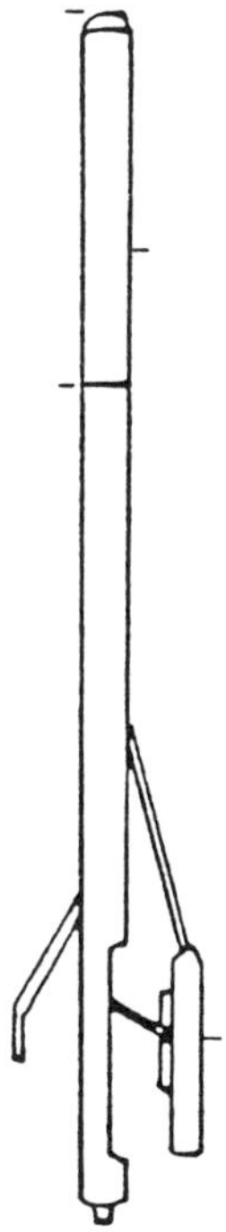

Service Company	Gearhart	Dresser	Schlumberger	Welex
Trademark Name	Sidewall Neutron (SNL)	Sidewall Epithermal Neutron Log	Sidewall Neutron Porosity Log (SNP)	Sidewall Neutron Log
Distance from bottom to measure point, Ft	2	2	2	2
Length, Ft	10.5	9.1		21
Diameter, Ins	4	4 7/8 (3*)	4 3/8	4 1/8
Weight, Lbs	275		275	
Max Temp, °F	300	300	350	350
Max Press, PSI	17500	20000	20000	20000
Minimum Hole Size, Ins	5	6	5.5	5.5
Maximum Hole size, Ins	18	16	16	20
Oil base Mud	Yes	Yes	Yes	Yes
Cased Hole	No	No	No	No
Air or Gas Filled Holes	Yes	Yes	Yes	Yes
Tools that can be combined	Density Gamma Ray Micro Electrical Log Caliper	Gamma Ray Caliper	Gamma Ray Caliper	Gamma Ray Caliper Density

* Available by Special Order

Compensated Density

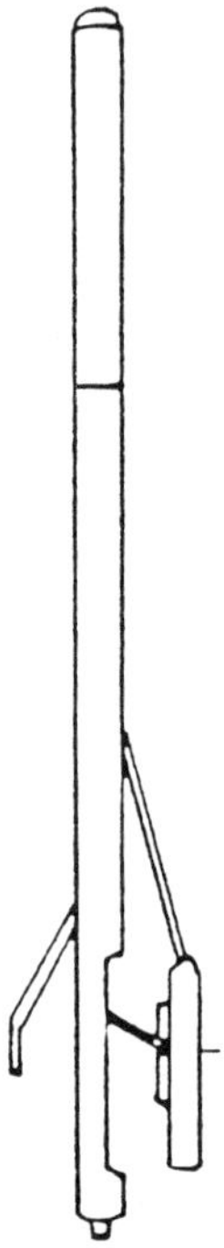

Service Company	Gearhart	Dresser	Schlumberger	Welex
Trademark Name	Compensated Density (CDL)	Compensated Densilog	Density Log Formation (FDC)	Density Log
Distance from bottom to measure point, Ft	2	2	2	2
Length, Ft	9	10		
Diameter, Ins	4	4 7/8 (3*)	4 5/8 (2.75*)	4 1/8
Weight, Lbs	275		275	
Max Temp, °F	350	300	350 (500*)	350
Max Press, PSI	17500	20000	20000 (25000*)	20000
Minimum Hole Size, Ins	5	6	6	5 1/4
Maximum Hole size, Ins	18	16	21	18
Oil base Mud	Yes	Yes	Yes	Yes
Cased Hole	No	No	No	No
Air or Gas Filled Holes	Yes	Yes	Yes	Yes
Tools that can be combined	Neutron Gamma Ray Sidewall Neutron Dual Induction Induction Electrical Micro Electrical	Gamma Ray Neutron Dual Induction Induction Electrical	Gamma Ray Neutron Induction SFL Dual Induction	Gamma Ray Neutron Contact Log

* Available by Special Order

Compensated Sonic

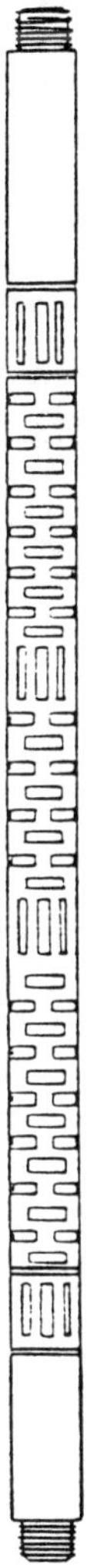

Service Company	Gearhart	Dresser	Schlumberger	Welex
Trademark Name	Borehole Compensated Sonic (BCS)	Borehole Compensated Acoustilog	Borehole Compensated Sonic Log (BHC)	Acoustic Velocity Log
Distance from bottom to measure point, Ft	8		5	
Length, Ft	16	19.5	19	
Diameter, Ins	3.5	3 3/8	3 3/8 (1 11/6*)	3 5/8
Weight, Lbs	255		250	
Max Temp, °F	350	350	350 (500*)	400
Max Press, PSI	18500	20000	20000 (25000*)	20000
Minimum Hole Size, Ins	4	4 1/4	4	4 1/4
Maximum Hole size, Ins	24	24	24	24
Oil base Mud	Yes	Yes	Yes	Yes
Cased Hole	No	No	No	No
Air or Gas Filled Holes	No	No	No	No
Tools that can be combined	Gamma Ray Dual Induction Induction Electrical Caliper	Gamma Ray Dual Induction Induction Electrical Caliper	Gamma Ray Caliper Induction Spherical Focused Log Dual Induction	Gamma Ray Dual Induction Neutron

* Available by Special Order

Repeat Formation Tester

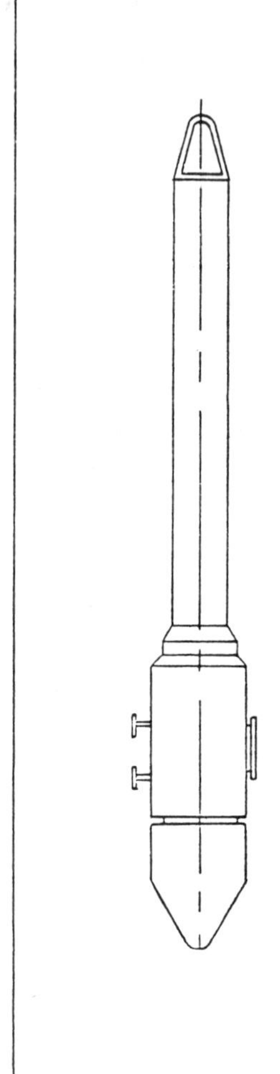

Service Company	Gearhart	Dresser	Schlumberger	Welex
Trademark Name	Selective Formation Tester (SFT)	Formation Multi-Tester (FMT)	Repeat Formation Tester (RFT)	Multiset Tester
Distance from bottom to measure point, Ft	1.5 *			
Length, Ft	8	35.5	33	
Diameter, Ins	5.5	6 1/4	5 1/8	6 1/8
Weight, Lbs				
Max Temp, °F	350	350	350	350
Max Press, PSI	15000	15000	20000	20000
Minimum Hole Size, Ins	6 1/4	7.5	6	7.5
Maximum Hole size, Ins	12.75	14	14.75	13
Oil base Mud	Yes	Yes	Yes	Yes
Cased Hole	Special Tool	No	Special Tool	No
Air or Gas Filled Holes	NA	NA	NA	NA
Tools that can be combined	Gamma Ray SP	Gamma Ray SP	Gamma Ray SP	Gamma Ray
Gallon Chambers available	2 1/4 & 5		1,2.75,6,& 12	1,2.5, & 7

* Distance from Test Point to bottom depend upon sample Chamber Size.
All distances given are minimum.

Core Gun

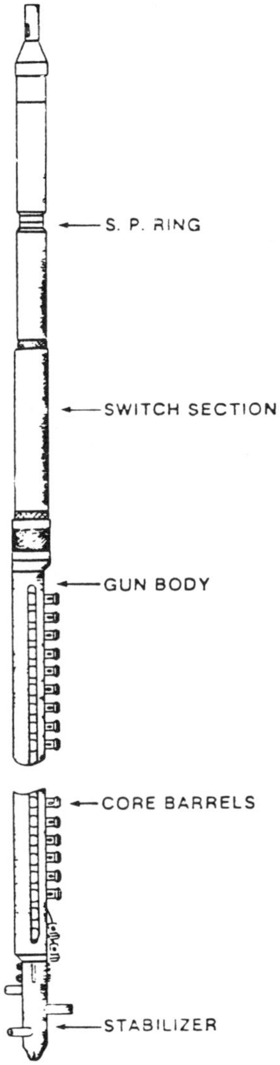

Service Company	Gearhart	Dresser	Schlumberger	Welex
Trademark Name	Sidewall Core Gun (SWC)	Corgun	Sidewall Sampler (CST)	Sidewall Coring
Distance from bottom to measure point, Ft				
Length, Ft	6.8	10.4		
Diameter, Ins	4	4	5 1/4	4 7/8
Weight, Lbs				
Max Temp, °F	350	350	280 (430*)	350
Max Press, PSI	20000	20000	20000	20000
Minimum Hole Size, Ins	6	5.5	6.5	6
Maximum Hole size, Ins			14	
Oil base Mud	Yes	Yes	Yes	Yes
Cased Hole	No	No	No	No
Air or Gas Filled Holes	No	No	No	No
Tools that can be combined	Gamma Ray SP	Gamma Ray SP	Gamma Ray SP	Gamma Ray SP
Maximum Number of Cores/trip	46	50	72	60
Diameter of Core Barrels, Ins	7/8, 1	1 1/8, 1 13/16, 11/16		3/4, 7/8, 1

* Available by Special Request

S. P. RING
SWITCH SECTION
GUN BODY
CORE BARRELS
STABILIZER

4-Arm Dipmeter

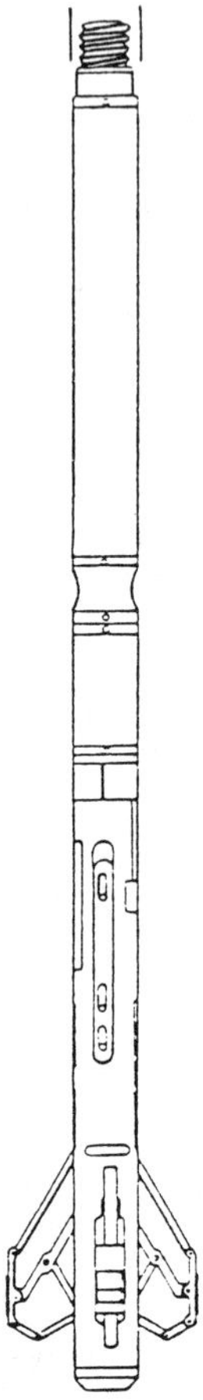

Service Company	Gearhart	Dresser	Schlumberger	Welex
Trademark Name	Four Electrode Dipmeter	Hi-Resolution 4-Arm Diplog	High Resolution Dipmeter Tool (HDT)	Dip Log 4-Arm
Distance from bottom to measure point, Ft	1	1	1	1
Length, Ft	15 9	26		
Diameter, Ins	5.5	4 9/10	4.5	4.5
Weight, Lbs				
Max Temp, °F	300	350	350	350
Max Press, PSI	20000	20000	20000	20000
Minimum Hole Size, Ins	6	5.5	5	5
Maximum Hole size, Ins	16	18	18	18
Oil base Mud	No	Yes, w/special electrodes	Yes, w/special electrodes	No
Cased Hole	No	No	No	No
Air or Gas Filled Holes	No	Yes, w/special electrodes	Yes, w/special electrodes	No
Tools that can be combined	Gamma Ray	Gamma Ray	Gamma Ray	

Caliper Log

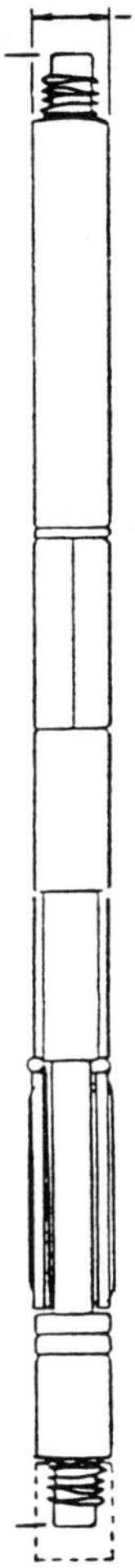

Service Company	Gearhart	Dresser	Schlumberger	Welex
Trademark Name	X-Y Caliper (4 arm)	4-Arm Dual Caliper	Borehole Geometry Tool (BGT)	4-Arm Single Service Caliper
Distance from bottom to measure point, Ft	2.5			
Length, Ft	8 11	9 9		
Diameter, Ins	4	2.75	4	3
Weight, Lbs	275			
Max Temp, °F	350	400	350	300
Max Press, PSI	18500	20000	20000	20000
Minimum Hole Size, Ins	5	3.5	5	4
Maximum Hole size, Ins	60		30	
Oil base Mud	Yes	Yes	Yes	Yes
Cased Hole	Yes	Yes	Yes	Yes
Air or Gas Filled Holes	Yes	Yes	Yes	Yes
Tools that can be combined	Dual Induction Induction Electrical Gamma Ray			

Gamma Ray

Service Company	Gearhart	Dresser	Schlumberger	Welex
Trademark Name	Gamma Ray	Gamma Ray	Gamma Ray	Gamma Ray
Distance from bottom to measure point, Ft	2		2	
Length, Ft	3		5	
Diameter, Ins	3.5	3 5/8	3 3/8 (2.75*)	3 5/8 (1 5/8*)
Weight, Lbs	50		50	
Max Temp, °F	300 (350*)	400	350 (500*)	350 (375*)
Max Press, PSI	20000	20000	20000 (25000*)	20000
Minimum Hole Size, Ins	4	4.5	4.5	4.5
Maximum Hole size, Ins				
Oil base Mud	Yes	Yes	Yes	Yes
Cased Hole	Yes	Yes	Yes	Yes
Air or Gas Filled Holes	Yes	Yes	Yes	Yes
Tools that can Combined	Gamma Ray can be combined with almost any other tool.			

* Available by Special Order.

APPENDIX B: GEOLOGIC TIME SCALE

ERA	PERIOD	EPOCH	YEARS BEFORE THE PRESENT
Cenozoic	Quaternary	Holocene (Recent)	11,000
		Pieistocene (Glacial)	500,000 to 2,000,000
	Tertiary	Pliocene	13,000,000
		Miocene	25,000,000
		Oligocene	36,000,000
		Eocene	58,000,000
		Paleocene	63,000,000
Mesozoic	Cretaceous		135,000,000
	Jurassic		180,000,000
	Triassic		230,000,000
Paleozoic	Permian		280,000,000
	Carboniferous: Pennsylvanian (Upper Carboniferous)		310,000,000
	Carboniferous: Mississippian (Lower Carboniferous)		345,000,000
	Devonian		405,000,000
	Silurian		425,000,000
	Ordovician		500,000,000
	Cambrian		600,000,000
Precambrian			

APPENDIX C: WELL SYMBOLS

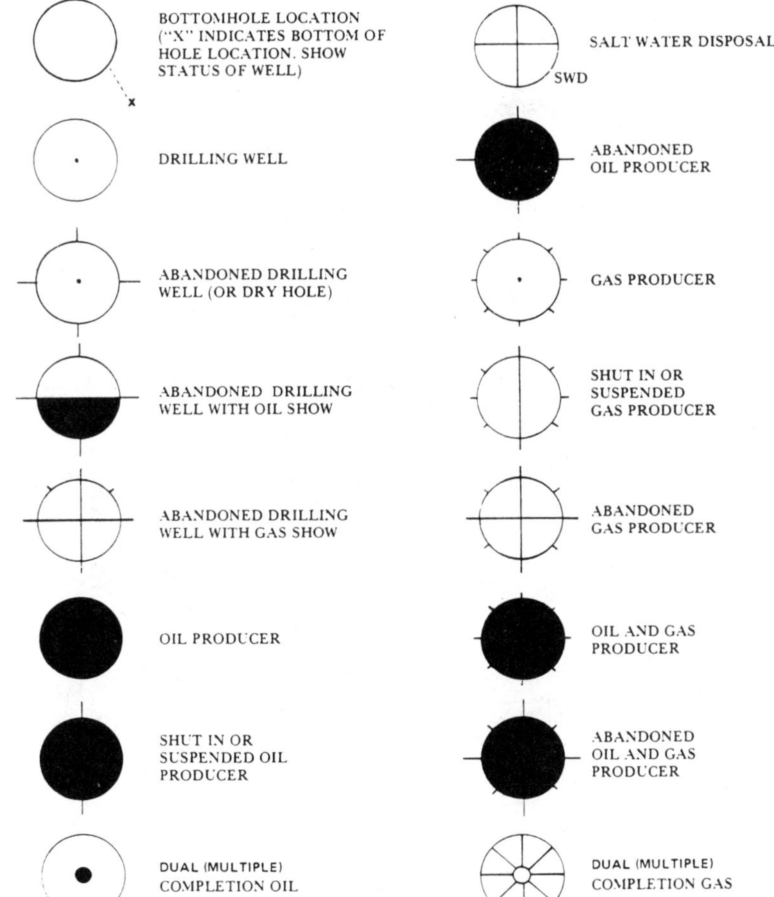

APPENDIX D: LITHOLOGY SYMBOLS

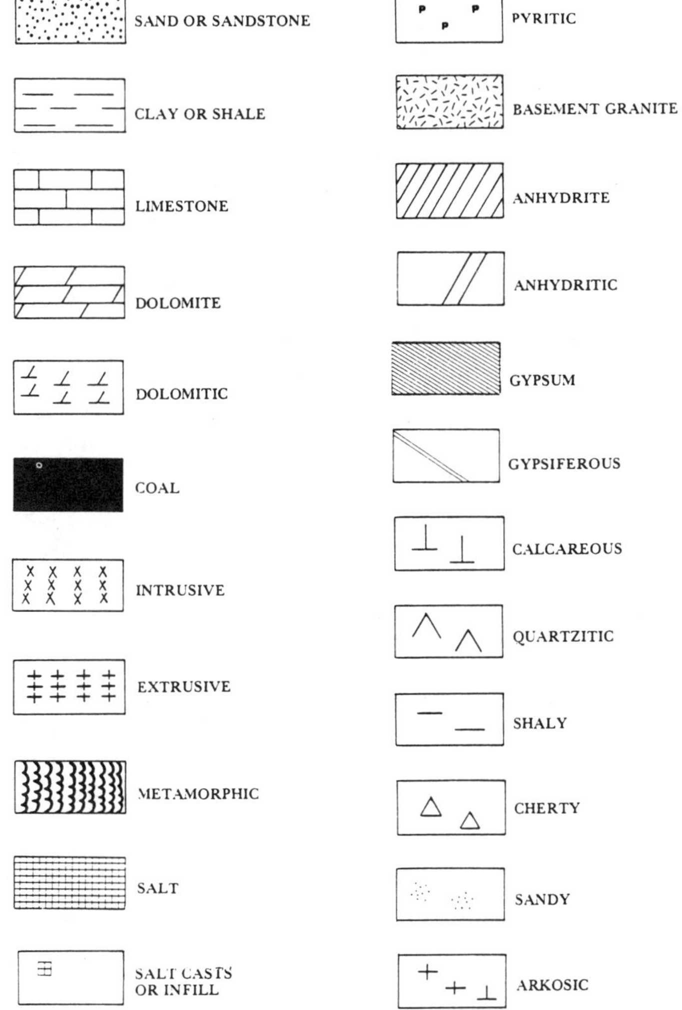

APPENDIX E: DESCRIPTIVE SYMBOLS USED ON WELL LOGS

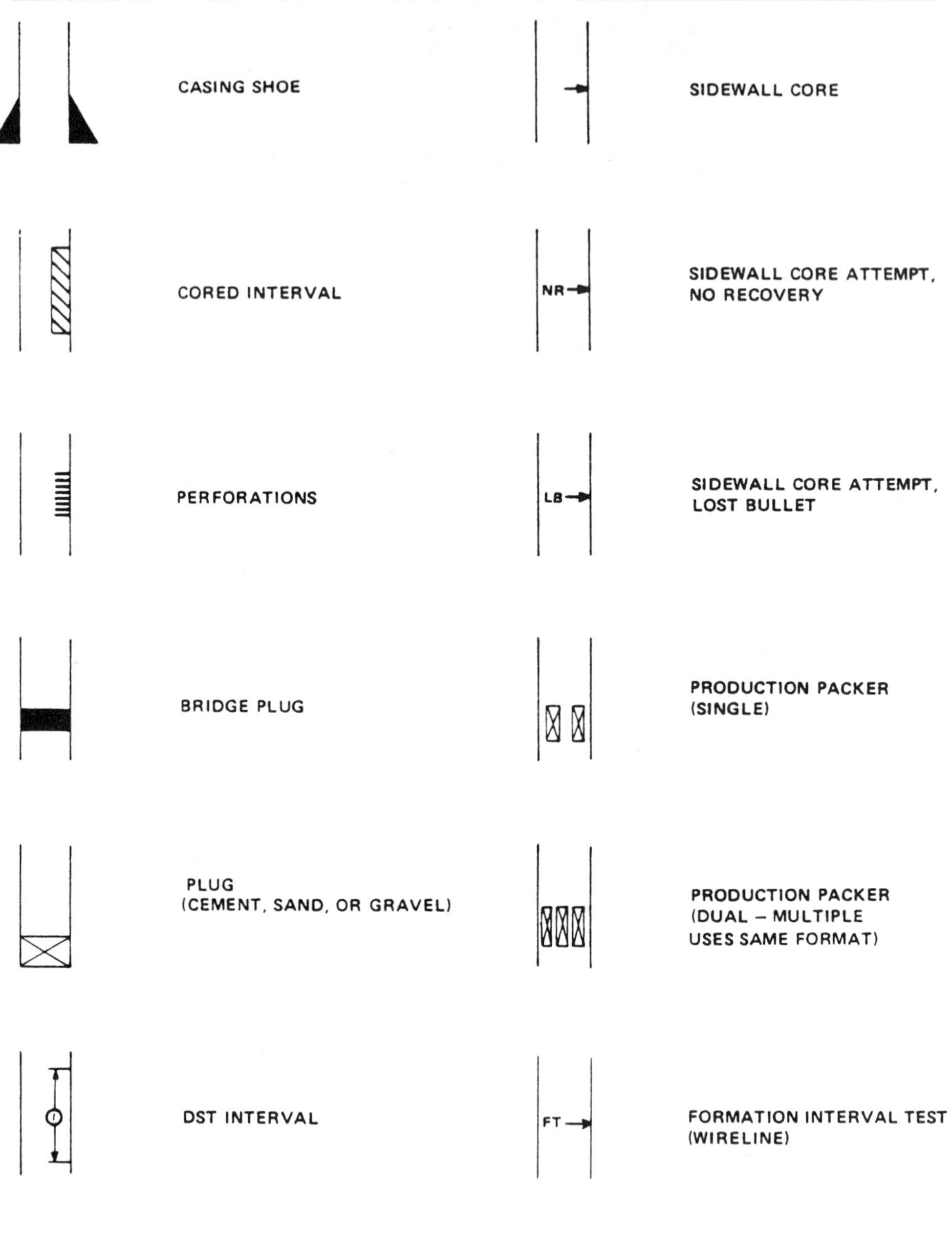
CASING SHOE
SIDEWALL CORE
CORED INTERVAL
NR
SIDEWALL CORE ATTEMPT, NO RECOVERY
PERFORATIONS
LB
SIDEWALL CORE ATTEMPT, LOST BULLET
BRIDGE PLUG
PRODUCTION PACKER (SINGLE)
PLUG (CEMENT, SAND, OR GRAVEL)
PRODUCTION PACKER (DUAL – MULTIPLE USES SAME FORMAT)
DST INTERVAL
FT
FORMATION INTERVAL TEST (WIRELINE)

APPENDIX F: WELL LOG DEPTH SCALE COMPARISON

The three basic log scale systems are:

FOOT–DUODECIMAL (inches per hundred feet)
FOOT–DECIMAL (ratio)
METRIC (ratio)

Common scales used:

FOOT-DUODECIMAL SYSTEM

1/1200 (1 inch of log per 100 feet of well)
1/600 (2 inches of log per 100 feet of well)
1/240 (5 inches of log per 100 feet of well)

FOOT-DECIMAL SYSTEM

1/1000 (1 foot of log per 1000 feet of well)
1/500 (1 foot of log per 500 feet of well)
1/200 (1 foot of log per 200 feet of well)

METRIC SYSTEM

/1000 (1 meter of log per 1000 meters of well)
/500 (1 meter of log per 500 meters of well)
1/200 (1 meter of log per 200 meters of well)

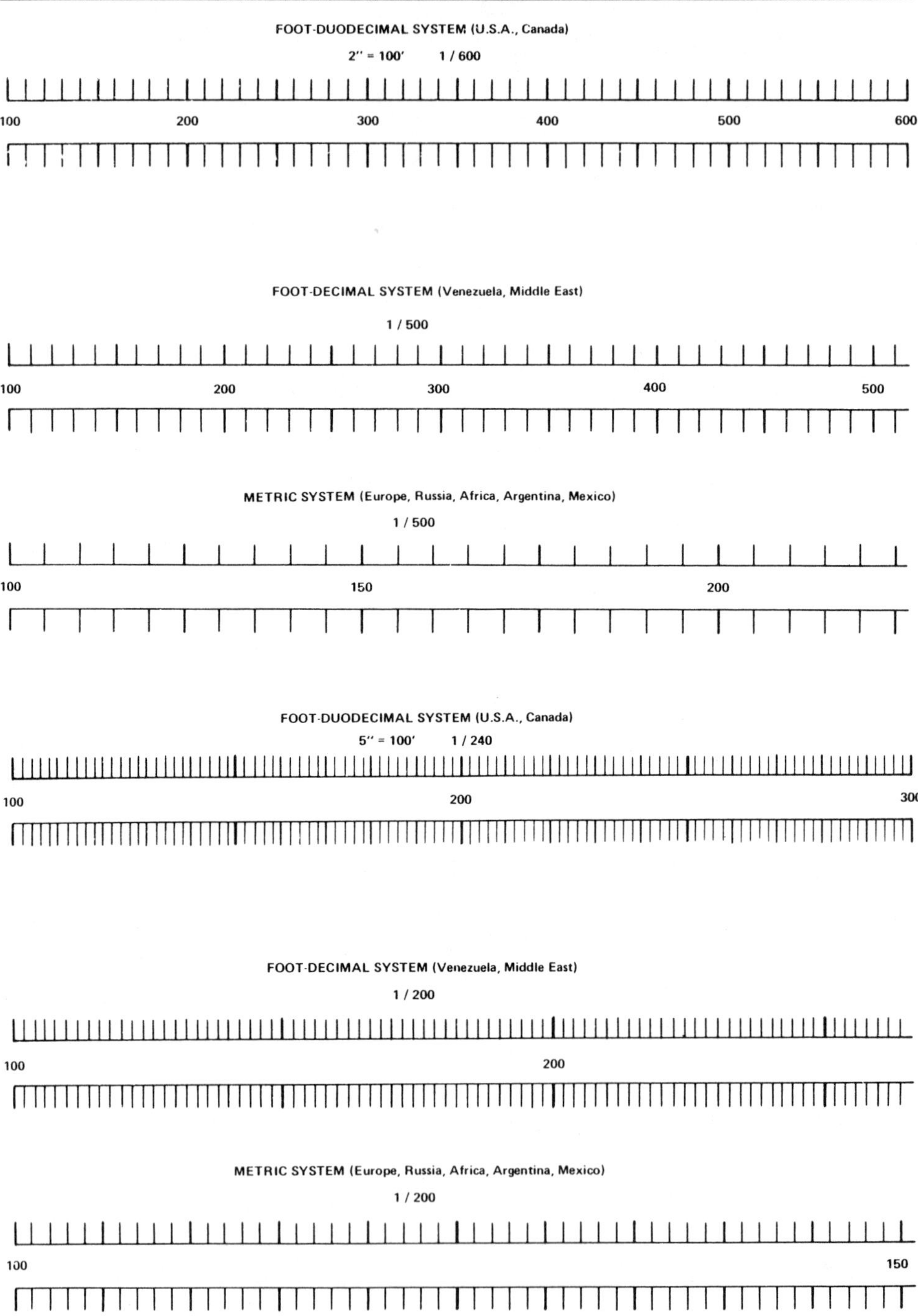

Note: These reproductions of common log depth grids are at approximately half scale.

APPENDIX G: DIVISION OF TOWNSHIP INTO SECTIONS

USA

6	5	4	3	2	1
7	8	9	10	11	12
18	17	16	15	14	13
19	20	21	22	23	24
30	29	28	27	26	25
31	32	33	34	35	36

6 MILES

6 MILES

CANADA

31	32	33	34	35	36
30	29	28	27	26	25
19	20	21	22	23	24
18	17	16	15	14	13
7	8	9	10	11	12
6	5	4	3	2	1

13	14	15	16
12	11	10	9
5	6	7	8
4	3	2	1

Division of Canadian Section into Legal Sub-divisions

APPENDIX H: BRITISH COLUMBIA, CANADA DIVISION OF "ZONE"

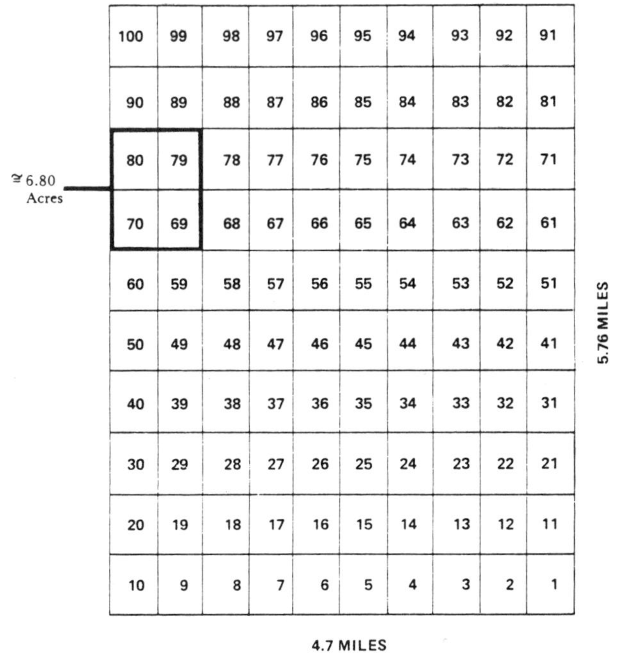

100	99	98	97	96	95	94	93	92	91
90	89	88	87	86	85	84	83	82	81
80	79	78	77	76	75	74	73	72	71
70	69	68	67	66	65	64	63	62	61
60	59	58	57	56	55	54	53	52	51
50	49	48	47	46	45	44	43	42	41
40	39	38	37	36	35	34	33	32	31
30	29	28	27	26	25	24	23	22	21
20	19	18	17	16	15	14	13	12	11
10	9	8	7	6	5	4	3	2	1

≅6.80 Acres

5.76 MILES

4.7 MILES

APPENDIX I: CONVERSION FACTORS BETWEEN PRACTICAL OILFIELD UNITS: METRIC, SI, AND OTHER MEASURES

(SI symbols shown in parentheses in the right-hand column)

MULTIPLY	BY		TO OBTAIN
	LENGTH		
angstrom	1.0	· E − 10	meter (m)
centimeter	3.280 840	E − 02	foot
centimeter	3.937 008	E − 01	inch
centimeter	1.0	· E − 02	meter (m)
centimeter	1.0	· E + 01	millimeter (m)
fathom	1.828 800	· E + 00	meter (m)
foot	3.048	· E − 01	meter (m)
foot	3.048	· E + 01	centimeter (cm)
foot	3.048	· E + 02	millimeter (mm)
foot	1.2	· E + 01	inch
inch	2.540	· E − 02	meter (m)
inch	2.540	· E + 00	centimeter (cm)
inch	2.540	· E + 01	millimeter (mm)
inch	8.333 333	E − 02	foot
kilometer	6.213 712	E − 01	mile [U.S. statute]
meter	3.937 008	E + 01	inch
meter	5.468 066	E − 01	fathom
meter	3.280 840	E + 00	foot
meter	1.093 613	E + 00	yard
meter	6.213 712	E − 04	mile [U.S. statute]
microinch	2.54	· E − 02	micrometer (μm) [micron]
micrometer [micron]	1.0	· E − 06	meter (m)
micrometer [micron]	3.937 008	E + 01	microinch
mile [U.S. statute]	1.609 344	· E + 03	meter (m)
mile [U.S. statute]	1.609 344	· E + 00	kilometer (km)
mile [U.S. statute]	5.280	· E + 03	foot
millimeter	3.280 840	E − 03	foot
millimeter	3.937 008	E − 02	inch
yard	9.144	· E − 01	meter (m)

MULTIPLY	BY	TO OBTAIN
	AREA	
acre	4.046 856 E−01	hectare (ha)
acre	4.356 · E+04	foot^2
acre	4.046 856 E+03	meter^2 (m^2)
centimeter^2	1.550 003 E−01	inch^2
centimeter^2	1.076 391 E−03	foot^2
foot^2	2.295 684 E−05	acre
foot^2	9.290 304 · E+02	centimeter^2 (cm^2)
foot^2	9.290 304 · E−02	meter^2 (m^2)
foot^2	1.44 · E+02	inch^2
foot^2	9.290 304 · E+04	millimeter^2 (mm^2)
hectare	1.0 · E+04	meter^2 (m^2)
hectare	2.471 054 E+00	acre
hectare	1.076 391 E+05	foot^2
hectare	1.0 · E−02	kilometer^2 (km^2)
inch^2	6.451 6 · E+00	centimeter^2 (cm^2)
inch^2	6.451 6 · E−04	meter^2 (m^2)
inch^2	6.451 6 · E+02	millimeter^2 (mm^2)
inch^2	6.944 444 E−03	foot^2
millimeter^2	1.076 387 E−04	foot^2
millimeter^2	1.550 003 E−03	inch^2
mile^2	6.4 · E+02	acre
mile^2	2.589 988 E+06	meter^2 (m^2)
meter^2	2.471 054 E−04	acre
meter^2	1.550 003 E+03	inch^2
meter^2	1.076 391 E+01	foot^2
meter^2	1.195 990 E+00	yard^2
yard^2	8.361 274 E−01	meter^2 (m^2)
	VOLUME	
acre-foot	7.758 368 E+03	barrel
acre-foot	4.356 · E+04	foot^3
acre-foot	3.258 515 E+05	gallon
acre-foot	1.233 482 E+03	meter^3 (m^3)
barrel [API]	1.288 931 E−04	acre-foot
barrel [API]	1.589 873 E+05	centimeter^3 (cm^3) (cc)[a]
barrel [API]	1.589 873 E+02	liter (l)
barrel [API]	1.589 873 E−01	meter^3 (m^3)
barrel [API]	5.614 583 E+00	foot^3
barrel [API]	4.2 · E+01	gallon
barrel [API]	9.701 999 E+03	inch^3
centimeter^3	6.289 981 E−06	barrel [API]
centimeter^3	3.531 466 E−05	foot^3
centimeter^3	2.641 721 E−04	gallon
centimeter^3	6.102 374 E−02	inch^3
foot^3	1.781 076 E−01	barrel [API]
foot^3	2.831 685 E+04	centimeter^3 (cm^3) (cc)[a]
foot^3	7.480 520 E+00	gallon
foot^3	1.728 · E+03	inch^3

MULTIPLY	BY	TO OBTAIN
VOLUME (CONTINUED)		
foot^3	2.831 685 E+01	liter (l)
foot^3	2.831 685 E−02	meter^3 (m^3)
gallon	2.380 952 E−02	barrel [API]
gallon	3.785 412 E+03	centimeter^3 (cm^3)
gallon	3.785 412 E+00	liter (l)
gallon	3.785 412 E−03	meter^3 (m^3)
gallon	2.310 001 E+02	inch^3
gallon	1.336 806 E−01	foot^3
gallon [U.K.]	1.200 950 E+00	gallon [U.S.]
inch^3	1.030 715 E−04	barrel [API]
inch^3	4.329 003 E−03	gallon
inch^3	1.638 706 E+01	centimeter^3 (cm^3) (cc)[a]
inch^3	1.638 706 E−02	liter (l)
inch^3	1.638 706 E−05	meter^3 (m^3)
inch^3	5.787 037 E−04	foot^3
inch^3	1.638 706 E+04	millimeter^3 (mm^3)
liter	1.0 · E+03	centimeter^3 (cm^3) (ml)
liter	1.0 · E−03	meter^3 (m^3)
liter	6.289 811 E−03	barrel [API]
liter	3.531 466 E−02	foot^3
liter	2.641 720 E−01	gallon
liter	6.102 373 E+01	inch^3
meter^3	1.0 · E+06	centimeter^3 (cm^3) (cc)[a]
meter^3	1.0 · E+03	liter
meter^3	8.107 131 E−04	acre-foot
meter^3	6.289 811 E+00	barrel
meter^3	3.531 466 E+01	foot^3
meter^3	2.641 720 E+02	gallon
meter^3	6.102 376 E+04	inch^3
meter^3	1.307 951 E+00	yard^3
millimeter^3	6.102 376 E−05	inch^3
yard^3	7.645 549 E−01	meter^3 (m^3)

[a]Not standard.

VELOCITY, ACCELERATION, TIME		
centimeter/second	1.181 102 E+02	foot/hour
centimeter/second	1.968 504 E+00	foot/minute
centimeter/second	3.280 840 E−02	foot/second
centimeter/second	6.0 · E−01	meter/minute
foot/hour	8.466 667 E−03	centimeter/second
foot/hour	8.466 667 E−05	meter/second (m/s)
foot/hour	5.08 · E−03	meter/minute
foot/hour	1.666 667 E−02	foot/minute
foot/minute	5.08 · E−01	centimeter/second
foot/minute	3.048 · E−01	meter/minute
foot/minute	5.08 · E−03	meter/second (m/s)
foot/minute	6.0 · E+01	foot/hour

MULTIPLY	BY		TO OBTAIN
VELOCITY, ACCELERATION, TIME (CONTINUED)			
foot/second	3.048	·E+01	centimeter/second
foot/second	1.828 8	·E+01	meter/minute
foot/second	3.048	·E−01	meter/second (m/s)
kilometer/hour	6.213 712	E−01	mile/hour
kilometer/hour	5.399	E−01	knots [International]
knots [International]	1.852	E+00	kilometer/hour
knots [International]	1.151	E+00	mile/hour
mile/hour	4.470 4	·E+01	centimeter/second
mile/hour	8.8	·E+01	foot/minute
mile/hour	1.466 667	E+00	foot/second
mile/hour	1.609 334	·E+00	kilometer/hour
mile/hour	4.470 4	·E−01	meter/second (m/s)
meter/minute	1.666 667	E+00	centimeter/second
meter/minute	3.280 840	E+00	foot/minute
meter/minute	1.968 504	E+02	foot/hour
meter/minute	5.468 067	E−02	foot/second
meter/second	3.280 840	E+00	foot/second
meter/second	1.181 102	E+04	foot/hour
meter/second	1.968 504	E+02	foot/minute
foot/second^2	3.048	·E−01	meter/second^2 (m/s^2)
meter/second^2	3.280 840	E+00	foot/second^2
day	2.4	·E+01	hour
day	1.44	·E+03	minute
day	8.64	·E+04	second (s)
minute	6.944 444	E−04	day
minute	1.666 667	E−02	hour
minute	6.0	·E+01	second (s)
second	1.157 407	·E−05	day
second	2.777 778	E−04	hour
VOLUMETRIC FLOW RATE			
barrel/day	1.840 131	E+00	centimeter^3/second
barrel/day	1.589 873	E−01	meter^3/day (m^3/d)
barrel/day	6.624 472	E−03	meter^3/hour
barrel/day	3.899 016	E−03	foot^3/minute
barrel/day	2.916 667	E−02	gallon/minute
barrel/day	6.624 472	E+00	liter/hour
foot^3/minute	1.699 011	E+03	liter/hour
foot^3/minute	4.719 474	E−04	meter^3/second (m^3/s)
foot^3/minute	2.564 749	E+02	barrel/day
foot^3/second	2.831 685	E−02	meter^3/second (m^3/s)
gallon/minute	3.428 571	E+01	barrel/day
gallon/minute	1.428 571	E+00	barrel/hour
gallon/minute	1.336 806	E−01	foot^3/minute
gallon/minute	2.271 247	E+02	liter/hour
gallon/minute	6.309 020	E−05	meter^3/second (m^3/s)
gallon/minute	5.450 993	E+00	meter^3/day

MULTIPLY	BY	TO OBTAIN
VOLUMETRIC FLOW RATE (CONTINUED)		
liter/hour	1.509 554 E−01	barrel/day
liter/hour	5.885 776 E−04	foot^3/minute
liter/hour	2.4 · E−02	meter^3/day
liter/minute	3.531 466 E−02	foot^3/minute
liter/minute	2.641 720 E−01	gallon/minute
liter/second	1.585 032 E+01	gallon/minute
meter^3/second	5.434 396 E+05	barrel/day
meter^3/second	2.118 880 E+03	foot^3/minute
meter^3/second	3.531 466 E+01	foot^3/second
meter^3/second	1.585 032 E+04	gallons/minute
meter^3/minute	2.641 720 E+02	gallons/minute
meter^3/day	6.289 811 E+00	barrels/day
meter^3/day	3.531 466 E+01	foot^3/day
meter^3/day	2.452 407 E−02	foot^3/minute
meter^3/day	1.834 528 E−01	gallon/minute
meter^3/day	4.166 667 E+01	liter/hour
meter^3/day	4.166 667 E−02	meter^3/hour
VISCOSITY, PERMEABILITY		
centipoise	1.0 · E−03	pascal-second (Pa · s)
centipoise	1.0 · E−02	dyne-second/centimeter^2
centipoise	2.088 543 E−05	pound-force-second/foot^2
centistoke	1.0 · E−06	meter^2/second (m^2/s)
centistoke	1.0 E+00	centipoise/(g/cm^2)
meter^2/second	1.0 · E+06	centistoke
meter^2/second	1.0 · E+04	stoke
pascal-second	1.0 · E+03	centipoise
pascal-second	1.0 E+01	poise
pound-force-second/foot^2	4.788 026 E+01	pascal-second (Pa · s)
poise	1.0 · E−01	pascal-second (Pa · s)
stoke	1.0 · E−04	meter^2/second (m^3/s)
darcy	9.869 23 E−13	meter^3 (m^2)
darcy	9.869 23 E−09	centimeter^2
darcy	9.869 23 E−01	micrometer^2 (μm^2)
darcy	1.0 · E+03	millidarcy
millidarcy-foot	3.008 142 E+04	micrometer^2-meter
MASS		
grain [avoirdupois]	6.479 891 E−02	gram (g)
gram	1.543 236 E+01	grain
gram	1.0 · E−03	kilogram (kg)
gram	3.527 397 E−02	ounce
gram	2.204 622 E−03	pound
kilogram	1.0 · E+03	gram
kilogram	3.527 397 E+01	ounce
kilogram	2.204 622 E+00	pound

MULTIPLY	BY	TO OBTAIN
	MASS (CONTINUED)	
kilogram	6.852 178 E−02	slug
kilogram	9.842 064 E−04	ton [long]
kilogram	1.102 311 E−03	ton [short]
kilogram	1.0 · E−03	ton [metric]
ounce	2.834 952 E+01	gram (g)
ounce	2.834 952 E−02	kilogram (kg)
pound	7.0 · E+03	grains
pound	4.535 924 E−01	kilogram (kg)
slug	1.459 390 E+01	kilogram (kg)
ton [long: 2240 pounds]	1.016 047 E+03	kilogram (kg)
ton [short: 2000 pounds]	9.071 847 E+02	kilogram (kg)
ton [metric]	1.0 · E+03	kilogram (kg)
	DENSITY	
gram/centimeter^3	1.0 · E+03	kilogram/meter^3 (kg/m^3)
gram/centimeter^3	6.242 797 E+01	pound/foot^3
gram/centimeter^3	8.345 405 E+00	pound/gallon
gram/centimeter^3	3.612 730 E−02	pound/inch^3
kilogram/meter^3	1.0 · E−03	gram/centimeter^3
kilogram/meter^3	6.242 797 E−02	pound/foot^3
kilogram/meter^3	1.002 242 E−02	pound/gallon
pound/foot^3	1.601 846 E−02	gram/centimeter^3
pound/foot^3	1.601 846 E+01	kilogram/meter^3 (kg/m^3)
pound/foot^3	1.336 805 E−01	pound/gallon
pound/gallon	1.198 264 E−01	gram/centimeter^3
pound/gallon	1.198 264 E+02	kilogram/meter^3
pound/gallon	7.480 520 E+00	pound/foot^3
	FORCE	
dyne	1.0 · E−05	newton (N)
dyne	2.248 089 E−06	pound-force
kilogram-force	9.806 650 · E+00	newton (N)
kilogram-force	2.204 622 E+00	pound-force
newton	1.0 · E+05	dyne
newton	1.019 716 E−01	kilogram-force
newton	2.248 089 E−01	pound-force
newton	3.596 942 E+00	ounce-force
ounce-force	2.780 139 E−01	newton (N)
pound-force	4.448 222 E+00	newton (N)
pound-force	4.535 925 E−01	kilogram-force

MULTIPLY	BY	TO OBTAIN
	TEMPERATURE	
TO CONVERT	SOLVE	TO OBTAIN
Celsius degree	K = °C + 273.15	Kelvin (K)
Celsius degree	F = 1.8°C + 32	Fahrenheit degree
Fahrenheit degree	C = (°F − 32)/1.8	Celsius degree
Fahrenheit degree	R = °F + 459.67	Rankine degree
Rankine degree	K = °R/1.8	Kelvin (K)

The SI unit of temperature is the degree Kelvin (written K without a degree symbol) and defined so the triple point of water is 273.16 exactly.

PRESSURE AND PRESSURE GRADIENT			
atmosphere, normal	1.013 25	E+00	bar
atmosphere, normal	3.389 95	E+01	feet of water, 4°C
atmosphere, normal	2.992	E+01	inches of Hg, 32°F
atmosphere, normal	7.6	·E+02	millimeters of Hg, 0°C
atmosphere, normal	1.013 25	E+05	pascal (Pa)
atmosphere, normal	1.469 60	E+01	pound/inch²
bar	9.869 23	E−01	atmosphere, normal
bar	1.019 716	E+00	kilogram-force/centimeter²
bar	1.0	·E+05	newton/meter² (N/m²)
bar	1.0	·E+05	pascal (Pa)
bar	1.450 377	E+01	pound/inch²
centimeters of Hg at 0°C	1.333 22	E+03	pascal (Pa)
centimeters of Hg at 0°C	1.933 67	E−01	pound/inch²
dyne/centimeter²	1.0	·E−01	pascal (Pa)
dyne/centimeter²	1.450 377	E−05	pound/inch²
inches of Hg at 60°F			pascal (Pa)
inches of Hg at 60°F	4.898	E−01	pound/inch²
feet of water at 4°C	2.988 98	E+03	pascal (Pa)
feet of water at 4°C	4.335 15	E−01	pound/inch²
kilogram-force/centimeter²	9.806 650	·E−01	bar
kilogram-force/centimeter²	9.806 650	·E+04	pascal (Pa)
kilogram-force/centimeter²	1.422 334	E+01	pound/inch²
kilogram-force/meter²	9.806 650	·E+00	newton/meter² (N/m²)
kilogram-force/meter²	9.806 650	·E+00	pascal (Pa)
kilogram-force/meter²	2.048 161	E−01	pound/foot²
kilogram-force/meter²	1.422 334	E−03	pound/inch²
newton/centimeter²	1.450 377	E+00	pound/inch²
newton/meter²	1.0	·E−05	bar
newton/meter²	1.0	·E+00	pascal (Pa)
newton/meter²	1.450 377	E−04	pound/inch²
newton/meter²	1.019 716	E−01	kilogram-force/meter²
newton/millimeter²	1.450 377	E+02	pound/inch²
pascal	9.869 23	E−06	atmosphere, normal
pascal	1.0	·E−05	bar

MULTIPLY	BY		TO OBTAIN
PRESSURE AND PRESSURE GRADIENT (CONTINUED)			
pascal	1.019 716	E−01	kilogram-force/meter^2
pascal	1.0	·E+00	newton/meter^2 (N/m^2)
pascal	2.088 543	E−02	pound/foot^2
pascal	1.450 377	E−04	pound/inch^2
pound/foot^2	4.882 429	E+00	kilogram-force/meter^2
pound/foot^2	4.788 026	E+01	pascal (Pa)
pound/foot^2	6.944 444	E−03	pound/inch^2
pound/inch^2	6.804 60	E−02	atmosphere, normal
pound/inch^2	6.894 757	E−02	bar
pound/inch^2	7.030 697	E−02	kilogram-force/ centimeter^2
pound/inch^2	6.894 757	E−01	newton/centimeter^2
pound/inch^2	6.894 757	E+00	kilonewton/ meter^2
pound/inch^2	6.894 757	E+03	newton/meter^2 (N/m^2)
pound/inch^2	6.894 757	E−03	newton/millimeter^2
pound/inch^2	6.894 757	E+03	pascal (Pa)
ENERGY AND WORK			
Btu [International table]	1.055 056	E+03	joule (J)
Btu [International table]	2.518 021	E−01	calorie [Kg, mean]
Btu [International table]	7.781 693	E+02	foot-pound
Btu [International table]	2.930 711	E−01	watt-hour
calorie [Kg, mean]	4.190 021	E+03	joule (J)
calorie [Kg, mean]	3.971 372	E+00	Btu [International table]
calorie [Kg, mean]	3.090 400	E+03	foot-pound
calorie [Kg, mean]	1.560 808	E−03	horsepower-hour
calorie [Kg, mean]	4.190 020	E−03	newton-meter
calorie [Kg, mean]	1.163 894	E+00	watt-hours
foot-pound	1.285 067	E−03	Btu [International mean]
foot-pound	1.355 818	E+00	joule (J)
foot-pound	3.235 826	E−04	calorie [Kg, mean]
joule (J)	9.478 170	E−04	Btu [International table]
joule (J)	2.386 623	E−04	calorie [Kg, mean]
joule (J)	7.375 621	E−01	foot-pound
joule (J)	2.777 778	E−04	watt-hour
joule (J)	1.0	·E+00	watt-second
watt-hour	3.6	·E−03	joule (J)
POWER			
Btu/HR [International table]	2.930 711	E−01	watt (W)
foot-pound/hour	3.766 161	E−04	watt (W)
foot-pound/minute	3.030 303	E−05	horsepower
foot-pound/minute	2.259 697	E−02	watt (W)

MULTIPLY	BY		TO OBTAIN
	POWER (CONTINUED)		
foot-pound/second	1.818 182	E−03	horsepower
foot-pound/second	1.355 818	E+00	watt (W)
horsepower [550 ft-lb/s]	7.456 99	E−01	kilowatt (KW)
horsepower	5.50 999 ·	E+02	foot-pounds/second
horsepower	3.3 ·	E+04	foot-pounds/minute
horsepower [electric]	7.46 ·	E+02	watt (W)
kilowatt	1.341 022	E+00	horsepower [550 ft-lb/s]
watt	2.655 224	E+03	foot-pound/hour
watt	4.425 372	E+01	foot-pound/minute
watt	1.341 022	E−03	horsepower [550 ft-lb/s]

MULTIPLY	BY		TO OBTAIN
TEMPERATURE GRADIENT			
°C/100 meters	5.486 4	E+01	°F/100 feet
°F/100 feet	1.822 7	E−02	°C/100 meters
CONCENTRATION			
grains/gallon	1.714	E−02	grams/liter
grains/gallon	1.712	E+01	parts per million
grains/gallon	1.429	E+02	pounds per million gallon
pounds/million gallon			grains/gallon
pounds/million gallon	1.198 2	E−01	parts per million
GAS-OIL RATIO			
m^3/m^3 at 15°C and 1 atm	1.801 175	E−01	scf/B at 60°F, 14.65 psia
scf/B at 60°F, 14.65 psia	5.551 931	E+00	m^3/m^3 at 15°C and 1 atm
GAS-VOLUME			
m^3 at 15°C and 1 atm	3.549 373	E+01	scf at 60°F and 14.65 psia
scf at 60°F and 14.65 psia	2.817 339	E−02	m^3 at 15°C and 1 atm

GIVEN	USE FORMULA	TO OBTAIN
OIL GRAVITY		
°API	$\frac{141.5}{131.5 + °API}$	sp. gr. @ 60°F
sp. gr. @ 60°F	$\frac{141.5}{\text{sp. gr. @ 60°F}} - 131.5$	°API
GAS GRAVITY		
gas density (g/cm^3) (g/cc)[a] at 60°F, 14.7 psia	$\frac{\text{gas density (g/cm}^3\text{)}}{0.00122}$	gas specific gravity
gas density (lb/cu ft) at 60°F, 14.7 psia	$\frac{\text{gas density (lb/cu ft)}}{0.762}$	gas specific gravity
	$\frac{\text{gas mol. weight}}{28.966}$	gas specific gravity

[a]Not standard.

APPENDIX J: COMPUTER CALIBRATIONS

INTRODUCTION AND ACKNOWLEDGMENT

Major wireline service companies now perform nearly all of their logging services using on-board computers. Calibrations performed with such logging units do not look like the conventional calibration tails recorded with analog systems. This section reproduces the pertinent data published by Schlumberger in their "Cyber Service Unit Wellsite Products and Calibrations Guide," CP 26, 1984. It is hoped that in future editions of this book similar information will be available from other service companies to keep it up to date.

CSU CALIBRATIONS GUIDE

Calibration Theory

A logging tool must produce a signal that is related in a known way to the parameter being measured. Calibration establishes that relationship.

The Schlumberger CSU unit is the surface instrumentation used in most tool calibrations. A review of the basic CSU calibration system will make individual calibrations easier to understand.

Measured	These values represent the raw or uncalibrated signals from the downhole tool. They are converted from signals to units of measurement and are then expressed in the engineering units the CSU system uses in computations.
Zero Measurement (ZMEA)	This is the uncalibrated signal from the downhole tool representing the low point for the calibration.
Plus Measurement (PMEA)	This is the uncalibrated signal from the downhole tool representing the high point for the calibration.
Zero Reference (ZREF)	This is the low point calibration standard. This is what the calibrated zero value should read.
Plus Reference (PREF)	This is the high point calibration standard. This is what the calibrated plus value should read.
Gain	Gain = (PREF − ZREF)/(PMEA − ZMEA).
Offset	Offset = ZREF − Gain (ZM) or Offset = PREF − Gain (PMEA).
Calibrated Value	Calibrated Value = Gain (Measured Value) + Offset.

Units — This is the engineering or physical unit used to express the data.

The CSU system uses a two-point calibration technique which assumes the tool response is linear. Because the response is linear, any two points on the tool response line can be compared to their associated calibration references to determine a gain and an offset. The gain and offset values are used in the calibrated value equation, which converts any measured value to its associated calibrated value.

The simple caliper device is a good illustration. The downhole tool sends to the surface a signal voltage that varies linearly with the diameter of the measured hole. In calibration, a two-point method, large ring and small ring, is used to establish both the slope and zero-intercept of the response line. Then, if the response is indeed linear and there is no drift in the electronics, the recorded values accurately reflect hole diameters over the range of the tool.

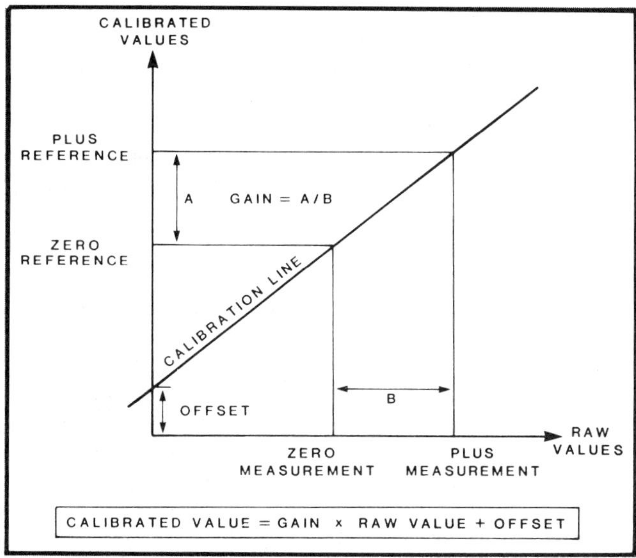

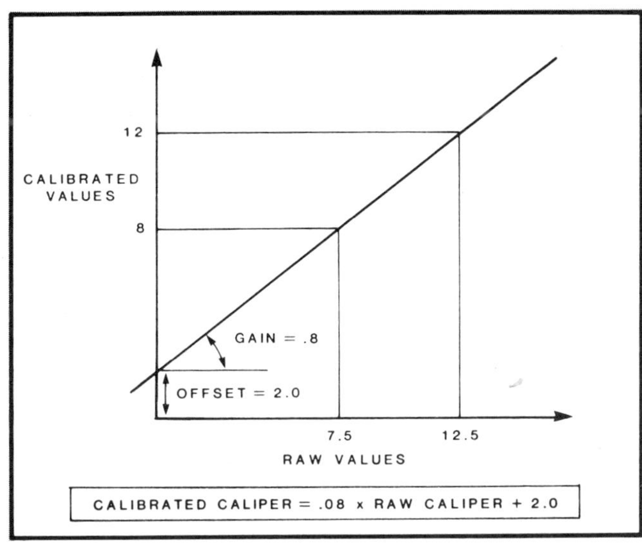

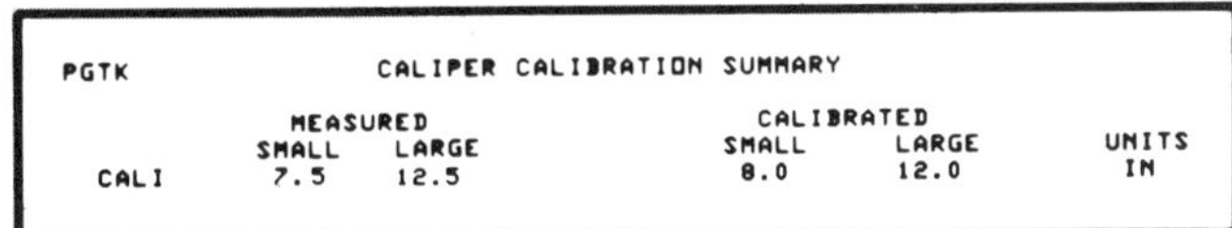

PGTK CALIPER CALIBRATION SUMMARY

	MEASURED		CALIBRATED		
	SMALL	LARGE	SMALL	LARGE	UNITS
CALI	7.5	12.5	8.0	12.0	IN

Schlumberger assures linearity in the caliper and most other tools by design and by periodic shop checks. The effects of possible drift are monitored by tool checks before and after logging.

Three types of calibration summaries appear on a log: the Shop, Before Survey, and After Survey.

The Shop Summary is used for tools which have field primary or shop standard calibrators which are too bulky or involved to be used at the wellsite. Some common tools which have Shop Summaries are the Dual Induction, FDC, and CNL. These Shop Calibrations are typically run once a month. The tool is calibrated to the shop standard and then compared to the secondary field standard which will be used at the wellsite to ensure the tool characteristics have not changed.

A Before Survey Calibration is run at the well. This calibrates the CSU unit for the tool being used. If the tool has a Shop Summary, the CSU unit is calibrated using the shop values. The secondary standard or jig is checked and its values can be compared to the values found on the Shop Summary to see that there has been no significant change in the tool response. If the engineer thinks it is necessary, an alternative calibration can be used which is based on the secondary or jig values rather than those of the primary standard. Some tools have no Shop Summary because the primary field standard is used at the wellsite. In these cases there are only Before and After Survey Summaries.

After the job, an After Survey Tool Check Summary is run to confirm that downhole tool drift has remained within tolerances while logging. Tolerances are chosen to represent realistic standards of tool performance under reasonable logging conditions; downhole temperature 300°F or less, logging time (between Before and After readings) of no more than about 8 hours, and no spudding or hard pulling with the tool.

If the values on the After Survey Summary are out of tolerance, a decision must be made. "Good" and "no good" are usually just questions of degree, but the calibration summaries will tell you the maximum error which can be expected on the log.

One of the demonstrated characteristics of the CSU system is its improved accuracy and consistency of tool calibrations. The quantitative nature of numeric printouts, as compared with analog traces on graphic film, serves to emphasize any out-of-tolerance calibrations, but these quantitative values are responsible for the very existence of valid, realistic tolerances.

Detailed Calibration Summaries

This section is designed to clarify the meaning of the various calibration summaries generated by the CSU system. Only one example will be given of each general log type; for instance, the Dual Induction SFL* printout illustrates the significance of all induction log calibration summaries.

The CSU programs which govern the log formats are regrouped under the term CYBERPACK. These programs are updated at regular intervals to include the latest improvements in acquisition and processing techniques.

Therefore, the calibration methods and summary format may change as the need arises. However, the principle will remain the same, and the various entries will still be recognizable.

Calibration of Resistivity Tools

```
                 SHOP SUMMARY

   PERFORMED:       83/09/23
   PROGRAM FILE: SHOP    (VERSION    26.15   83/08/13)

 DITD                  ELECTRONICS CALIBRATION SUMMARY

                 TEST LOOP CALIBRATION               TOOL CHECK
             MEASURED             CALIBRATED          CALIBRATED
           ZERO      PLUS      ZERO       PLUS      ZERO       PLUS    UNITS
    ILD     -.5      584.3      0.0       500.0      0.0       495.7   MSIE
    ILM    -1.1      534.2      0.0       500.0      0.0       499.5   MSIE
 ILD  SONDE ERROR CORRECTION :            5.1 MSIE
 ILM  SONDE ERROR CORRECTION :            4.7 MSIE

                     (IS:291 , IC:1010)

                   FILE      0        23-SEP-83 13:39
                   DATA ACQUIRED    01-JAN-80 00:00
```

```
                 BEFORE SURVEY CALIBRATION SUMMARY

  PERFORMED:       83/09/23
  PROGRAM FILE: ISN     (VERSION     26.15     83/08/13)

DITD                 ELECTRONICS CALIBRATION SUMMARY

              MEASURED                       CALIBRATED
            ZERO      PLUS              ZERO          PLUS          UNITS
   ILD       4.3      586.5              0.0          495.9         MSIE
   ILM       4.5      539.9              0.0          499.7         MSIE
   SFL        .3      531.5              0.0          500.0         MSIE

ILD   SONDE ERROR CORRECTION :          5.1 MSIE
ILM   SONDE ERROR CORRECTION :          4.7 MSIE

                     FILE      1        23-SEP-83  15:05
                     DATA ACQUIRED      01-JAN-80  00:00
```

```
                 AFTER SURVEY TOOL CHECK SUMMARY

  PERFORMED:       83/09/23
  PROGRAM FILE: ISN     (VERSION     26.15     83/08/13)

DITD                       TOOL  CHECK

                    ZERO                        PLUS
            BEFORE          AFTER        BEFORE       AFTER       UNITS
   ILD       0.0             -.0          495.9       495.0       MSIE
   ILM       0.0              .0          499.7       499.2       MSIE
   SFL       0.0              .0          500.0       500.0       MSIE

ILD   SONDE ERROR CORRECTION :          5.1 MSIE
ILM   SONDE ERROR CORRECTION :          4.7 MSIE

                     FILE      4        23-SEP-83  15:07
                     DATA ACQUIRED      01-JAN-80  00:00
```

DUAL INDUCTION TOOL (DIT). The Dual Induction tool requires a shop check. The primary calibration standard is a precision test loop which must be used in a zero-conductivity medium, and this is impractical at the wellsite.

The tool is calibrated in the district to the primary standard by taking three readings. A measurement is made first with the sonde in a zero-conductivity environment and then with the precision test loop installed on the sonde to represent a 500-mmho formation. The calibration gain is computed using these two values. Next, the sonde is disconnected electrically, using a relay in the cartridge to get an electronic or tool check zero measurement. A calibration offset is computed using this value and the computed gain. With the gain and offset, the CSU unit is calibrated to the primary standard. The secondary standard is a precision resistor in the sonde which can be connected to the cartridge by a relay. The CSU system will measure the signal received uphole when the resistor in the sonde is connected. It will then apply the gain and offset to this value to determine the effective resistivity this resistor represents. Typically, this calibrated tool check value is 500 ± 5 mmho. However, if it is not, this calibrated tool check plus value can be substituted for the plus reference used in the Before Survey Calibration to ensure the tool will be calibrated as it was during the Shop Calibration. Next, a sonde error correction will be determined using the zero-conductivity measurement and the electronic zero measurement.

Basically, a sonde error correction is the amount of correction needed to make the sonde read zero in a zero-conductivity medium. This varies from sonde to sonde. The SFL measurement does not require a sonde calibration because the primary standard is a precision resistor network in the cartridge.

All of this information is translated on the log by a Shop Summary. The Shop Summary begins, as do all sections of the calibration summary, with the date performed, the program used, and what version of program tape was used. Next, the tool type (DITD) and the type of calibration are defined (Electronics Calibration Summary). There are two induction measurements made with a Dual Induction tool; thus, there are two calibrations. One is for the deep induction (ILD) measurement and the other is for the medium induction measurement (ILM). However, the tolerances of the two measurements are the same and will be discussed together. The tolerances for the different Shop Summary values are listed.

DIT Shop Summary

Item	Explanation	Tolerance
Test Loop Measured Zero	Tool reading in zero-conductivity medium.	0 ± 15
Test Loop Measured	Tool reading with test loop.	455 – 560
Test Loop Calibrated Zero	Set at zero.	0
Test Loop Calibrated Plus	Value obtained by applying gain and offset to item 2.	500
Tool Check Calibrated Zero	Value obtained by applying an offset to signal measured with zero relay activated.	0
Tool Check Calibrated	Value obtained by applying gain and offset to signal measured with calibrate relay activated.	500 ± 15
Sonde Error Correction	Negative value obtained by applying gain and offset to item 1 (negative of value this sonde would read in a zero-conductivity environment).	± 15

The numbers of the cartridge and sonde used in the Shop Calibration are listed at the bottom of the Shop Summary. These should agree with the tool numbers listed on the log heading.

DIT BEFORE SURVEY CALIBRATION. For calibration, the CSU system uses the internal tool check values obtained from the calibrate plus and zero relays for the induction and SFL measurements. The zero reference is 0. The plus reference for the SFL is 500 mmho, and it is typically 500 mmho for the ILD and ILM measurements. However, to refine the calibration, the ILD and ILM 500 mmho plus references can be substituted at the beginning of the Before Survey Calibration with the tool check calibrated values from the Shop Summary. The table explains each item in the summary as well as tolerances.

DIT Before Survey Calibration Summary

Item	Explanation	Tolerance
Performed Program file	Date of survey and program and version used.	Agrees with date on log.
Measured Zero	Signal measured with zero relays activated.	0 ± 15
Measured Plus	Signal measured with plus relay activated.	450 − 565; SFL = 500 ± 100.
Calibrated Zero	Value after gain and offset applied to item 8.	0
Calibrated Plus	Value after gain and offset applied to item 9.	ILD, ILM = ± 0.5 of item 6 Shop Summary; SFL = 500 ± 0.5.
Sonde Error Correction	See Shop Summary.	Same as item 7 in Shop Summary, but may be changed by engineer to correct for borehole effect. If different, there should be an explanation on log heading.

DIT AFTER SURVEY CALIBRATION. After the log is run, a calibration identical to the Before Survey Calibration is performed. This checks for any significant drift in the downhole tool that might have occurred during logging. The After Survey Tool Check Summary just compares the Before and After calibrated zero and plus values. The table explains the After Survey Summary.

DIT After Survey Tool Check Summary

Item	Explanation	Tolerance
Zero Before	—	Same as item 10 of BSC.
Zero After	—	Zero Before ± 2 mmho.
Plus Before	—	Same as item 11 of BSC.
Plus After	—	Plus Before ± 20 mmho.
Sonde Error Correction	—	Same as item 12 of BSC or an explanation in the remarks section of the log heading.

Dual Laterolog (DLL–MSFL)*

```
                BEFORE SURVEY CALIBRATION SUMMARY

     PERFORMED:      83/09/16
     PROGRAM FILE: MILL    (VERSION    26.1    83/05/23)

DSTD                 ELECTRONICS CALIBRATION SUMMARY

             MEASURED                      CALIBRATED           UNITS
     LLS       32.40                          31.6094           OHMM
     LLD       31.76                          31.6094           OHMM

DST                  ELECTRONICS CALIBRATION SUMMARY

             MEASURED                      CALIBRATED
           ZERO     PLUS                 ZERO       PLUS        UNITS
    CMSF     .1   1000.                  0.0      1000.0        MMHO
      I1    -.0    163.7                 0.0       200.0        MMHO

DST                  CALIPER CALIBRATION SUMMARY

             MEASURED                      CALIBRATED
          SMALL    LARGE                SMALL      LARGE        UNITS
    CALI   7.5      11.4                 8.0        12.0        IN
                        FILE 0             16-SEP-83 00:33
```

AFTER SURVEY TOOL CHECK SUMMARY

PERFORMED: 83/09/16
PROGRAM FILE: MILL (VERSION 26.1 83/05/23)

DSTD TOOL CHECK

	PLUS		
	BEFORE	AFTER	UNITS
LLS	31.60	31.5789	OHMM
LLD	31.60	31.6249	OHMM

DST TOOL CHECK

	ZERO		PLUS		
	BEFORE	AFTER	BEFORE	AFTER	UNITS
CMSF	0.0	.0	1000.0	1001.5	MMHO
I1	0.0	-.1	200.0	200.4	MMHO

FILE 0 16-SEP-83 00:27

The formats of the Before Survey and After Survey Summaries are the same as those for the induction tools. However, a Shop Summary is not used because the primary calibration standard is an internal resistor in each measuring circuit, permitting primary calibration at the wellsite. The geometry of the sensors is very stable so an external reference is unnecessary.

Dual Laterlog tool calibrations are straightforward. The plus values are produced by downhole resistors which represent formation resistivities of 31.6 ohm-m. The actual readings usually differ from this figure because of downhole circuit variables, but they are adjusted to read exactly 31.6 by the CSU calibration computations. The zero values are measured with the downhole tool electronically disconnected.

The MSFL tool is calibrated similarly, but uses conductivity units (mmho). The zero end (3) is read with the downhole circuits disconnected, but the calibrating resistor gives a plus response equal to that of a 1000-mmho formation (4). Deviations caused by circuit variables are adjusted by the CSU unit to give precise calibrated readings (5 and 6).

The After Survey Tool Check Summary is self-explanatory. The After Survey tolerances are given.

DLL/MSFL Before Survey Calibration Summary

Item	Explanation	Tolerance
DSTD		
Measured LLS, LLD	Uncalibrated resistivity measured with calibrate relay.	29 − 40
Calibrated LLS, LLD	Computed gains applied to item 1.	31.6
DST		
Measured Zero CMSF, I1	Uncalibrated conductivity measured with zero relay.	CMSF = 0 ± 2; I1 = 0 ± 36
Measured Plus CMSF, I1	Uncalibrated measurement with plus relay.	CMSF = 1000 ± 80; I1 = 200 ± 80
Calibrated Zero CMSF, I1	Computed gains and offsets applied to item 3.	0
Calibrated Plus CMSF, I1	Computed gains and offsets applied to item 4.	CMSF = 1000; I1 = 200

DLL–MSFL After Survey Tool Check Summary

Item	Explanation	Tolerance
DSTD		
Plus Before	Same as item 2 of BSC.	31.6
Plus After	After survey calibrated plus values.	31.6 ± 0.7
DST		
Zero Before	Same as item 5 of BSC.	0
Zero After	After survey calibrated zero values.	0 ± 2
Plus Before	Same as item 6 of BSC.	
Plus After	After survey calibrated plus values.	± 20

Microlog, Microlaterolog, Proximity (ML, MLL-PROX) Tools

```
                BEFORE SURVEY CALIBRATION SUMMARY

  PERFORMED:     83/09/23
  PROGRAM FILE: BMS     (VERSION    26.15    83/08/13)

 MPT              ELECTRONICS CALIBRATION SUMMARY

           MEASURED                      CALIBRATED
         ZERO     PLUS                  ZERO      PLUS       UNITS
  MINV     .0      5.2                   0.0       5.0       OHMM
  MNOR    -.0      3.5                   0.0       5.0       OHMM
  PROX     .3    514.5                   0.0     500.0       MMHO

 MPT              CALIPER CALIBRATION SUMMARY

           MEASURED                      CALIBRATED
         SMALL    LARGE                 SMALL     LARGE      UNITS
  CALI    5.9     10.3                   8.0      12.0       IN

                     FILE     3       23-SEP-83 16:24
                     DATA ACQUIRED    01-JAN-80 00:00
```

```
                AFTER SURVEY TOOL CHECK SUMMARY

  PERFORMED:     83/09/23
  PROGRAM FILE: BMS     (VERSION    26.15    83/08/13)

 MPT                      TOOL   CHECK

                 ZERO                        PLUS
           BEFORE       AFTER          BEFORE      AFTER     UNITS
  MINV      0.0          -.0             5.0        4.8      OHMM
  MNOR      0.0           .1             5.0        5.0      OHMM
  PROX      0.0           .0           500.0      497.3      MMHO

                     FILE     6       23-SEP-83 16:27
                     DATA ACQUIRED    01-JAN-80 00:00
```

The Microresistivity tool runs a Microlog with either a Proximity or Microlaterolog tool; each has a caliper. It is calibrated by means of a calibrating resistor, and the operation is identical to that described for the Dual Laterolog tools. The After Survey Summary is self-explanatory; the reasonable tolerances are given.

MPT Before Survey Calibration Summary

Item	Explanation	Tolerance
Measured Zero MINV, MNOR, PROX	Uncalibrated zero relay values.	0 ± 10
Measured Plus MINV, MNOR, PROX	Uncalibrated plus relay values.	MINV/MNOR = 5 ± 2; PROX = 500 ± 100.
Calibrated Zero	Computed gains and offsets applied to item 1.	0 ± 0.1
Calibrated Plus	Computed gains and offsets applied to item 2.	MINV/MNOR = 5 ± 0.1; PROX = 500 ± 0.5.

MPT After Survey Calibration Summary

Item	Explanation	Tolerance
Zero Before	Same as item 3.	—
Zero After	After survey calibrated zero measurement.	Item 5 ± 1 for PROX; 7 ± 0.1 for MINV, MNOR
Plus Before	Same as item 4.	—
Plus After	After survey calibrated plus measurement.	Item 7 ± 15 for PROX; Item 7 ± .2 for MINV, MNOR

Calibration of Density Tools

Density tool calibration consists of a primary calibration standard, a shop calibration standard that is referenced to the primary standard, and a wellsite calibration which uses a portable source to check proper tool operation and sensitivity.

A cesium source emits gamma radiation at a constant level of 662 keV. The two detectors receive impinging gamma rays which have undergone both Compton scattering and photoelectric effect interactions with the formation.

By design, the PGT Density tool responds almost exclusively to Compton scattering. Therefore, its response is a function of electron density, ρ_e.

$$\rho_e = No\ (Z/A)\rho_b$$

where
No = Avogadro's number
Z = Electrons per atom
A = Atomic weight

Elements in common reservoir rock have a Z/A value very close to 0.5. For those elements ρ_e is directly proportional to their bulk density ρ_b. In formations such as salt or gypsum where Z/A is not 0.5, the measured density will differ from the true bulk density.

The primary calibration standard for density tools is a series of water-filled, limestone, laboratory blocks whose densities are known precisely. The tool response is adjusted to read those density values.

Since detector sensitivity and source strength vary from tool to tool, a shop calibration standard is required to normalize the response of each tool to the primary standard. The shop calibration standard is a special, extremely pure aluminum block that is very stable and repeatable. After the density tool is calibrated to the laboratory limestone blocks, count rates are measured in the laboratory aluminum block. These count rates are the reference values used in shop aluminum block calibrations to normalize all PGT tools to the primary calibration standard. The block values have no geological significance. In the PGT calibration:

$$\rho_b = A - B \ln N$$

where
A = f (normalization)
B = f (spacing)
N = Normalized count rates

Because the source-detector spacing is tightly controlled in manufacturing, the value for constant B is determined from measurements in the laboratory limestone blocks. The value for A is determined by measuring the count rates in the laboratory aluminum block.

The shop calibration for the PGT tool consists of measuring the count rates in the aluminum block and computing a gain to normalize the tool response to the primary standard.

$$N = N_{log} \frac{N_{ref}}{N_{al}}$$

where
N_{log} = Log counts
N_{al} = Shop aluminum block measurement
N_{ref} = Laboratory aluminum block measurement

A portable jig containing two small gamma ray sources is used at the wellsite to verify proper tool operation and to check for changes in detector sensitivity.

The jig is placed on the detector skid during the shop calibration and is adjusted to approximate the count rates from the aluminum block. The same gain used to normalize the tool in the aluminum block is applied to calibrate the count rates from the jig in the shop and at the wellsite.

LDT CALIBRATION. With the Litho-Density tool, the gamma rays are sorted at the detector level by using a proportional counter. This allows us to accumulate and count gamma rays within an energy window. Gamma rays with energy above 180 keV originate from Compton scattering; those with an energy level below 80 keV are a result of the photoelectric effect.

Since proportional counters are used to measure the energy flux of gamma rays within a particular energy range, the stability of the detection system is extremely important. This stability is maintained by a continuous monitoring of the counts from tiny sources implanted in the crystals and the location of the counts within the energy spectrum. These count rates are extremely stable and repetitive, and a measurement of them is an excellent indicator of proper tool performance and stability.

The calibration of the density measurement of the LDT tool is similar to that for the PGT tool. The response in the shop aluminum block is referenced to the primary limestone block standards, but the technique is slightly different. Also, a practical difference exists in the calibration software although the mathematical results are the same.

During the Shop Calibration of the LDT tool, the measured count rates from the aluminum block are normalized by subtracting the counts from the built-in sources:

$$N_{log} = (N_{meas} - N_{internal})$$

The normalized counts from the high energy window are then used in an algorithm to calibrate the tool response to the primary standard.

$$\rho_b = A_1 - B_1 \ln \frac{N_{log}}{N_{al}}$$

where A_1 and B_1 are values for the shop aluminum block.

When the LDT tool density calibration is compared to the calibration for the PGT tool:

$$\rho_b = A - B \ln N_{log} \frac{N_{ref}}{N_{al}}$$

$$= A - B \ln N_{ref} - B \ln \frac{N_{log}}{N_{al}}$$

$$= A' - B \ln \frac{N_{log}}{N_{al}}$$

The calibration results are exactly the same with both tools, only the procedure is different.

The photoelectric absorption index (P_e) requires two calibration points. One is from the aluminum block and the other is from an iron sleeve inserted in the aluminum block.

PGT TOOL

```
                SHOP SUMMARY

   PERFORMED:      83/09/23
   PROGRAM FILE: SHOP   (VERSION     26.15    83/08/13)

 PGTK               DETECTOR CALIBRATION SUMMARY

              BLOCK                           JIG
       MEASURED   CALIBRATED     MEASURED   CALIBRATED     UNITS

   FFDC   518          337          519          337         CPS
   NFDC   914          528          915          528         CPS

         (PGS:310 , PDH:394 , GSR:5562, SFT :354 )

                       FILE     0       23-SEP-83 13:46
                       DATA ACQUIRED   01-JAN-80 00:00
```

```
               BEFORE SURVEY CALIBRATION SUMMARY

   PERFORMED:      83/09/23
   PROGRAM FILE: ISM     (VERSION     26.15    83/08/13)

 PGTK               DETECTOR CALIBRATION SUMMARY

               BLOCK                             JIG
          INPUT  CALIBRATED           MEASURED   CALIBRATED     UNITS
   FFDC    518       337                 515          335
   NFDC    914       528                 912          526

 PGTK               CALIPER CALIBRATION SUMMARY

            MEASURED                         CALIBRATED
          SMALL    LARGE                   SMALL      LARGE         UNITS
   CALI    194.0    261.1                   200.0      300.0          MM

                       FILE     1       23-SEP-83 15:36
                       DATA ACQUIRED   01-JAN-80 00:00
```

```
               AFTER SURVEY TOOL CHECK SUMMARY

   PERFORMED:      83/09/23
   PROGRAM FILE: ISM     (VERSION     26.15    83/08/13)

 PGTK                        TOOL   CHECK

                    JIG
              BEFORE       AFTER      UNITS
   FFDC        335          337        CPS
   NFDC        526          529        CPS

   RHOB   CHANGE (WATER): -2.118 K/M3

                       FILE     2       23-SEP-83 15:37
                       DATA ACQUIRED   01-JAN-80 00:00
```

PGTK (FDC) Shop Summary

Item	Explanation	Tolerance
Performed Program File	Date Shop Calibration was performed and version and program used.	—
Block Measured FFDC (Far Detector CPS); NFDC (Near Detector CPS)	Counts measured with tool in A1 block. Gains computed with these values.	FFDC − 320 to 450; NFDC − 500 to 1400
Block Calibrated FFDC, NFDC	Values derived by applying computed gains to item 2. Should equal PREFs.	FFDC = 337 NFDC = 528
Jig Measured FFDC, NFDC	Counts measured with wellsite jig on. Should be close to item 2, but not necessary.	Item 2 ± 10 percent.
Jig Calibrated FFDC, NFDC	Computed gains applied to item 4.	Item 3 ± 10 percent. Depends on how close jig is set to the block.
Tool Numbers	Should agree with log heading.	—

For tool calibration at the wellsite, the engineer enters the block measured count rates from the Shop Summary. The jig is used to check proper tool operation and stability.

PGTK (FDC) Before Survey Calibration Summary

Item	Explanation	Tolerance
Block Input	Shop Summary FFDC/NFDC block counts.	—
Block Calibrated	Computed gains are applied to item 7.	If block input counts used to compute gains, FFDC = 337; NFDC = 528. If jig used, FFDC = 337 ± 7; NFDC = 528 ± 11.
Jig Measured	FFDC/NFDC counts measured with wellsite jig.	Jig measured counts on shop summary ± 4 percent.
Jig Calibrated	Computed gains are applied to item 9.	Item 5 ± 4 percent.

After the survey the calibrated count rates from the jig are compared to the before survey count rates to check for stability.

PGTK After Survey Tool Check Summary

Item	Explanation	Tolerance
Jig Before FFDC/ NFDC	Same as item 9 of BSC.	—
Jig After FFDC/ NFDC	Calibrated FFDC/NFDC counts with jig after log.	Item 11 ± 4 percent.

LDT TOOL

```
                    SHOP SUMMARY

   PERFORMF         83/09/29
   PROGRA     _E: CCSHOP (VERSION    26.15   83/08/13)

 LDTC                 DETECTOR CALIBRATION SUMMARY

         DRS SONDE NUMBER                          :    3834
         NUCLEAR SERVICE CARTRIDGE NUMBER          :    991
         POWERED DETECTOR HOUSING NUMBER           :    2754
         POWERED GAMMA-GAMMA DETECTOR NUMBER       :    2768
         LDT LOGGING SOURCE NUMBER                 :    6996
         LDT CALIBRATION MODE                      :    AIR
         MUD DENSITY                               :    8.00000

           MASTER  CALIBRATED
           BKGD    AL+FE     AL    UNITS
     LL    18.6     92.0   103.5    CPS
     LU    72.5    143.0   160.8    CPS
     LS    54.9    163.6   183.8    CPS
   LITH     5.5     41.3    62.2    CPS
    SS1    15.6    166.2   188.3    CPS
    SS2    10.6    263.2   292.2    CPS
```

```
                 BEFORE SURVEY CALIBRATION SUMMARY

   PERFORMED:       83/09/29
   PROGRAM FILE: LEP     (VERSION    26.15   83/08/13)

 LDTC                 DETECTOR CALIBRATION SUMMARY

         DRS SONDE NUMBER                          :    3834
         NUCLEAR SERVICE CARTRIDGE NUMBER          :    991
         POWERED DETECTOR HOUSING NUMBER           :    2754
         POWERED GAMMA-GAMMA DETECTOR NUMBER       :    2768
         LDT LOGGING SOURCE NUMBER                 :    6996
         LDT CALIBRATION MODE                      :    AIR
         MUD DENSITY                               :    8.00000

             M E A S U R E D
           BKGD    AL+FE     AL    UNITS
     LL    18.2     92.0   103.5    CPS
     LU    72.1    143.0   160.8    CPS
     LS    54.4    163.6   183.8    CPS
   LITH     5.6     41.3    62.2    CPS
    SS1    15.1    166.2   188.3    CPS
    SS2    10.4    263.2   292.2    CPS
```

```
                AFTER SURVEY TOOL CHECK SUMMARY

    PERFORMED:       83/09/29
    PROGRAM FILE: LEP     (VERSION     26.15    83/08/13)

  LDTC                        TOOL  CHECK

          DRS SONDE NUMBER                           :   3834
          NUCLEAR SERVICE CARTRIDGE NUMBER           :   991
          POWERED DETECTOR HOUSING NUMBER            :   2754
          POWERED GAMMA-GAMMA DETECTOR NUMBER        :   2768
          LDT LOGGING SOURCE NUMBER                  :   6996
          LDT CALIBRATION MODE                       :   AIR
          MUD DENSITY                                :   8.00000

              BACKGROUND MEASURED
              BEFORE         AFTER         UNITS
      LL      18.29          18.37          CPS
      LU      72.18          72.25          CPS
      LS      54.43          54.53          CPS
    LITH       5.656          5.70          CPS
     SS1      15.15          15.07          CPS
     SS2      10.46          10.46          CPS
```

The shop calibration consists of count rate measurements from the internal sources and Al and Al + Fe measurements made with the logging source inserted.

LDT Shop Summary

Item	Explanation	Tolerance
Performed Program File	Date tool was calibrated and program and version used.	—
BKGD	Count rates in detector windows from internal sources.	—
LL	Long spacing lower.	17-21
LU	Long spacing upper.	72-88
LS	LL + LU/2.	55-62
LITH	Lithology.	4.5-6.5
SS1	Short spacing 1.	13-16
SS2	Short spacing 2.	8-12
AL + FE	Count rates in aluminum block and iron sleeve.	—
AL	Count rates in aluminum block.	LS > 150 SS1 > 155

During the wellsite calibration the engineer records the count rates from the internal sources. They should match those from the shop calibration, within tolerances. This assures the tool is working properly and the detector is stable.

The engineer enters values Al and Al + Fe from the Shop Summary to calibrate the tool.

LDT Before Survey Calibration Summary

Item	Explanation	Tolerance
BKGD	Count rates in detector windows from internal sources.	± 1CPS for LL, LU; ± .3 CPS for LITH; .5 CPS for SS1, SS2
AL + FE	Shop Summary values for aluminum block with iron sleeve.	—
AL	Shop Summary values for aluminum block.	—

After the survey, count rates from the internal sources are compared to those before the survey to check for proper tool operation.

LDT After Survey Tool Check Summary

Item	Explanation	Tolerance
Background Measured Before	Same values as 5.	—
Background Measured After	Count rates from internal sources after survey.	± 1CPS for LL, LU; ± .3 CPS for LITH; .5 CPS for SS1, SS2

Calibrations of Neutron Tools

The primary calibration standard for neutron tools is a series of water-filled limestone blocks. The porosities of these carefully controlled laboratory blocks are known within ± 0.5 porosity units.

Since source strength and detector sensitivity may vary from tool to tool, a field calibration standard is required to normalize the response of all tools to the primary standard. The field calibration standard is the neutron calibration tank (NCT).

COMPENSATED NEUTRON TOOL (CNTA). The ratio of the counting rates from the two thermal neutron detectors is observed in the limestone calibration blocks and is related to the porosity in each block.

The CNT-A tool is then placed in the water-filled NCT tank and the ratio is measured. This is the value used to normalize all CNT-A tools during shop calibration.

```
                SHOP SUMMARY

  PERFORMED:       83/09/23
  PROGRAM FILE: SHOP    (VERSION    26.15    83/08/13)

CNTA                DETECTOR CALIBRATION SUMMARY

               TANK                            JIG
        MEASURED   CALIBRATED        MEASURED    CALIBRATED

  NRAT     2.212        2.158            2.197        2.143

                        (CNC:258 , CNB:1039)

                        FILE     0        23-SEP-83 11:08
                        DATA ACQUIRED   01-JAN-80 00:00
```

```
                BEFORE SURVEY CALIBRATION SUMMARY

  PERFORMED:       83/09/23
  PROGRAM FILE: ISN     (VERSION    26.15    83/08/13)

CNTA                DETECTOR CALIBRATION SUMMARY

              TANK                              JIG
          INPUT   CALIBRATED             MEASURED    CALIBRATED
  NRAT    2.21200   2.158                2.207          2.153

                        FILE     1        23-SEP-83 15:00
                        DATA ACQUIRED   01-JAN-80 00:00
```

```
                AFTER SURVEY TOOL CHECK SUMMARY

  PERFORMED:       83/09/23
  PROGRAM FILE: ISN     (VERSION    26.15    83/08/13)

CNTA                        TOOL   CHECK

                         JIG
               BEFORE          AFTER
  NRAT         2.153           2.153

  POROSITY CHANGE  (LIME):    .000

                        FILE     0        23-SEP-83 14:59
                        DATA ACQUIRED   01-JAN-80 00:00
```

CNT-A Calibration. The detector calibration summary for the CNT-A tool recaps the calibration of the ratio of the count rates from the near and far detectors in the water-filled NCT calibration tank. The measured ratio (NRAT) is calibrated to the value observed during the primary calibration in the laboratory.

The count rate ratio from a portable jig is also recorded during the Shop Calibration. The jig is then used at the wellsite to check for changes in detector sensitivity and to verify proper tool operation.

CNT-A Shop Summary

Item	Explanation	Tolerance
Performed Program File	Date Shop Calibration was performed and the version and program used.	—
Tank Measured NRAT (Normalized Ratio)	Uncalibrated ratio measured with the tool in calibration tank. Used to compute gain.	2.158 ± 0.25
Tank Calibrated NRAT	Value derived by applying computed gain to item 2. Should equal PREF.	2.158
Jig Measured NRAT	Uncalibrated ratio measured with wellsite jig on.	2.2 ± 0.3 usually within ± 0.04 of item 2, but not necessarily.
Jig Calibrated NRAT	Value derived by applying gain to item 4.	Depends on how close jig is set to the tank ratio.
Tool Numbers	Should agree with tool numbers on log heading.	—

For tool calibration at the wellsite, the engineer enters the NCT tank-measured ratio. The jig is used to check proper tool operation and stability.

CNT-A Before Survey Calibration Summary

Item	Explanation	Tolerance
Tank Input NRAT	Same as tank measured NRAT on Shop Summary. Normally used to compute gain.	—
Tank Calibrated NRAT	Computed gain is applied to item 3.	If tank NRAT used to compute gain, 2.158. If jig used, 2.158 ± 0.04.
Jig Measured NRAT	NRAT measured with wellsite calibrator on.	Jig measured NRAT on Shop Summary ± 0.03.
Jig Calibrated NRAT	Computed gain is applied to item 5.	Item 3 ± 0.04.

After the survey the calibrated ratio from the jig is compared to the Before Survey ratio to check for stability.

CNT-A After Survey Tool Check Summary

Item	Explanation	Tolerance
Jig Before NRAT	Same as item 5 of BSC.	—
Jig After NRAT	Calibrated NRAT with jig after logging.	Item 13 ± 0.04.
Porosity Change	Maximum amount of error in porosity due to Before Survey and After Survey tool drift.	± 1 porosity unit.

DUAL POROSITY NEUTRON TOOL (CNTG). The Dual Porosity Neutron tool (CNT-G) has two thermal and two epithermal detectors for separate porosity measurements. The epithermal porosity measurement is processed using individual count rates from each detector. The technique is similar to the "spine-and-ribs" analysis used for density logs. The thermal neutron porosity is calibrated by the ratio of the near and far detector count rate.

Count rates and ratios from the thermal and epithermal detectors are recorded in the water-filled, limestone laboratory blocks and are related to the porosity in each block.

The tool is then placed in the NCT tank and the individual count rates and ratios are recorded. These values are used to normalize all CNT-G tools during shop calibration.

```
                    SHOP SUMMARY

   PERFORMED:       83/03/22
   PROGRAM FILE: CCS18  (VERSION     24.1     83/01/26)

 CNTG                  DETECTOR CALIBRATION SUMMARY

 NSSN(NEUTRON SOURCE SERIAL NUMBER): 157

 MEAS ID         PLUS REF            SHOP VALUE            GAIN

   TNRA            2.158                 2.208             0.977
   ENRA            2.845                 2.320             1.226
   NECN         3304.     CPS         2572.     CPS        1.284
   FECN         1235.     CPS         1102.     CPS        1.120
```

```
                 BEFORE SURVEY CALIBRATION SUMMARY

   PERFORMED:       83/03/03
   PROGRAM FILE: CNTGQC (VERSION     24.1     83/01/26)

 CNTG                  DETECTOR CALIBRATION SUMMARY

 NSSN(NEUTRON SOURCE SERIAL NUMBER): 157

 MEAS ID         PLUS REF            SHOP VALUE            GAIN

   TNRA            2.158                 2.210             0.976
   ENRA            2.845                 2.332             1.220
   NECN         3304.     CPS         2572.     CPS        1.284
   FECN         1235.     CPS         1102.     CPS        1.120
                         FILE
```

CNT-G Calibration. The Shop Summary for the CNT-G tool includes calibrations for the thermal and epithermal neutron detectors.

The individual count rates for the epithermal neutron detectors are measured in the NCT tank and calibrated to the reference laboratory values. These measured values are used as an input for the wellsite calibration of the epithermal detectors.

The thermal detectors are calibrated by normalizing the measured ratio of the tank count rates from the near and far detectors to the laboratory value. This measured ratio is the input for the wellsite calibration.

A new feature of the CNT-G operation is a wellsite test phase to assure proper tool performance. During the test phase the detector plateaus and all digital circuitry are checked and verified for stability. Therefore a test check with an external source or jig is not performed.

Calibrations for Gamma Ray Tools

The primary calibration standard for gamma ray tools is the API test facility in Houston. By definition, the difference between the two levels of radiation in the API pit is equal to 200 gamma ray API units.

The count rate response in the API test pit is recorded and calibrated in gamma ray API units.

Because the sensitivity of tools may vary, a field calibration standard is used to normalize the response of each tool to the calibration standard. The field standard is a small, stable, radioactive source placed a fixed distance from the GR counter by means of a special jig.

```
                BEFORE SURVEY CALIBRATION SUMMARY

PERFORMED:       83/02/25
PROGRAM FILE: DUAL    (VERSION     24.1      14/20/97)

SGTL                   DETECTOR CALIBRATION SUMMARY

              MEASURED
           BKGD      JIG          CALIBRATED          UNITS
   GR       19       185             165              GAPI
```

```
                AFTER SURVEY TOOL CHECK SUMMARY

PERFORMED:       83/02/25
PROGRAM FILE: DUAL    (VERSION     24.1      14/20/97)

SGTL                          TOOL  CHECK

                BEFORE        AFTER       UNITS
   GR            165           164        GAPI
```

The tool response to the field calibration standard is recorded and related to the API standard. Each different size and type of GR tool is calibrated in this manner. The increase in radiation field strength due to the field calibration source is 165 API units for the standard $3\frac{3}{8}$-inch tools and 135 API units for the standard $1\frac{11}{16}$-inch tools. These are the values used to normalize the GR tool response during field calibrations.

NATURAL GAMMA RAY SPECTROMETRY TOOL (NGT)

```
                      SHOP SUMMARY

   PERFORMED:        83/09/26
   PROGRAM FILE: NGSHOP (VERSION     26.1     83/05/23)

NGTC                   DETECTOR CALIBRATION SUMMARY

                 MEASURED
               BKGD     JIG          CALIBRATED          UNITS
   SGR          29      183              160             GAPI
                   SHOP       MEASURED
                 BKG      JIG       UNITS
 W1NG           68.8     430.0       CPS
 W2NG           27.7     190.7       CPS
 W3NG            9.1      29.4       CPS
 W4NG            2.0      15.9       CPS
 W5NG            1.8      25.0       CPS
                             FILE 0             26-SEP-83 00:45
```

```
                  BEFORE SURVEY CALIBRATION SUMMARY

   PERFORMED:        83/09/26
   PROGRAM FILE: TL8      (VERSION      26.15    83/08/13)

NGTC                   DETECTOR CALIBRATION SUMMARY

                 MEASURED
               BKGD     JIG          CALIBRATED          UNITS
   SGR          31      187              160             GAPI
                MEASURED BEFORE SURVEY
                 BKG      JIG       UNITS
 W1NG           79.0     444.5       CPS
 W2NG           28.1     197.7       CPS
 W3NG            8.8      32.0       CPS
 W4NG            2.2      16.3       CPS
 W5NG            2.3      25.6       CPS

                        FILE      0        26-SEP-83 09:07
                        DATA ACQUIRED      01-JAN-80 00:00
```

```
                   AFTER SURVEY TOOL CHECK SUMMARY

   PERFORMED:        83/09/26
   PROGRAM FILE: TL8      (VERSION      26.15    83/08/13)

NGTC                          TOOL   CHECK

                    BEFORE         AFTER       UNITS
   SGR               160            158        GAPI
                MEASURED AFTER SURVEY
                 BKG      JIG       UNITS
 W1NG           78.3     442.0       CPS
 W2NG           28.4     196.6       CPS
 W3NG            8.1      32.0       CPS
 W4NG            2.2      15.7       CPS
 W5NG            2.5      25.0       CPS

                        FILE      0        26-SEP-83 09:08
                        DATA ACQUIRED      01-JAN-80 00:00
```

The primary calibration standard for NGT tools is a laboratory test pit where the values for uranium, thorium, and potassium are precisely known. Each NGT tool is calibrated in this test pit and delivered to the field with a specification sheet.

The detectors are stabilized by using a small internal source. The amplitude/energy ratio is continuously calibrated providing an extremely accurate measurement that is free from drift.

Calibration includes a Shop Summary where the background count rates are recorded. Then, the increased count rates due to a special test source are measured in each of the 5 energy windows. This is not a calibration; it is a check to ensure the tool is operating properly and that the count rates are within specifications.

The wellsite procedures are similar. The special test source is used to ensure the tool is operating properly and that the count rates are within tolerances to those measured at the shop.

The standard GR curve is calibrated to give a 160-API unit increase from the test jig.

NGT-C Shop Summary

Item	Explanation	Tolerance
Measured Background	Background reading in the 5 windows.	—
Jig Measurement	Increased count rates due to test source.	—

NGT-C Survey Calibration Summary

Measured Background	Background reading in the 5 windows.	—
Jig Measurement	Increased count rate due to test source.	Item 2 ± 6 percent for W1, W2, W3; ± 20 percent for W4, W5.

NGT-C After Survey Calibration Summary

Jig Measurement	Increased count rate to test source.	Item 4 ± 6% for W1, W2, W3; ± 20% for W4, W5.

GAMMA RAY SPECTROMETRY TOOL (GST)

```
               BEFORE SURVEY CALIBRATION SUMMARY

PERFORMED:       83/10/01
PROGRAM FILE: GST      (VERSION     26.15    83/08/13)

GST                 CALIBRATION SUMMARY

                       NEUTRON BURST PROFILE

              FRACTION         FRACTION         FRACTION
              OF COUNTS       OF COUNTS        OF COUNTS
             BEFORE GATE     DURING GATE      AFTER GATE    SQUARENESS
MEASURED       0.0              .9726            .0273           .694
TYPICAL        0.01             0.97             0.02            0.75

TOTAL NET COUNTS      7446.
ACCUMULATION TIME     516.   SEC
```

```
GST                 CALIBRATION SUMMARY

             GSTA ZINC RESOLUTION

                             ZINC PEAK       ZINC
             RESOLUTION     COUNT RATE     CENTROID
MEASURED        5.95          1008.9         153.3
TYPICAL        5.5-6.9       1000-1700       70-240

                     FILE      0        01-OCT-83 13:31
                     DATA ACQUIRED     01-JAN-80 00:00
```

```
GST                 CALIBRATION SUMMARY

                        GSTA TANK

  INELASTIC
                                           PUR
              SADC      CHI-SQUARE     CHI-SQUARE
MEASURED     3309.9        1.466          2.270
TYPICAL      4000.0        1.500          2.000

                IBSF =    .750

  CAPTURE
              SCDC      CHI-SQUARE
MEASURED     1269.1        1.623
TYPICAL      1700.0        1.500

                          YIELDS

             INEL O     INEL FE     CAPT H     CAPT FE
MEASURED      .518        .316        .479       .276
TYPICAL       .550        .500        .680       .290

   GAIN          OFFSET        RESOLUTION
CORRECTION     CORRECTION     DEGREDATION      NEW       NEW
  FACTOR         FACTOR          FACTOR        PPR       ZPC
  1.000          -.039            0.0         3.473    35.101

ACCUMULATION TIME     300    SEC
```

```
GST              CALIBRATION SUMMARY
                      GSTA CHECK
  INELASTIC
                                           PUR
             SADC      CHI-SQUARE     CHI-SQUARE
MEASURED    3855.9        1.561          2.304
TYPICAL     4000.0        1.500          2.000

                 IBSF =    .693

  CAPTURE
             SCDC      CHI-SQUARE
MEASURED    1558.0        1.883
TYPICAL     1700.0        1.500

                       YIELDS

            INEL O     INEL FE    CAPT H     CAPT FE
MEASURED     .520        .308      .468        .279

   GAIN         OFFSET        RESOLUTION
CORRECTION    CORRECTION     DEGREDATION      NEW        NEW
  FACTOR        FACTOR          FACTOR        PPR        ZPC
  1.001         -.028            0.0         3.471     35.185

ACCUMULATION TIME    300    SEC
```

There are 4 calibration tasks for the GST-A tool.

Zinc resolution
Spectral check calibration
Water tank calibration
Neutron burst profile test

For the zinc test, the important parameters are the count rate and zinc resolution. The resolution calibration is used to determine the resolution of the detector and its temperature rating for logging. The count rate is important to ensure that the zinc peak can be seen above background for proper gain-offset regulation. The gain is controlled by maintaining the zinc peak centroid in the gain spectrum at the proper value. The offset is controlled by maintaining the proper ratio between the upper pulser peak centroid and the lower pulser peak centroid in the offset spectrum. The hardware gain and offset are checked every 15 seconds while logging and adjusted as necessary.

The check and tank tests are performed to determine the overall operation of the GST system. Either can be run in both capture mode and inelastic mode. In both tests the position of the capture iron and capture hydrogen are used to determine the zinc peak location and pulser peak ratio reference values for the particular tool.

The check test is performed in the borehole before logging and is used to determine the gain, offset, and degradation factor for the GST analysis. These software values are updated every 5 minutes while logging.

The tank test is conducted at the shop in a water tank. It reports the elemental yields that should be observed. The typical yields listed on CP26 summaries do not match the new standards and will be updated in CP28. The limits listed on the calibration specifications include values for tank tests with and without fluid excluder sleeves.

GST Calibration Specifications Neutron Burst Profile

Item	Limits
Fraction of counts before gate	< 0.03
Fraction of counts during gate	0.92 <
Fraction of counts after gate	< 0.05
Squareness	0.65 <

Zinc Resolution

Item	Limits
Resolution	< 6.4 @ AMB
Zinc peak count rate	1000–1700
Zinc Centroid	70–240

Tank Test—Capture Mode

Item	Limits
SADC count rate	< 5500
CHI-square (goodness of fit)	< 4.0
PUR CHI-square	< 4.0
Yields	
Capture H	.49–.60
Capture FE	.22–.29
TAU	180–200
GTQ	.95–1.05
Gain correction factor	.98–1.02
Offset correlation factor	+ /.25
Resolution degradation factor	< 2.0 @ AMB
New PPR	3.4–3.5
New ZPC	34–36

Tank Test—Inelastic Mode

Item	Limits
Inelastic	
SADC count rate	< 5500
CHI-square	< 4.0
PUR CHI-square	< 4.0
Capture	
CHI-square	< 4.0
Yields	
INEL O	.47–.66
INEL FE	.25–.52
CAPT H	.45–.89
CAPT FE	.07–.31
Gain correction factor	.98–1.02
Offset correction factor	+/− .25
Resolution degradation factor	< 2.0 @ AMB
New PPR	3.4–3.5
New ZPC	34–36

The neutron burst test is made before logging in the inelastic mode. This test is made to assure that there are minimum capture background counts in the inelastic gate.

Limits for the check test and tank test are the same except for the yields. Yields listed on the check test are actual formation values.

Thermal Decay Time Logging

Thermal Decay Time tools measure the neutron capture cross section of the formation. The parameter is derived from the decay in time of the formation's activation after a neutron burst, created by a minitron source. The TDT* tool contains two detectors.

TDT-K TOOL

```
                 BEFORE SURVEY CALIBRATION SUMMARY

  PERFORMED:       83/10/13
  PROGRAM FILE: TDT     (VERSION     26.1     83/07/15)

SGTG                  DETECTOR CALIBRATION SUMMARY

              MEASURED
            BKGD     JIG           CALIBRATED           UNITS
    GR       48      205              135               GAPI

TDTK                  DETECTOR CALIBRATION SUMMARY

   BKGRND CORR  (CPS)     MEAS    REF
   NEAR DET    FAR DET    RATIO   RATIO
    597.53      902.04     .66     .66

                            FILE 2              13-OCT-83 14:43
```

In the TDT-K tool each detector has three gates corresponding to three time intervals after the neutron burst. The first two, after background subtraction, are used to compute sigma (near counts) and a TDT ratio (near to far counts in gates 1 + 2). The third one measures background after complete decay.

TDT-K Before Survey Calibration Summary. Counts are not normalized, and a sigma measurement will not depend on the magnitude of the counts but on their distribution. On the contrary, the ratio has to be calibrated because it depends on detector sensitivity. A calibrating source gives us a static measurement where the counts of all the gates of the near and far detectors are added and the ratio is computed.

The ratio is then normalized to the reference ratio of the tool.

TDT-M TOOL. Sixteen time gates are selected for each detector of the TDT-M tool. This allows more elaborate decay time computation. Counts are not calibrated. Only the ratio is normalized, and this is done with a static measurement: a source is placed near the detectors and counts in all the gates are added. Background is measured first and subtracted from the jig counts. The calibration summary looks the same as that for the TDT-K tool.

There is no Shop Calibration for TDT tools. The static calibration with the source is made at the wellsite just before the job.

Electromagnetic Propagation Logging

The Electromagnetic Propagation tool measures attenuation and propagation time of an ultra-high-frequency electromagnetic wave.

The CSU system continuously calibrates the attenuation during logging by comparing the formation attenuation to that caused by a calibrated resistor network. The transit time is obtained from a phase shift measurement and the frequency of the wave. The accuracy is ensured by the use of a high-stability oscillator and a crystal clock. These components need only periodic checks. There is no CSU calibration summary for the EPT tool because no engineer intervention is needed.

Sonic Logging

There are no CSU system calibration summaries for sonic logs. Sonic transit time logs are direct measurements of a basic parameter: time. The only calibrations of conventional analog sonic recordings are of the galvanometers and associated circuitry, not of the measurement itself. In the CSU system, time intervals are keyed to a crystal-controlled oscillator and processed in digital form; the surface equipment does not vary in sensitivity. The accuracy of the crystal clock is verified during the general system check before each job.

Wellsite Seismic Logging

The Wellsite Seismic tool records waveforms digitally and computes transit times between the surface and downhole.

Amplitudes are normalized to give an even response throughout the hole, but are not quantitized. Time measurement is similar to that of the sonic log. This tool does not require any CSU calibration, except for the caliper.

Summary of Tolerances Between Before and After Summaries

Log	Curve	Tolerance Zero	Tolerance Plus	Units
ISF	ILD	± 2	± 20	MHO/M
	SFL	± 2	± 20	MHO/M
DIT	ILM	± 2	± 20	MHO/M
	ILD	± 2	± 20	MHO/M
IES	ILD	± 2	± 20	MHO/M
	16″ N	± 0.08	± 1.05	ΩM
DLL	LLS		± 0.7	ΩM
	LLD		± 0.7	ΩM
MSFL	CMSF	± 2	± 20	MMHO/M
	I1	± 2	± 20	MMHO/M
Proximity	PROX	± 1	± 15	MMHO/M
MLL	MLL	± 1	± 15	MMHO/M
GR	GR		± 15	CPS
CNL	NRAT		± 0.04	RATIO
Dual Porosity	NRAT		± 0.04	RATIO
Neutron	NECN			CPS
	FECN			CPS
SNP	SNP		± 13	CPS
FDC	FFDC		± 14	CPS
	NFDC		± 22	CPS
LDT	LL		± 1	CPS
	LU		± 1	CPS
	LITH		± 0.3	CPS
	SS1		± 0.5	CPS
	SS2		± 0.5	CPS
NGS	W1NG		± 6	%
	W2NG		± 6	%
	W3NG		± 6	%
	W4NG		± 20	%
	W5NG		± 20	%

NOTES

This section is intentionally left blank for the user to file any additional log quality control material acquired from other sources.

BIBLIOGRAPHY

Belknap, W. B., Dewan, J. T., Kirkpatrick, C. V., Mott, W. E., Pearson, A. J., and Rabson, W. R.: "API Calibration for Nuclear Logs," *Drill. and Prod. Prac.*, API (1959).

Bosworth, A. F.: "Log Calibration, Surface and Downhole," *Canadian Well Logging Society J.* (December 1972) **5**, 39–68.

Brown, R. J. S. and Neuman, C. H.: "The Nuclear Magnetism Log—A Guide for Field Use," *The Log Analyst* (Sept.–Oct. 1982) **23**, 4–9.

Cochrane, J. E.: "Principles of Log Calibration and Their Application to Log Accuracy," *J. Pet. Tech.* (July 1966) 817–826.

Holt, O. R.: "Log Quality Control," paper BB presented at the SPWLA 15th Annual Logging Symposium, 1975.

Knox, C. C.: "Quality Control of Well Logs," paper A presented at the SWPLA 15th Annual Logging Symposium, 1974.

Lang, W. H., Jr.: "Porosity Log Calibration," SPWLA ad hoc Calibration Committee report, *The Log Analyst* (March–April 1980).

McCulloh, T. H., Kandle, J. R., and Schoellhamer, J. E.: "Application of Gravity Measurements in Wells to Problems of Reservoir Evaluation," paper presented at the SPWLA 9th Annual Logging Symposium, 1968.

Neinast, G. S. and Knox, C. C.: "Normalization of Well Log Data," paper I presented at the SPWLA 14th Annual Logging Symposium, 1973.

Patchett, J. G. and Coalson, E. B.: "The Determination of Porosity in Shaly Sandstones," Part One: Quality Control. Paper QQ presented at the SPWLA 20th Annual Logging Symposium, volume II, 1979.

Schultz, A. L., Bell, W. T., and Urbanosky, H. J.: "Advancements in Uncased-Hole Wireline-Formation-Tester Techniques," paper SPE 5035 presented at the SPE 49th Annual Meeting, Houston, Oct. 6–9, 1974, fig. 6.

Scott, A. C.: "A New Generation Directional Survey System Using Continuous Gyro Compassing Techniques," paper SPE 11169 presented at the SPE 57th Annual Technical Conference and Exhibition, New Orleans, Sept. 26–29, 1982.

Stephen, W. and Walter, F.: "Participatory Log Quality Control," paper D presented at the SPWLA 23rd Annual Logging Symposium, July 1982.

Waller, W. C., Cram, M. E., and Hall, J. E.: "Mechanics of Log Calibration," presented at the SPWLA 16th Annual Logging Symposium, June 4–7, 1975.

COMPANY PUBLICATIONS

Dresser Atlas, Houston, Texas

"Computerized Logging Service," 9410 (1980).

"Formation Multi-Tester Interpretation Manual," 9404 (1980).

"Production Services," 9307 (1981).

"Spectralog," 3334 (1981).

"Wireline Services Catalog," 82-02-96-2 9102 (1982).

Gearhart Industries, Inc., Fort Worth, Texas

"Formation Evaluation Chart Book," WS–226 4–78 R 7–83 (1983).

Schlumberger Well Services, Houston, Texas

"Calibration and Quality Standards," (March 1974).
"Cased Hole Applications," C–11976 (1975).
"Data Processing Services Catalogue," M 0821020 (1981).
"Dipmeter Interpretation. Volume I—Fundamentals" (1981).
"Engineered Open Hole Services."
"The Essentials of Cement Evaluation," 261–56–11 (1976).
"The Essentials of Wireline Formation Tester," 261–56–11.
"Formation Tester Interpretation Methods and Charts" (1966).
"Fundamentals of Dipmeter Interpretation" (1970).
"Interpretation of TDT Logs" (1974).
"Log Interpretation Charts," SMP–7006 (1984).
"Log Interpretation Principles," C–11759 (1972).
"Production Log Interpretation," C–11811 (1973).
"Production Services" (1975).
"RFT Essentials of Pressure Test Interpretation," M 081022 (1982).
"Services Catalog," C 12001 (1977).
"Well Evaluation Developments—Continental Europe" (1982).
"Wellsite Products and Calibration Guide CP 26," M 081027 (1983).

Welex, a Halliburton Company, Houston, Texas

"Open Hole Services," G–6003.
"Thermal Multigate Decay Logging," 1YD (1983).

INDEX